The pre-glacial Pleistocene of the Norfolk and Suffolk coasts

TO HAZEL

The pre-glacial Pleistocene of the Norfolk and Suffolk coasts

R. G. WEST

WITH CONTRIBUTIONS BY P.E.P. NORTON, B.W. SPARKS AND D.G. WILSON

CAMBRIDGE UNIVERSITY PRESS

CAMBRIDGE

LONDON NEW YORK NEW ROCHELLE

MELBOURNE SYDNEY

Published by the Press Syndicate of the University of Cambridge
The Pitt Building, Trumpington Street, Cambridge CB2 1RP
32 East 57th Street, New York, NY 10022, USA
296 Beaconsfield Parade, Middle Park, Melbourne 3206, Australia

First published 1980

Printed in Great Britain at the University Press, Cambridge

ISBN 0 521 21962 0 hard covers

CONTENTS

Index. The reader will find there is no index. The contents list, however, supplies in detail the page numbers of the major heads of subject matter, so doing duty which would otherwise be served by an index. Thus reference can be made via the contents list to particular places and subjects. Likewise, references to flora and fauna, in general or at particular localities, can be obtained by use of the relevant tables, as well as via the contents list.

PREFACE

The following account is the result of many years' observations on the receding cliffs of north and north-east Norfolk and north-east Suffolk. The first field studies and collections were made at the time of the Centenary Excursion of the Geologists' Association in East Anglia in 1958, directed by D. F. W. Baden-Powell and myself. At that time it seemed amazing that no substantial work on the stratigraphy of the pre-glacial Pleistocene had been carried out since the end of the last century, in spite of the miles of excellent coastal sections. The word of Clement Reid, who wrote the Cromer Memoir of the Geological Survey (published 1882), still held sway, a tribute to the thoroughness of his work and to his interpretative abilities. Reid surveyed the coastal area from 1876 to 1879, just a hundred years ago, and produced his ideas on the stratigraphy in the Cromer Memoir and subsequently in his Pliocene Memoir (1890). He retained his interest in the Cromer Forest-bed after he left the area to work elsewhere, later making further contributions on the flora.

Reid's work on the Cromer Forest-bed flora stimulated his interest in the history of the British Flora, an interest which subsequently resulted in his book on the Origin of the British Flora (1899). He studied many other Pleistocene floras, and he and his wife made a classic contribution in 1915 to the Pliocene and Early Pleistocene flora of north-west Europe in their study of plant remains from the Dutch–German border area of Reuver, Brunssum and Tegelen. In working on the Cromer Forest-bed flora I therefore have a sense of going back to the beginning of Pleistocene floristic studies in Britain, and it has been stimulating to rediscover many of Reid's sequences and to realise the excellence of his descriptions, even though the results have meant a re-interpretation and replacement of some of his conclusions.

Although cliff recession through coastal erosion has given ideal opportunities for the study of new sections, to the residents of the cliff tops it is a menace, and protection from it must be secured by local authority coastal defence work. So in the course of the work much of the observed sections have become obscured by the building of concrete walls or by talus protected by groyne systems. As a result many of the sections are now lost, until the ravages of the sea again gain control. In many instances the local authorities helpfully delayed their work until observations and collections had been made. However, it is sad to report that there has been little positive result of efforts to protect type sections for future students of the Pleistocene, particularly because some of the sections must be considered as fundamental for studies of environmental history in Europe. Thus the term Cromerian attained wide usage in Europe as a stratigraphical or biostratigraphical unit. In addition the sections are extraordinary in showing evidence of great vegetational change, faunal change, climatic change and sea-level change, compressed into a few metres thickness, and they provide ideal opportunities for teaching.

The work could be regarded as an ideal scientific challenge to a field observer. Continual new sections, great variations in lithology, visible and invisible fossils, and a lack of publication for a hundred years, have all provided a great stimulus, sharpened by the present coastal environment, best appreciated at dusk on a cold winter's day. Like Clement Reid, I shall probably return to the coast, though officially finished in the area.

ACKNOWLEDGEMENTS

I have been greatly assisted in the work by Mrs D. G. Wilson, who over a long period has contributed data on plant macroscopic remains. I gratefully acknowledge the help and advice I have also received concerning different aspects of the work from many colleagues and organisations, including Dr R. W. Hey on geological matters; Miss R. Andrew for pollen identification and certain analyses; Mr B. W. Sparks in relation to non-marine mollusc faunas; Dr P. E. P. Norton in relation to marine mollusc faunas; Dr A. J. Stuart in relation to vertebrate faunas; Dr H. J. B. Birks on numerical analysis and computer work; Mr B. Belson, Mr C. W. English, Mr K. Gardner, Mr R. Mallett, Mr R. A. D. Markham, Mr S. W. Mobbs and Mrs H. V. West for assistance in the field and for details of exposures and boreholes; officers of the Erpingham and Lothingland Rural District Councils and Messrs Pontins Ltd for access and permissions; Professor J. K. St Joseph for aerial photographs; and Dr W. H. Zagwijn for discussion and assistance with problems of vegetational and environmental history of the southern North-Sea area. I also acknowledge the support of the Natural Environment Research Council in providing the research assistantship held by Mrs D. G. Wilson from 1964–70 and funding boreholes, excavations and assistance with production of pollen diagrams. This support greatly helped the completion of the work.

THE ABBREVIATIONS

GR	Grid Reference
OD	Ordnance Datum
HWM	High Water Mark
LWM	Low Water Mark
AP	Arboreal pollen
NAP	Non-arboreal pollen
p.a.b.	Pollen assemblage biozone

Ho	Hoxnian
An	Anglian
Cr	Cromerian
Be	Beestonian
Pa	Pastonian
Pre-Pa	Pre-Pastonian
Ba	Baventian
Lu	Ludham, Ludhamian

1: Introduction

The Cromerian as a name for a stage or other unit of the Pleistocene has had a varied and long history. It has been used to denote a temperate climatic stage, a faunal assemblage and a supposed Palaeolithic culture. These various usages are described in the Lexique Stratigraphique International (Oakley & Baden-Powell, 1963). In particular, the term has come to be used for the period of time of temperate climate immediately preceding the first major glaciation of northern Europe and Britain, in the earlier part of the Middle Pleistocene, and with such a generalised meaning that it has achieved wide circulation in stratigraphical, climatic and faunal tables of the Pleistocene in many parts of Europe. The reason for this is that there is obvious stratigraphical evidence in the Cromer area of Norfolk for a distinct immediately pre-glacial fauna and flora. Since these had been studied for over a hundred years, the area became regarded as a type-area.

However, the absence of detailed stratigraphical and palaeontological data from the pre-glacial sediments has prevented the characterisation of the Cromerian in more precise terms. The work described here was undertaken to provide a substantial basis for reconstructing environmental history in the immediately pre-glacial period of time, especially important because of the wide use of the term Cromerian in Europe. In the work a number of climatic stages, including the Cromerian, have been recognised and defined, and it has been possible to relate these to the history of sedimentation, vegetation and sea-level change.

The sediments concerned are highly fossiliferous and are found beneath the glacial drift of the Norfolk and Suffolk coastal cliff sections. In the past, they have been grouped together as the Cromer Forest-Bed Series and have been placed in the Middle and Lower Pleistocene. The sections present, perhaps better than other sections in Europe, an opportunity for detailed palaeoecological reconstruction of the immediately pre-glacial period of time in northern Europe.

The pre-glacial sediments have been observed from Weybourne on the north coast of Norfolk, east to Happisburgh, and on the Suffolk coast from Hopton and Corton to Kessingland (Fig. 1). Few inland occurrences have been recorded. The classic description of the sections was made by Clement Reid (1882*a*, *b*; 1890) in Memoirs of the Geological Survey. Previously there had been a long history of observation of the pre-glacial deposits, summarised by Reid. A bibliography of relevant publications up to 1890 was given by Reid (1890) and to 1897 by Harrison (1897).

Reid classified the pre-glacial sequence into two major units, the Weybourne Crag and (the younger) Cromer Forest-bed Series, in which he included three units, the Lower Freshwater Bed, the 'Forest-bed' (estuarine), and the Upper Freshwater Bed. He considered the Forest-bed Series to have been formed under a temperate climate. Since Reid's time, no further substantial work has been done on the stratigraphy of his Cromer Forest-bed Series, even though the term 'Cromerian' attained wide use in the literature of European Pleistocene stratigraphy, as indicating an immediately pre-glacial temperate stage. However, an important contribution on the palynology was published by Duigan (1963), but the pollen assemblage biozones could not be satisfactorily related to the sequence established by Reid.

The contributions described here concern principally the stratigraphy and palaeobotany. In the work I was greatly assisted by D. G. Wilson, who identified the macroscopic plant remains and analysed all the macroscopic samples.

The present study began in 1958, and observations of the stratigraphy and samples for palaeobotanical analysis were obtained up to 1977. Provisional stratigraphical results of the project were briefly reported by West & Wilson (1966). It was demonstrated that two temperate stages and a cold stage were included in the Cromer Forest-bed Formation (for definition see Chapter 18). Since 1966 many additional data have been obtained, and it is now evident that the sequence is even more complex than was thought then.

During the period 1958–75 samples were taken for analysis of marine molluscs and non-marine molluscs, studied by P. E. P. Norton and B. W. Sparks respectively. The marine mollusc analyses by P. E. P. Norton are listed and discussed in Appendix 2. The non-marine mollusc analyses by B. W. Sparks are listed in the tables, or, if brief, in the text, with his comments on their significance. He has contributed sections in the text on the West Runton Freshwater Bed fauna and on the Cromerian non-marine mollusc fauna. The authorities for the mollusc nomenclature are given in the tables. R. W. Hey has made analyses of the gravels encountered in the field work, results of which have been published (Hey, 1976). Certain sections have been sampled by A. J. Stuart for vertebrate remains and results of his investigations have also been published (Stuart, 1975).

The presentation of the results dealing with such numerous sections over long coastal stretches is complex. Two sets of data are to be given, one dealing with lithostratigraphy, the second with biostratigraphy. Introductions to these two subjects are given first. The biostratigraphy is intimately related to the lithostratigraphy and it is convenient to divide the coast into stretches and treat each stretch as a unit, giving details of lithostratigraphy and biostratigraphy. As the stratigraphical basis rests primarily on the West Runton and Beeston stretches of coast, these two areas are dealt with before the other areas studied.

But before dealing with West Runton, since the underlying sediments of the whole area have a basic similarity and are included in a Pre-Pastonian a substage, it is appropriate to give details of this Pre-Pastonian a substage before continuing with the detailed descriptions of the successive coastal stretches.

The description and interpretation of each stretch of coast entails the following heads: lithostratigraphy, periglacial stratigraphy, biostratigraphy, sedimentary and vegetational history. A table summarises the conclusions for each stretch of coast. The sequence of stages identified on the coast as a result of this work is: Pre-Pastonian (cold), Pastonian (temperate), Beestonian (cold), Cromerian (temperate), Anglian (cold, glacial).

After the description and interpretation of each stretch of coast, the southern and north-western limits of the various sediments are described. The nature of the marine sediments is then considered and relative land/sea-level changes are estimated. The environmental history of the whole succession is then summarised stage-by-stage, and the final chapter draws conclusions about the succession, flora and fauna and suggests possible correlations with sequences elsewhere.

(a) Lithostratigraphy

The best sections were seen after scours during north-west gales, but even then rapid talus accretion quickly obscured the sections. Another difficulty resulted from the building of sea defences along the coast from 1953 onwards. Fortunately, many sections were recorded and sampled before protection works were built, and I am grateful to the local authorities of the coastal areas for their cooperation in this aspect of the work.

The stratigraphical data stem from repeated observation of the cliff sections from 1958–77, from boreholes carried out on behalf of the coastal defence authorities, from specially commissioned boreholes (at Happisburgh and Mundesley), and from excavations made by excavating machines. The data so obtained are presented in the form of coastal sections, while the lithological details of the most important sites are given in Appendix 1. The symbols used in the sections are shown in Fig. 3. The observations cover a period of 19 years, in which time cliff recession occurred. Where sections were observed at the same point in a receding cliff, the dates of observation are mentioned in the sections. The positions of sample series and single samples for pollen and macroscopic plant remains analysis are shown in the diagrams of the sections. The sites are labelled by letters denoting the stretch of coast concerned, then letters denoting the particular site, allocated alphabetically, in order of description. The drawn sections also show the sites where R. W. Hey has analysed gravels (Hey, 1976).

The sections were drawn in the field at a scale of 4 cm to a metre, a scale promoting the study of detail. The location of the sections was fixed by identification on large-scale oblique aerial photographs of the cliffs and then reference to 6-inch and 25-inch maps, and by taping or pacing to a datum point. Such datum points were established at Beeston, West Runton, Sidestrand/Overstrand, Trimingham, Mundesley (two), Bacton, Happisburgh, Corton and Pakefield. Their positions are shown in Fig. 2. The exact positions of these points are as follows:

Beeston: at the termination of the concrete revetment to the east of Sheringham in 1964. GR 16824339.

West Runton (a levelled datum, 4·90 m OD): on a flat concrete surface at the south-west corner of the concrete slopeway at Woman Hithe. GR 18434317.

Sidestrand/Overstrand: opposite the final groyne to the south-east of Overstrand in 1964. GR 25484058.

Trimingham: at the northern boundary to the minefield area, as marked in 1964. GR 26034020.

Mundesley 1: Opposite the final zig-zag groyne to the south-east of Mundesley in 1964. GR 31783639.

Mundesley 2: Near the end of the iron revetment to the north-west of Mundesley, at the point where it turns towards the cliff, in 1964. GR 30953714.

Bacton: 69 m north-west of the point where the road meets the cliff. GR 34083430.

Happisburgh: At the gangway next to the Lifeboat Station. GR 38343113.

Corton: Opposite where the southern boundary of the radio station field meets the cliff. GR 53879903.

Pakefield: Opposite the centre of Crazy Mary's Hole (Old Lighthouse Gap). GR 53678868.

Heights at West Runton, Skelding Hill and Pakefield were obtained by levelling from bench-marks. Heights at Corton were estimated from known tidal range. Elsewhere heights were obtained from data on levels of coast defence works and boreholes provided by local authority engineers.

The sections studied covered several kilometres of coastline (Figs. 1, 2), and, as mentioned previously, they can be grouped for convenience into the stretches listed below. The type area for Reid's Cromer Forest-bed Series is at West Runton, where he first gave a detailed description (Reid, 1882*a*, *b*) of his Upper Freshwater Bed, which has since become the type section for the Cromerian temperate stage.

After a description of the Pre-Pastonian a substage, the sections first to be described will therefore be in the West Runton area, and since the sections at Beeston provide the best evidence of pre-Cromerian stratigraphy, these will be next in order to be described. The sequence of description of the groups of sections will then be as follows, with the tables indicated summarising the environmental history of each stretch of coast.

Pre-Pastonian a substage around the coast
West Runton (Table 8)
Beeston (Table 11)
East Runton
Cromer-Overstrand (Table 14)
Sidestrand (Table 17)
Trimingham (Table 20)
Paston-Mundesley (Table 23)
Bacton (Table 26)
Ostend (Table 26)
Happisburgh (Table 29)
Corton (Table 32)
Pakefield-Kessingland (Table 34)
Covehithe
Skelding Hill, Sheringham

A separate chapter (Chapter 15) considers the southern boundary of the Pre-Pastonian and younger pre-glacial sediments, using evidence from the cliff section and boreholes at Covehithe, 4 km south of Kessingland. The north-western boundary on the coast is to the east of Weybourne, where the low cliff shows glacial drift covering pre-glacial Pleistocene sediments. Reid (1882*a*, *b*) described the Sheringham to Weybourne sections in some detail. Unfortunately, except for a number of sections under Skelding Hill, 1 km west of Sheringham, the sediments on this part of the coast are weathered, and pollen analysis has not proved possible, so that the stratigraphy of this most western part of the area remains problematical. The stretch of coast between Happisburgh and Great Yarmouth (28 km) is at present barren of good stratigraphical and biostratigraphical data, so that this area also remains for future investigation.

The sediments in each stretch of the coast considered are grouped into beds, labelled by lower-case letters in alphabetical order of decreasing age. The beds are allocated to stages identified by biostratigraphy. In Chapter 18, a formal nomenclature (Table 43) is suggested for certain well-distributed beds, and these are grouped into members and formations. The two formations are the Norwich Crag Formation, including the Pre-Pastonian a substage, and the Cromer Forest-bed Formation, including Pre-Pastonian b, c and d substages, the Pastonian, Beestonian and Cromerian stages, and the Early Anglian substage. In the subsequent descriptions Cromer Forest-bed Formation replaces Reid's Cromer Forest-bed Series as a stratigraphical term. The terms put forward in Table 43 are used throughout the descriptions, rather than Reid's terms. The relation of

Reid's, Gunn's and Prestwich's sequences to the succession proposed here is explained in Chapter 18.

In the sections on periglacial stratigraphy, ice-wedges and involutions are described. Under the heading involution, I include disturbed bedding seen at clay or silt/sand interfaces. The origin of such disturbances may lie in load-casting as well as freeze-thaw, but the presence of involution horizons is included in these sections without prejudice to their origin.

(b) Biostratigraphy

The biostratigraphy of each stretch of coast listed above is described separately. Pollen assemblage biozones are described and related to the stratigraphy, and correlations are made via the pollen stratigraphy with the successions at West Runton and Beeston. Analyses of macroscopic plant remains by D. G. Wilson, presented as tables, provide a very substantial basis for elucidating local vegetational conditions and amplifying the pollen-analytical evidence of vegetational history. Notes on the identification and behaviour of particular taxa are given in Chapter 18.

(i) *Palynology*

Slides for pollen analysis were prepared by standard methods used in the Sub-department of Quaternary Research (West, 1977), using hydrochloric acid or sodium hydroxide to disperse calcareous and organic sediments respectively, sieving, removing inorganic fractions by hydrofluoric acid and hydrochloric acid treatment, acetolysis with glacial acetic acid and sulphuric acid, oxidising with glacial acetic acid and sodium chlorate, and mounting in safranin-stained glycerine jelly.

The Cambridge pollen reference collection was used as the basis for identification of pollen and spores and the taxa identified; their nomenclature is in accordance with the taxa given by Andrew (1970). The identification of certain taxa is discussed in Chapter 18.

The basis for calculation of all pollen diagrams is as follows. The percentage base is the sum of land plant pollen, shown in the diagrams. In the diagrams, the percentage of each land taxon is given on this basis, with the percentage of aquatics, pteridophytes and dinoflagellates based on the total land plant pollen plus the total of the respective group. A + in the pollen diagram indicates the taxon has been observed in scanning the pollen slide, not in the count. *Azolla* totals refer to massulae in whole or part.

The degree of preservation of pollen and spores was very variable, being generally best in freshwater organic sediments and worst in weathered inorganic marine sediments. With the pollen and spore counts are included counts of types of dinoflagellates and derived spores. Cysts with hooked appendages are placed as hystrichospheres in the pollen diagrams. Three genera of dinoflagellates are also recorded, *Operculodinium*, *Tectatodinium* and *Leptodinium*. These taxa, and their distribution in marine sediments of early Pleistocene age in the Ludham, Norfolk, borehole have been described by Wall & Dale (1968). The occurrence of these genera in the pre-glacial Pleistocene is discussed in Chapter 18.

(ii) *Macroscopic plant remains*

Samples for analysis, of size between 500 and 1000 ml depending on the volume available, were mostly broken down in water or sodium hydroxide. Hydrochloric acid or detergent was used where necessary for intractable clays and clay-muds. The sediment was then sieved (to 100 mesh per inch (152 μm)), and the residues examined and sorted under a binocular microscope ($\times$25). Identifications at higher magnifications (to $\times$1000) were made as needed. As well as seeds and fruits, the analyses gave information on the presence of many other kinds of plant and animal remains, including moss remains, foraminifers and derived megaspores.

The identifications were based on detailed comparisons with the Cambridge reference collections, which include British and a wide selection of non-British European and other taxa, as well as named fossil material. The nomenclature, apart from *Potamogeton*, is for British taxa based on that of Dandy (1958); for non-British European taxa on the *Flora Europaea* (Tutin *et al.*, 1964–76). Where the taxon is dealt with in these floras, the authority for the name is not given in the text. Where it is not dealt with, the authority is given.

The *Potamogeton* macroscopic remains will be described in more detail elsewhere by M. Aalto and D. G. Wilson. D. G. Wilson, who made the identifications listed in the macroscopic analyses, has already published an account of the more interesting macroscopic records (Wilson, 1973). She has in hand further contributions on a review of old records from the Cromer Forest-bed Formation and on further interesting records. Dickson (1973) has described certain of the moss remains isolated from the pre-glacial sediments.

The analyses of macroscopic plant remains for each stretch of the coast considered, are presented in a table showing the chronostratigraphical position of each sample analysed. Three figures (Figs. 15, 28, 41) show the results in the form of the frequency of physiognomic/ecological groups in the various samples: trees, shrubs, maritime taxa, heath taxa, montane taxa, dry-ground herbs, damp-ground herbs, plants of muddy substrates or shallow water, fen and reedswamp helophytes, aquatics, water ferns, *Carex* and unclassified. Such a grouping is of course an ideal one, since more than one ecological category may be covered by a particular taxon. However, the grouping does convey information on major changes of local vegetation, as reflected in varying frequencies of the groups. In these figures, the percentage occurrences shown use as a basis the total number of remains excluding those of water ferns and in the case of Fig. 15, excluding *Equisetum* remains. In the case of the water ferns the basis used is the total including remains of water ferns.

(iii) *Taphonomy: pollen*

With the variety of sedimentary environments seen in the pre-glacial sequence it is obvious that pollen assemblages in various sediments will originate in different ways. Changes in pollen assemblages which may be a result of changes in the taphonomy of the assemblages are discussed in the sections on vegetational history. The most obvious changes are those associated with changes from freshwater to marine sedimentation and vice versa, partly dependent on the wider pollen catchment via flowing water of the marine sediments. Thus a horizon of humified mud with *Phragmites* sandwiched between tidal sediments shows a mainly locally derived pollen assemblage, in contrast to the assemblages of the tidal sediments below and above (Cr IV, Mundesley, MAM). There are also abrupt changes in freshwater muds which can be related to changes in water level and related changes of taphonomy of the assemblages (Cr II, Mundesley, MF, MU).

Marine sediments containing pollen occur at several horizons in the pre-glacial sequence. Fine-grained tidal laminated sediments are present in the Pre-Pastonian a, Pastonian and in the later part of the Cromerian. Coarser marine sands with silt/clay laminae are present in the Pre-Pastonian a and in the Pastonian.

The interpretation of the pollen spectra from these sediments in vegetational and climatic terms is not straightforward. In offshore and estuarine areas the pollen content of sediment will be related to the source of pollen and to the sinking velocity of pollen and its transport by currents and turbulence. The processes involved have been discussed in many papers (e.g. Muller, 1959; Cross, Thompson & Zaitzeff, 1966; Groot, 1966; Groot & Groot, 1966; Koreneva, 1966; Traverse & Ginsburg, 1966; Stanley, 1969).

Source of pollen

Groot (1966) showed a relationship between the number of pollen grains and the quantity of suspended mineral matter in estuarine waters (Delaware river), indicating that pollen and sediment were transported together by water currents. Likewise Muller (1959) showed that pollen in sediments offshore to the Orinoco delta was transported by water currents, in contrast to sediments within the delta, where the local swamp flora is much better represented. The pollen content of the estuarine and offshore sediments will therefore contain pollen taxa transported from the catchment area of the neighbouring area, and will give some idea of the broad vegetational types of neighbouring lands. Thus Koreneva (1957) was able to divide the Sea of Okhotsk, a nearly enclosed sea, into palynological provinces related to neighbouring phytogeographic provinces, and Groot (1966) showed that pollen spectra in the Delaware river were representative of the vegetation of the whole catchment rather than the local communities bordering the river.

If much of the pollen in marine sediments has its source in large catchment areas, then we must consider how the catchments are contributing pollen to the river. Investigations of the type of Koreneva (1957) and Groot (1966) show the relationship at the present time. In regions of low relief, such as East Anglia, pollen input into rivers will be largely related to airborne pollen rain onto lake, river and soil surfaces, with rivers aggraded to present sea level. If, however, rivers are in a state of more active transport of sediment at times of falling sea-level and/or cold climate, then a large contribution of pollen may be made to rivers by destruction of soils and reworking through erosion of sediments. In a region of low relief, therefore, we might expect pollen assemblages of the temperate stages (Pastonian, Cromerian) found in marine sediments to be generally representative of the catchment vegetation, but in cold periods a large element of reworked pollen might be expected to be present.

In the temperate stages marine and freshwater sediments forming a transgression series will be likely to contain different pollen assemblages since the transgression will be accompanied by changes in pollen source. Such changes are identified later. It will be clear that it is of prime importance to relate sedimentary regime to pollen assemblages in order to identify the pollen source and therefore the vegetation types which supplied the pollen.

Sinking velocity of pollen and transport by currents and turbulence

Theoretical considerations (Stoke's Law), experiments on settling times and buoyancy (Hopkins, 1950; Brush & Brush, 1972) and field observations (Muller, 1959; Davis & Brubaker, 1973) indicate that sinking velocities and buoyancy of pollen in water vary with size and morphology of pollen grains. For example, Hopkins (1950) showed that the settling times of *Pinus* and *Quercus* were very different. Muller (1959) showed that size selection of pollen grains takes place off the Orinoco delta during the process of gradual settling of the pollen load transported by marine currents. The best-known example of differential rate of sinkage is that of *Pinus* compared with pollen of deciduous trees such as *Quercus*. *Pinus* becomes widely distributed in offshore regions because of its buoyancy. Thus in Koreneva's (1957) study of the Sea of Okhotsk, *Pinus* pollen is very widely distributed compared with other arboreal pollen taxa, though in part this may be due to the excellence of the long-distance airborne transport performance of the *Pinus* pollen grain. The relation of *Pinus* pollen frequencies to the sedimentation pattern on Great Bahama Bank (Traverse & Ginsberg, 1966) showed that *Pinus* pollen remained suspended for long periods and its deposition was related to sedimentation of fine-grained sediments and absence of turbulence.

The size of most pollen grains and spores is that of coarse to medium silt, but the change of density on loss of protoplasm means that exines have a hydraulic equivalent of perhaps fine silt or clay, and morphological variations will also contribute to variation in settling rate. Fine-grained sediments show the highest concentrations of pollen, indicating that their transport and deposition behaviour follows that of silt and clay. A consideration of silt and clay deposition in the estuarine and offshore regions will help the interpretation of pollen assemblages in them. Fine-grained sediments may be formed offshore, seaward of the coastal high-energy environment, or closer inshore if the influx of fine sediment is sufficient, as it may be in estuaries and tidal regions. In the former situation, the more buoyant and the smaller grains will be selectively transported seaward, forming assemblages in the offshore silts and clays. In the inshore region, silts and clays may contain more representative assemblages which have not been subjected to such a degree of selective transport as the offshore assemblages. Thus in these inshore regions the heavier pollen grains will be sedimented. There may still be some selective transport in these inshore regions, however, since Postma (1967) has shown how the constitution of suspended matter varies from a tidal inlet to the shore, with the finer fractions (silt, 8 μm–fine sand, 256 μm) concentrating at the shore and coarser sand at the mouth of the inlet. The effect of this on pollen transport remains to be investigated.

Postma (1967) has also described how fine-grained sediments accumulate at estuary mouths and in tidal flats by settling and scour lags. He also describes the turbidity maxima found in rivers with an estuarine circulation, with suspended matter attaining values higher than in the river water and in the sea, often with consequent high deposition rates of sediment at slack tide. The region with stratified river and sea water acts as a sediment trap, with particles caught in a circulating system of river water flowing downstream, and sea water upstream at a lower level. The turbidity maxima may work as an effective selective pollen trap, since the more buoyant grains will sediment to the sea-water level and may be recirculated before final deposition.

Thus we have here two mechanisms for selective transport. One is the selection of smaller or lighter grains in the shore part of a tidal system; the second is the extraction of heavy grains by the processes accompanying a turbidity maximum in a river mouth.

Notes on particular pollen taxa

Pinus. The buoyancy and consequent wide marine dispersal of *Pinus* have already been mentioned. Where pollen influx is low (as in offshore regions) there may be high *Pinus* percentage frequencies (a comparable situation to the high *Pinus* frequencies found in terrestrial sediments of the high Arctic). An expression of pollen influx in terms of absolute pollen frequency would assist interpretation of such pollen spectra, but in the absence of a knowledge of sedimentation rates this cannot be determined. High *Pinus* frequencies in fine-grained sediments may thus indicate deposition in non-turbid waters in areas where selective transport has favoured *Pinus*. If the fine sediments are found as laminae in coarse sand, we can envisage short tranquil periods for deposition, interrupted by a higher energy regime with sand deposition. Such conditions might be found below mean tide level in estuaries or offshore. If the fine sediments are massive, the more continuous low energy regime of deeper offshore areas may be indicated.

However, in the present study marine sediments with high *Pinus* frequencies occur in both fine-grained tidal sediments (e.g. Skelding Hill WM) and in coarser sands (e.g. Sidestrand SSV 12) in a similar stratigraphical position, suggesting that selective transport of *Pinus* is a factor which is subordinate to regional pollen production.

Ericales. The tetrads of Ericales are present in high frequencies in the marine tidal laminated silt and clay of the Pre-Pastonian a, as they are in the Lu 4b horizon (Baventian) at Ludham (West, 1961), Easton Bavents (Funnell & West, 1962) and Aldeby (Norton & Beck, 1972). At Ludham and Aldeby, in zone Lu 4b frequencies of 50% of total pollen were reached and frequencies of about 40% at Easton Bavents and Covehithe in similar sediment. At all these sites *Pinus* pollen reached 10–15%. Lower frequencies of Ericales pollen and higher frequencies of *Pinus* pollen are found in Pre-Pa a sediments at Pakefield, Happisburgh and Mundesley and elsewhere. The sediments at the two former sites are mainly sands with thin silt/clay laminae. A possible explanation of this difference in the pollen content is that the high Ericales–low *Pinus* assemblages were deposited with tidal silt/clay inshore or in estuaries, whereas the high *Pinus*–low Ericales assemblages were deposited under conditions where more selective transport had taken place and where a higher energy environment prevailed, i.e. in channels between barrier islands or more seaward. But the Pre-Pa a type of assemblage is found in both massive laminated silty clays (e.g. Mundesley MAT) as well as in more sandy facies. It is therefore concluded that again selective transport is subordinate to regional pollen input.

The origin of the high Ericales pollen frequencies in the Pre-Pa a and Lu 4b pollen spectra is of especial interest. In the discussion on the high Ericales frequencies of the Ludham borehole (West, 1961), I suggested that they were the result of the development of oceanic heath vegetation in the colder parts of the sequence (Lu 2 (Thurnian) and Lu 4 (Baventian)). It is also possible that the Ericales frequencies are raised by the same processes that promote sediment traps in estuaries, as discussed above, with concentration of the heavy Ericales tetrads. A third point is that, at times of lowering base level, soil erosion will result and input of Ericales tetrads, preserved in the soil, may increase as a result. Ericales tetrads are not widely dispersed in the air and tend to be dispersed in run-off.

It is known that in the early Pleistocene and Pliocene Ericales frequencies are consistently high at other sites of continental deposition in north-west Europe. Thus Menke (1975) has described sediment sequences from north-west Germany, associated with salt structures, which show high Ericales frequencies. Heath was thus obviously an important component of the vegetation in certain areas in north-west Europe in early Pleistocene times.

Gramineae. The frequencies of Gramineae pollen in the Pre-Pastonian and Pastonian marine sediments is variable. In general, high *Pinus* frequencies are accompanied by lower Gramineae frequencies, and vice versa. In addition, the higher Gramineae frequencies may be accompanied by a wider representation of arboreal taxa. Observations on Gramineae pollen in recent marine sediments tend to demonstrate that the representation of Gramineae pollen falls off strongly seaward. Zagwijn & Veenstra (1966) remarked on the unexpectedly poor representation of Gramineae pollen in surface sediments of the Outer Silver Pit (North Sea). Koreneva (1966) noted that Gramineae pollen was of very little importance in the Sea of Okhotsk sediments, and Cross *et al.* (1966) demonstrated a decrease in Gramineae pollen westwards from the Pacific shore of Mexico. In the Delaware estuary considerable Gramineae pollen percentages were present and were related to marsh vegetation bordering the estuary (Groot, 1966).

In the light of these observations, we may consider that the marine spectra with higher Gramineae percentages were formed nearer the shore, and this would also explain the better representation of arboreal taxa in many of the samples with high Gramineae percentages, as in the Cromerian I and II assemblages at Sidestrand (SSTB) and Trimingham (BHT 14A) respectively. We may note that temperate pollen spectra with substantial frequencies of Gramineae and Chenopodiaceae pollen have been described from Flandrian and recent tidal sediments in the western Netherlands by Hartman (1968).

(iv) *Taphonomy: macroscopic plant remains*

Whereas the taphonomy of pollen assemblages has been studied in some detail, the taphonomy of macroscopic assemblages remains a largely uninvestigated field. Seed and fruit productivity, the processes of dispersal and the processes of sedimentation must vary very much from species to species and from environment to environment. There are data on productivity (e.g. Salisbury, 1942) and dispersal (Praeger, 1913), but little evidence is available on the relation of fossil macroscopic plant remains to the vegetation yielding them in different sedimentary environments (Birks,

1973; McQueen, 1969). There is a great range of sedimentary environments represented in the pre-glacial sequence of the coast and we can briefly comment on variations in the macroscopic assemblages in respect of the following heads: the taxa most commonly represented, the numbers of taxa in the assemblages, the characteristics of assemblages from particular environments, the representation of the different physiognomic/ecological groups in various environments of deposition and the distribution of derived megaspores.

Taxa most commonly represented

Of the trees, *Alnus* and *Betula* are most frequently recorded. This must be related both to the occurrence of these genera in communities associated with high water levels and to their dispersal capabilities. *Alnus glutinosa* has a wind-disseminated fruit of poor efficiency, but its corky floats permit wider distribution by flowing water and by wind-drift on water. *Betula* fruits are more efficiently wind-distributed and may also disperse by wind-drift on water, though their floatability is lower than *Alnus*.

The most commonly found taxa are some in the fen and reedswamp group and in the aquatic group. In the former group the seeds or fruits of *Lycopus europaeus*, *Ranunculus sceleratus* and *Urtica dioica* are very well represented. These species have high seed/fruit production with, in the case of the first two, high floatability, due to the corky surround of the achene or nutlet. The fruits of *Alisma plantago-aquatica* and *Rumex maritimus*, common in many samples, likewise have properties of floatability which promote their dispersal by water. All these taxa also grow on mud banks, fringes of water bodies or in fens so that their contribution of fruits or seeds to the water body may be fairly direct.

The most common aquatics are *Azolla filiculoides*, *Stratiotes aloides* (leaf spines), *Ranunculus* subg. *Batrachium*, *Scirpus lacustris*, *Typha* and *Zannichellia*. Some of these have propagules or fragments with a high floatability (e.g. *Azolla*, *Stratiotes*, *Typha*, *Zannichellia*), while the others have achenes (*R.* subg. *Batrachium*) or seeds (*Scirpus*) which sink quickly, and their abundance is likely to be related to the abundance of the plant at the place of sedimentation.

These remarks can do no more than illustrate the difficulty of interpreting macroscopic assemblages. Each taxon has a dispersal unit of unique character, whose behaviour in dispersal and sedimentation depends on many biotic and environmental factors.

Size of assemblages

Though the quantitative data cannot be related to sample size or to time, the number of taxa in each assemblage has some relation to the sediment. The highest number is found in sediments of temperate stages, either in freshwater organic sediments with a considerable silt-clay content (Cr I, Sidestrand SSTC, 60 taxa; Cr II, West Runton WRU, 58 taxa) or in a similar sediment showing a marine influence (Pa III, West Runton CNA, 59 taxa; Cr III, Pakefield PD, 62 taxa, PH, 63 taxa). The sedimentary environments are here such that plant remains from a variety of habitats are concentrated. Lower numbers are found in wood peat, presumably reflecting the very local origin of the remains, and, as would be expected, in the highly inorganic tidal sediments, though as mentioned below these contain remains from a wide variety of habitats. In the cold-stage floras the numbers are variable, at their highest (35 taxa) in the e An silty mud of Mundesley ME 2. The variability is well shown by the three samples from the ME channel, indicating that the factors controlling input of plant remains can act very locally.

Assemblages from particular environments

Freshwater. Wood peats show a fairly low number of taxa with a good representation of trees. Presumably the majority of taxa are from very local plant communities, heath in the case of Pa IVa sediments at Beeston BR–S, and alder fen carr indicated by the Cr II assemblage at Corton CA.

Detritus muds with a low inorganic fraction show a high content of the fen and reedswamp helophyte group and the aquatic and water-fern groups (Mundesley MU) and very restricted frequency or absence of other groups. Fringing reedswamp may filter remains from a wider variety of communities.

Where there is evidence for water flow in the presence of a proportion of inorganic sediment and particular faunas (Cr I, Sidestrand SSTC; Cr II, West Runton WRU, Trimingham TB), there is better representation of the dry-ground herb group and the muddy substrates shallow water group, indicating a wider source of macroscopic remains.

The alluvium of the Corton Rootlet Bed shows a flora with a predominant fen and reedswamp helophyte group and surprisingly little indication of distantly derived plant remains.

Apart from the Rootlet Bed, the most inorganic freshwater sediments are those of the cold stages. They usually show an assemblage with a diversity of physiognomic/ecological groups indicating a wide catchment of plant remains, with the sediment deposited in channels (Be, Beeston BA–B; e An, Mundesley ME, Paston PC) or in sheets resembling alluvium (e An, Mundesley MAJ). In the latter case they show a richness in derived megaspores.

Brackish. The organic seams of mud associated with tidal sediments show floras with a maritime component, as well as good representation of freshwater groups (Pa II, Sidestrand SSTB, Happisburgh HC). In these assemblages, much material must have originated by drift.

Marine. The temperate tidal laminated sediments contain a characteristic flora which includes a wide representation of the physiognomic/ecological groups, and a number of taxa with remains with good powers of flotation. These include *Dryopteris*-type sporangia, *Sphagnum* leaves and shoots, *Stratiotes* leaf-spines, Ericales leaves, coal fragments and derived megaspores. The floras thus contain a local element of maritime taxa, as well as much drifted material. In the deeper water marine sands (Happisburgh borehole HC), the remains are much scarcer and must all be drifted.

Representation of the groups in different environments of deposition

Here we consider the frequencies of groups shown in Figs. 15, 28 and 41.

Trees. The frequencies are generally under 10%, except in certain sites at West Runton (CGB), Mundesley (MAR, MAK, MPA), Ostend, Corton (CA) and Pakefield (PH). The highest frequencies are, as would be expected, in wood peat, and they are mostly provided by *Alnus*, *Betula*, *Pinus* and *Picea* remains. But they also occur in silty muds in the narrow channel at Pakefield, in tidal sediments at Mundesley (Cromerian) and Beeston (Pastonian).

Shrubs. These are poorly represented in all the samples, *Salix* (buds and bud-scales) being the most abundant member of the group.

Maritime. This group is best represented, as expected, in the tidal sediments of Pre-Pastonian, Pastonian and Cromerian age, and in silty muds closely associated with such tidal sediments. Very low frequencies also occur in Beestonian and early Anglian sediments with high NAP as members of the cold stage flora and in temperate freshwater sediments.

Heath. The heath group has its highest frequency in the late Pastonian wood peat at Beeston (BR–S), and is also well represented in the immediately underlying tidal sediments. Substantial frequencies seem characteristic of marine or near-marine sediments, probably a result of fluviatile transport of the seeds and leaves of Ericales and of *Eriophorum* spindles, also included in this group.

Montane. This group has a very limited representation, being recorded in the Pre-Pastonian, late Pastonian (Mundesley MPA 2), Beestonian and early Anglian.

Dry-ground herbs. The frequencies of this group are usually below 10%, except at West Runton (WRU) where the higher frequencies are associated with fluviatile deposition, and probably related to partial derivation of the assemblages by bank erosion, as with the vertebrate remains (Stuart, 1975). Frequencies around 10% are also found in the Cr Ib freshwater sediments at Sidestrand (SSTC) where a relatively high frequency of NAP is present. Otherwise low frequencies of this group are found in other freshwater sediments of temperate and cold stages and in tidal sediments.

Damp-ground herbs. This group is poorly represented. It is most frequent (*Selaginella*, *Juncus*) in cold-stage floras of Pre-Pa, Be and e An age, perhaps a reflection of poorly drained soils marginal to the basins of deposition.

Muddy substrates, shallow water. This group shows consistent fairly low frequencies in most of the analyses, but higher frequencies in both cold stage floras (Pre-Pa c, Beeston BR–S; l Be, West Runton WRAQ) and temperate stage floras, both in freshwater sediments (Cr II, Mundesley MAK) and marine sediments (Pa II, West Runton WRF, Paston PG, Happisburgh HC). The highest frequencies are due to the abundance of *Alisma* (*plantago-aquatica*) and *Eleocharis*. In the case of l Be and Cr II freshwater sediments the assemblage clearly results from deposition from very local plant communities, but with the marine sediments indicates successful transport from more distant communities.

Fen and reedswamp helophytes. This group is strongly represented in freshwater sediments of the Cromerian, both in the organic muds of West Runton (l Be, Cr I, II) and Mundesley (MF, MU) and the alluvium of the Rootlet Bed at Corton (CA). Apart from the high frequency (*Osmunda*) in Pa II at Happisburgh (HE), in organic sediment associated with tidal sediments and in tidal sediments the frequencies are much lower, depressed by increased frequencies of aquatics. Low frequencies are also recorded in cold stage sediments: Pre-Pa (Beeston BR–S), Be (Beeston BV) and e An (Trimingham TL, Mundesley ME, MAJ). But in the l Be at West Runton (AQ) a rise in frequency takes place just before the beginning of the Cromerian.

Aquatics. This group is well represented in the majority of samples. The lower frequencies are found in the l Be at West Runton (WRAQ) where the muddy substrate, shallow-water group is prevalent, in tidal sediments of Cr and Pa age, in the alluvium of the Rootlet Bed at Corton (Cr II, CA), of the wood-peat horizons at Beeston (Pa III, BR–S), Ostend (Cr IV, OA), Happisburgh (Pa II, HE) and Corton (Cr II, CA).

Water ferns. High frequencies occur in freshwater sediments of Cromerian age and in both freshwater and tidal sediments of Pastonian age. The megaspores are obviously widely dispersed into the marine environment. They also occur at low frequency in marine sediments of Pre-Pa a age (Mundesley MAT), but it cannot be determined whether they are primary or secondary in such a situation.

Derived megaspores. The numbers of derived pre-Pleistocene megaspores recorded in the macroscopic analyses are shown in the analytical tables given later. Apart from single records, megaspores are most frequent in the tidal marine sediments of Pre-Pa a (Mundesley MAT), Pa II, III (many sites) and less frequently Cr (Cr III, Mundesley MAK, MAR; Cr IV, Mundesley MAR) age. They are much less frequent in temperate freshwater sediments (e.g. Pa I, Beeston BR–S), except in the inorganic sediment of Pa IV at Mundesley MPA 2. On the other hand, in freshwater sediments of the cold stages, they attain higher frequencies, as Mundesley (e An, Mundesley ME, MAJ: Paston PC).

In the marine sediments, the megaspores must be part of the marine drift reworked over a long period of time and sedimenting with the fine-grained inorganic tidal sediments. In the freshwater sediment, they are again associated with inorganic sediments, reworked and redeposited with the fine-grained inorganic fraction. Their presence presumably indicates erosional regimes such as might be expected in periods of regression, in channel formation and in cold climates.

(v) *Zonation*

The pollen spectra are grouped into pollen assemblage biozones, the grouping being based on the frequency of tree genera and of major non-tree taxa. A sequence of biozones is established for each stretch of coast, with each biozone identified by a letter. There is a table of biozones for each stretch of coast, giving the site, location, depth or sample number, pollen characteristics and relation to lithostratigraphy and chronostratigraphy of each p.a.b. distinguished in the pollen diagrams.

Variations resulting from very local developments of vegetational history are apparent in some of the pollen diagrams and local p.a.b.s are described which are subsequently related to biozones of wider application. These local p.a.b.s are given abbreviations prefixed by (l).

The nature of a pollen assemblage in a sediment is dependent on the constitution of the regional and local vegetation and on the action of transporting agencies and processes on the dispersal of pollen and spores (West, 1973). In a sequence of homogeneous sediment, there may be minimum variation in the effects of transporting agencies and processes, so that variation in time of pollen assemblages may be clearly related to vegetational history. But in sequences which show great variation in facies from freshwater to marine, such as those of the pre-glacial sequence, assemblages change with changes of the sedimentary regime. Zonation will itself take no account of these variations in the taphonomy of the assemblages, but the biozonal sequences have to be interpreted with the variations in mind. It is not always possible to distinguish the effects of vegetational change from those of change in the taphonomy in the explanations of the differences between successive pollen assemblages. The variations in sediment and assemblages mean that it is necessary to distinguish biozones based on single samples as well as on successive samples.

The p.a.b.s distinguished are grouped into substages according to the system of division of temperate stages described by Turner & West (1968). The substages (Cr I, II, III, IV, etc.) are either equated to p.a.b.s or they are subdivided into lesser units with a letter suffix, with these units equated to p.a.b.s. A suffix to a substage title (e.g. Cr IIIa) is only used where a particular unit is firmly recognised along the coast.

While the subdivision of the Cromerian temperate stage on these lines is clear, the subdivision of the Pastonian temperate stage is not so easy. Pa I, II and IV are distinguishable, but Pa III is not so well developed as an entity as the substage III of younger temperate stages. Several assemblages have been tentatively placed in Pa III by reason of their showing a constitution intermediate between those of Pa II and Pa IV, with stratigraphy supporting the intermediate position (*see p.* 100). But it is not always clear whether these changes in the late Pastonian are associated with changes of vegetation or of taphonomy of the assemblages.

(vi) *Treatment of zonation and correlation*

In the following accounts of the biozones, after considering the Pre-Pastonian a substage, the basal stratigraphical unit recognised widely along the coast, we shall first deal with West Runton, where the most complete sequence of Cromerian biozones is found. In the pre-Cromerian at West Runton a temperate forest biozone (*Pinus–Quercus–Ulmus–Carpinus* p.a.b.) is identified. It is similar to a pre-Cromerian biozone at Paston–Mundesley, originally giving the name Pastonian to the temperate stage concerned. This biozone is again identified in the Beeston sections, where its detail and stratigraphic context is much clearer. The biozones at Beeston are therefore next described after those at West Runton, so that the relationships and details of the Cromerian and Pastonian sequences can be set out clearly from the first. Thereafter the biozones of each stretch of the coast are described, related to the stratigraphy and correlated with the regional sequence.

At West Runton and Beeston the biozones are given a regional notation related to chronostratigraphical subdivision, but in the other stretches of coast a local notation is first given and a correlation of this with the regional chronostratigraphical sequence is then made. Thus the sections at West Runton and Beeston are the basis for the major biostratigraphical divisions and may be considered 'type' sections in this respect.

Table 42 summarises the biozone sequences for each stretch of coast and shows correlations with the regional sequence. Very local biozones or those of uncertain provenance are omitted from this table.

2: The Pre-Pastonian a substage

The sediments of the Pre-Pastonian cold stage are principally marine, with freshwater horizons found only in the later part of the stage to the north-west of Cromer at the margin of the Crag basin (Fig. 1). The pollen of the stage assemblages is distinguished by considerable NAP frequencies with a restricted variety of AP taxa, principally *Pinus*, *Betula*, *Alnus* and *Picea*.

The most widely distributed of the substages of the Pre-Pastonian stage is substage Pre-Pastonian a. This is the basal stratigraphical unit recognised over most of the study area. The Mundesley borehole (MAT, Fig. 35) and the Happisburgh borehole (HC, Fig. 42) show what is regarded as the characteristic pollen assemblage of Pre-Pa a, with the NAP total showing substantial frequencies of Gramineae and Ericales pollen (*Pinus*–Gramineae–Ericales p.a.b.). Sediments of this substage, thickening from the margin to the centre of the Crag basin, have been identified round the coast from Beeston to Pakefield. The later substages, Pre-Pa b to d, are represented in the more restricted area of Beeston and East Runton.

We will first consider the Happisburgh borehole (HC, Plate 33), since this sequence enables us to place the Pre-Pa sediments in a wider setting, and then proceed to consider the correlation of Pre-Pa a before going on to present details of the environmental history in the Pre-Pa a substage.

(a) **The Happisburgh borehole** (HC)

(i) *Succession, biozones and correlation with the Ludham borehole*

The sediments of the Pre-Pa a substage, showing a *Pinus*–Gramineae–Ericales p.a.b. (biozone d), are underlain by further brackish or freshwater and marine sediments, giving evidence for the sequence of p.a.b.s listed at the base of Table 28.

Preceding the Pre-Pa a sediments are a Gramineae–Ericales p.a.b. (biozone c), a *Pinus*–Gramineae–Ericales p.a.b. (biozone b) and a *Pinus–Picea* p.a.b. (biozone a). Comparing this sequence with that of the Ludham borehole (West, 1961), 11 km to the south, the Pre-Pa a biozone (*Pinus*–Gramineae–Ericales p.a.b.) is similar to the Ludham pollen spectra in substage Lu 4c. The O.D. depths of the Pre-Pa a sediment at HC is −2·8 to −17 m, and of Lu 4c −8·2 m to −9·4 m.

In the Happisburgh borehole, a marked change in the sedimentation regime is indicated below Pre-Pa a by laminated clayey sediments with maritime, freshwater and terrestrial herb-plant remains and very high NAP percentages (bed c, biozone c, Gramineae–Ericales p.a.b.). The NAP is principally Gramineae, Cyperaceae and Compositae Liguliflorae. The ecological grouping and the short list of macroscopic plant remains are given in Fig. 41 and Table 2 respectively. The flora indicates regional grassland communities, with plant remains deposited in shallow water. The presence of fragments of *Mytilus edulis* (see Appendix 2) indicate a marine influence in the environment. At the present time the distribution of *M. edulis* in the Atlantic province is north to Iceland and the west and south-east coasts of Greenland and to the west coast of Novaya Zemlya, so its occurrence with a high NAP assemblage is reasonable. Probably the environment is a tidal one, with opportunity for sedimentation of a variety of plant remains.

Below the gravels of bed b, sands with laminae of silty clay indicate a resumption of deeper water marine conditions, as in bed e of Pre-Pa a. Two biozones are present between the gravels and the Chalk.

The flora of the next of these lower biozones (biozone b, *Pinus*–Gramineae–Ericales p.a.b.) shows again high NAP frequencies. The AP is predominantly *Pinus*, with *Picea*, *Alnus* and *Betula*. The spectra are thus similar to those of Pre-Pa a. The contained macroflora is limited and is listed in Table 2. Samples from 23·8 m and 27·5 m gave only fragmentary plant remains, but a number of species were found at 25·6 m, including a seed of the arctic–montane species *Arabis alpina* (Wilson, 1973). By comparison with Pre-Pa a conditions, it is likely that the flora indicates cold stage conditions. Some amelioration is indicated by the basal biozone of the borehole (biozone a, *Pinus–Picea* p.a.b.), where the AP frequencies are higher and where there is a low representation of thermophilous genera (*Quercus*, *Pterocarya*).

These two biozones are not similar to any of the Lu biozones before Lu 4c. They show too little thermophilous tree pollen (especially *Quercus*, *Tsuga*) to be correlated with Lu 1 (Ludhamian) and too low Ericales frequencies to be comparable to Lu 2 (Thurnian). The basal samples contain higher AP frequencies with thermophilous tree pollen more in evidence. It is possible that these two biozones lie near the boundary between Lu 1 and Lu 2 of the Ludham borehole.

The regression indicated by the laminated clays of bed c may be related to the non-sequence in the Ludham borehole, observed at a similar depth (*c.* −18 m OD, between stages Lu 2 and Lu 3 (Antian)), or to the possible non-sequence between Lu 4b and 4c at *c.* −9·3 m OD. The Gramineae–Ericales p.a.b. of the Happisburgh borehole may then be correlated with part of Lu 2 or Lu 4 cold stages at Ludham or some other cold stage not yet defined.

(ii) *Regional relationships*

The transgression of the Pre-Pa a sediments over the freshwater or tidal Gramineae–Ericales p.a.b. sediments at Happisburgh seems likely to be that which transgressed the Chalk surface in the Cromer–Beeston area, to produce the stone bed and its associated marine fauna. This correlation is supported by pollen analyses of sediments associated with the stone bed in this area. The deposition of the coombe rock at the surface of the Chalk in the West Runton and Beeston area will then be related to the cold period of the Pre-Pa a substage or an earlier cold episode (Lu 2 ?, Lu 4b ?).

If we now turn to the southern part of the study area, we have at Covehithe, in clays similar to those of the Easton Bavents Baventian type site (Funnell & West, 1962) 3 km south of Covehithe, pollen spectra very similar to those of the Easton Bavents clay and of its correlative sediments at Ludham and Aldeby (Norton & Beck, 1972) in substage Lu 4b. At Covehithe the laminated clays underlie Westleton Beds and overlie crag in the cliff section, as shown in Fig. 48. The pollen spectra show the presence of an Ericales–Gramineae p.a.b. and an Ericales p.a.b., with *Pinus*, *Picea*, *Alnus* and *Betula* contributing to a low AP total. These spectra are correlated with Lu 4b.

At Pakefield, 5 km north of Covehithe, clay laminae in sand contain spectra of Lu 4c type with higher *Pinus* frequencies and considerable Ericales frequencies. These are similar to Pre-Pa a assemblages further north and it seems reasonable to suppose that these should be placed in that substage. We may then conclude on present evidence that Pre-Pa a sediments form a recognisable and substantial part of the filling of the Crag basin in the Kessingland–Pakefield area and in the Happisburgh–Besston area. In the former area, the Pre-Pa a sediments along the cliff base incline northwards according to Blake (1890) and the Covehithe section (Fig. 48) shows a similar direction of apparent dip for the more massive clays of Lu 4b. We conclude that Pre-Pa a is later than Lu 4b in the Covehithe to Pakefield area, a conclusion tallying with the superposition of Lu 4c (≡ Pre-Pa a) on Lu 4b in the Ludham borehole. We may note a further similarity of the Ludham and coastal sequence in the presence of a temperate stage (Lu 5 in the Ludham borehole, Pastonian on the coast), with similar pollen assemblages, overlying the Lu 4c/Pre-Pa a assemblages.

(iii) *Pre-Pastonian and Baventian*

At Ludham, Lu 4b and Lu 4c were included in the Baventian stage, and if this grouping holds, then Pre-Pa a may be considered a late part of the Baventian stage and described as Baventian. However, it may be more prudent to retain the name Pre-Pa a, since it is possible that there is a non-sequence between Lu 4b and Lu 4c. In the Ludham borehole, a horizon of flint gravel was observed between sediments of Lu 4b and c at −9·3 m OD and though clayey sediments were present, exceptionally no identifiable pollen was countable from −9·3 m to −13·1 m OD, the depth of the sediment between Lu 4b and c. This may be a result of weathering associated with a non-sequence. The pollen was heavily corroded and scarce compared with much better preservation above the flint gravel in Lu 4b and below −13·1 m OD in Lu 4b.

It seems that further evidence will be required to

prove that Pre-Pa a is part of the Baventian stage, and therefore in the following account the term Baventian will be restricted to Lu 4b, and Lu 4c will be described as Pre-Pa a.

(b) Characterisation of Pre-Pastonian a

As mentioned above, pollen assemblages with high *Pinus* frequencies and a high NAP total showing substantial frequencies of Gramineae and Ericales pollen are grouped into this substage. Within the substage assemblages, there are variations in *Pinus* and Ericales frequencies, which may, as discussed on *pp.* 5ff, be related to facies variation. There is considerable danger here of subjective argument: first making correlations, then giving explanations for differences between correlated biozones. However, stratigraphical considerations provide support and an additional basis for the grouping of the Pre-Pa a biozones. There are a number of facies represented in this substage – thick-bedded laminated tidal clays (e.g. Mundesley MAT), clay laminae in sands, sometimes shelly (e.g. Sidestrand SSB), and silts associated with, or closely above, the stone bed on the Chalk (e.g. Beeston BE). This variation is likely to cause variation in taphonomy of the pollen assemblages, as discussed on *pp.* 5ff. In addition, the presence of erratics of freshwater mud in marine sediments of the substage gives the possibility of differentiating the pollen assemblages resulting from mainly aerial transport from those resulting from water transport.

(c) Distribution of Pre-Pa a sediments

The distribution of pollen assemblage biozones included in substage Pre-Pa a is shown in Table 1. The sediments concerned are all marine, many showing association with *Macoma balthica* faunas and with the stone bed on the Chalk, itself containing a marine fauna, except for pollen spectra from erratics of coarse detritus mud at Sidestrand (SSF2, SSJ) found in the Pre-Pa a sediments.

At the northern end of the crag basin (see Harvey *et al.* (1974) for the contours of the Chalk surface), at Beeston, the clays associated with the stone bed contain the Pre-Pa a pollen spectra. Further south-east, as the crag basin deepens, the spectra occur in the stone bed and in the overlying crag, as at West Runton, Overstrand and Sidestrand. South-east of Overstrand, to Happisburgh, the marine sands thicken as the crag basin further deepens. At Happisburgh the Pre-Pa a sediments reach a thickness of 11 metres. There is then a long gap and Pre-Pa a sediments are again recognised at Pakefield, where they can be related, as mentioned above, to the Lu 4b clays seen southwards in the cliffs at Covehithe and Easton Bavents.

(d) Environmental history

The high NAP content and scarcity of thermophilous tree pollen in the marine assemblages indicate the presence of a park–tundra type of landscape, as discussed by West (1961) in relation to similar pollen spectra in the Ludham borehole. *Pinus*, *Picea*, *Betula* and *Alnus* appear to have been present, but accompanied by high relative herb-pollen production. The high Gramineae assemblages of the mud erratics support this conclusion of a cold-stage type of flora. Sidestrand SSF2 in particular shows some variety of NAP taxa, with Compositae, *Rumex* and Umbelliferae. The same abundance of Gramineae and Compositae pollen is seen in Sidestrand SSJ. These spectra show Ericales in low frequency or absent. It is possible that the high Ericales frequencies of the marine sediments are a result of redeposition through upstream erosion or selective transport under fluvial conditions. The heavy tetrad of the Ericales is more likely to travel far by soil wash and fluvial transport than by aerial means (*see p.* 7).

The existence of cold conditions during substage Pre-Pa a is supported by evidence from foraminifera and molluscs. Funnell (1970) reported that the foraminiferal assemblages from the section at Sidestrand SSB indicated semi-glacial conditions, with the occurrence of arctic species. Norton (1967) reported that the mollusc assemblages from a section next to SSB (his SSK) was a very impoverished one, indicating very shallow tidal conditions (see Appendix 2). The occurrence of mud erratics in such an environment is to be expected. They also occur in later, Pastonian, sediments of similar provenance at Sidestrand (SSV).

Surface textures of sand grains from the Pre-Pa a sediments of Sidestrand SSB show evidence of glacial action followed by estuarine or beach conditions (Krinsley & Funnell, 1965). This is conformable with the evidence for environmental conditions derived from the sediment and its contained fossils, but more importantly, it indicates a period of glaciation before the deposition of the Pre-Pa a sediments.

(e) Notes on pollen spectra assigned to substage Pre-Pa a

As mentioned previously there is considerable variation in the pollen spectra assigned to Pre-Pa a. At Sidestrand the three sites on the Chalk anticline 1

(Fig. 22), SSB, SSD, SSH in the same sediment series (bed b) in the cliff section show considerable variations in frequencies of *Betula*, *Alnus*, Ericales and Gramineae. For example, SSH 5, taken from a horizon of drift mud in sand, shows a partial replacement of *Pinus* by *Alnus*, though the NAP frequency is higher. This section shows much lower Ericales frequencies than SSB or SSD. At Pakefield the spectra show rather lower frequencies of Gramineae pollen. These variations may reflect a real difference in age or variation in the taphonomy. The latter is much more likely, especially in view of the local variation in shallow water sediments seen at Sidestrand.

(f) Macroscopic plant remains

Macroscopic plant remains have been recorded from Pre-Pa a at Beeston, Mundesley and Happisburgh. At Beeston a 60 cm-long log of cf. *Pinus* was found in the stone bed on the foreshore opposite BTA (Fig. 12). At Mundesley MAT, in the massive laminated clays of bed b, together with marine molluscs and foraminifera, the plant remains listed in Table 3 were recorded. The frequency of ecological groups is shown in Fig. 28. The flora is a mixed one with elements of aquatic and reed-swamp floras, open-ground taxa, maritime or salt-marsh taxa, heath taxa and derived spores. Fragments of the thallus of *Sphacelaria* (Wilson, 1973) were found at one level (MAT 890 cm). Apart from one *Betula pendula/pubescens* fruit in the topmost sample (MAT 700 cm), no other remains of trees were recorded.

At Happisburgh a single sample (HC 710 cm) of silty clay yielded only few fragmentary plant remains with a nut of cf. Cyperaceae.

The origin of macroscopic remains in these marine sediments must be diverse. The maritime/marine species are presumably local and contemporary, but others may be delivered as sediment after reworking in fluvial conditions. The paucity of tree remains presumably reflects the herbaceous nature of the vegetation. The few remains of Ericaceae, despite the abundance of Ericales pollen, may indicate that the Ericales pollen is reworked.

(g) Conclusions

The flora and sediments of Pre-Pa a indicate the presence of a marine embayment in the area between Beeston and Covehithe, with beach conglomerate (stone bed) forming at the margin of the embayment in the West Runton–Beeston area. The presence of shallow-water sediments below the Pre-Pa a sediments at Happisburgh suggests a major marine transgression at the beginning of Pre-Pa a times, which resulted in the formation of the stone bed of the West Runton–Beeston area. The exact nature of the vegetation of the time cannot be discerned because of the difficulty of interpreting marine pollen spectra, but it seems likely that it was of a park–tundra type, with an important herbaceous element, together with *Pinus*, *Picea*, *Alnus* and *Betula*, but no certain representation of thermophilous trees. The Pre-Pa a substage is correlated to the Lu 4c substage of the Ludham borehole. It is thought best to regard it for the present as distinct from the Lu 4b substage represented at Ludham, Aldeby and Easton Bavents, restricting the term Baventian to the Lu 4a and b substages, typically developed at Easton Bavents.

3: West Runton

The sections along this stretch of coast (Fig. 2, 4–11) include a long sequence at the cliff base from Wood Hill, 745 m east of Woman Hithe (Figs. 4–8), westward to 1066 m west of Woman Hithe (Figs. 8–11), where till cuts down to the stone bed on the Chalk. Sections were also observed between tide marks.

The sections at West Runton were levelled to the levelled datum point at Woman Hithe (4·90 m OD).

Chalk is exposed at low tide all along this stretch of coast. Mean low water neaps is −0·88 m OD at Cromer (MLWS −2·01 m OD), giving an approximate maximum thickness of pre-glacial Pleistocene deposits in the area of 8 to 9 m.

The stratigraphy of the pre-glacial Pleistocene at West Runton is complex. It is simplest to make first a threefold division, the middle unit of which is constituted by the organic muds of Reid's Upper Freshwater Bed, from now on called the West Runton Freshwater Bed, and correlated organic sediments, a readily identified horizon at about HWM and slightly above (*c.* 3–4 m OD). Above this organic horizon occur clays, silts, sands and gravels of the Cromerian marine transgression, and clays, silts and sands of early Anglian age immediately below the till. Below the organic horizon is a series of clays, silts, sands and gravels of pre-Cromerian age, mainly marine, with only minor organic horizons.

The middle unit, the West Runton Freshwater Bed, will be described first, then the transgression and other sediments above it, and lastly the pre-Cromerian sediments of the area.

Table 8 summarises the environmental history described in this chapter.

(a) The West Runton Freshwater Bed

The West Runton Freshwater Bed, represented mainly by organic muds, richly fossiliferous, is seen in a series of three shallow depressions in the underlying deposits. The central depression, a channel filled with muds and fluviatile sediments, is the deepest and most fossiliferous of the three, with its best exposures east of Woman Hithe (West Runton Gap) (Figs. 5, 6). A shallow mud-filled swale (50 cm mud) occurs further eastward, about 25 m wide and 500 m east of Woman Hithe (Fig. 4); a second mud-filled swale (30 cm mud), about 50 m wide, occurs 245 m west of Woman Hithe (Figs. 10, 11). As will be seen from the figures, the organic muds of these three swales or channels lie at *c.* 1·80–4·20 m OD. The organic level indicates a land surface during the Cromerian temperate stage, with mud accumulation in two shallow wet swales and the central fluviatile channel. Where the underlying sediment rises the organic muds thin out, until above about 4·20 m OD no organic sediments are found, and a palaeosol has been reported from these underlying sediments by Valentine & Dalrymple (1975).

(i) *Central channel: lithostratigraphy*

The freshwater deposits lie in a broad, shallow channel cut in sand and gravel. The channel filling is up to about 2 m deep and at least 400 m wide, and it constitutes the type section of the Cromerian stage.

Fig. 6 shows 25 sections observed to the east of Woman Hithe in the period 1961–8, and Fig. 5 a more recently exposed section at the east end of the channel. The positions of the sections are shown in relation to the datum point at Woman Hithe, and they have been levelled to

OD. The western margin of the channel is a few metres to the west of Woman Hithe (Fig. 7); the eastern margin has not been seen but must lie between the section WRCT (368 m east of Woman Hithe and the exposure of gravels 443 m east of Woman Hithe (Fig. 4).

The section in Fig. 6 covers 222 m of the channel filling, and this is extended some 20 m east in Fig. 5 and 20 m west of Woman Hithe in Fig. 7. The following sequence of beds fills the channel, with the beds labelled by the letters in the figured sections.

a. The basal sediment unit of the channel varies from a light brown, stiff, sandy, calcareous, detritus mud with shells at the east end of the exposure (308–318 m) to a grey marl with shells at the west end (270–300 m). At the section WRAQ (285 m), the grey marl is much darker in the basal 25 cm. The section at WRAQ is described in detail in Appendix 1. This horizon reaches a maximum observed thickness of 90 cm at WRAQ. It has an irregular surface on which the later deposits lie unconformably.

b. This bed, 5–7 cm thick, was only seen in sections WRAR and WRCI, 340–345 m along the section. It is a brown stony mud, more clayey at WRCI, overlying the basal gravels at both sites. Pollen analyses indicate a younger age for this sediment than bed a, though the two are not seen in contact.

c.–e. These three beds form a series of fluviatile shelly sands associated with mud containing brecciated material from bed a.

c. This sediment is mainly a shelly sand, but may be gravelly or streaked with grey silt and clay. It reaches a maximum observed thickness of 75 cm at WRCJ and WRCT at the east end of the section. In this region it overlies bed b, and underlies the brecciated muds of d. In the centre of the section (177–261 m) it is a thin horizon between the brecciated muds and the basal gravel.

d. This bed (Plate 1) is characterised by the presence of brecciated sediments, largely composed of fragments of the marl of bed a, set in a matrix of dark clay-mud. It reaches a maximum observed thickness of 85 cm (at WRCH), and it may overlie beds a, b or c, or rest directly on the basal gravel (WRBZ, WRBY). The sediment is richest in breccia in its basal parts; upwards the content of apparently autochthonous mud increases and the brecciated marl fragments become scarcer.

e. This sediment resembles that of bed c. It overlies the breccia of bed d and is seen in the western part of the section (163–186 m), reaching a maximum observed thickness of about 70 cm between WRCA and WRBZ.

f. The uppermost bed of the channel filling in the major part of its width is a dark-brown, humified, detritus mud with some silt and fine sand, containing wood fragments, and shelly in its lower half. It reaches a maximum observed thickness of 134 cm at WRBR. The shell content and sandiness of the mud vary considerably along the section. The top 10–20 cm of the detritus mud is stained red by iron oxide. In WRCT (Fig. 5), this bed rises, lying on bed c, and thins out to become a brown-red oxidised horizon.

g. At the east of the section (WRCT, Fig. 5), the mud of bed f is overlain by grey and red laminated silt with freshwater shells.

The west end of the channel is seen in Fig. 7. Here its base rises from 3·80 m OD at the datum point to 4·80 m OD at the point (16 m west of datum) where the channel filling is cut out by the basal gravel of the marine transgression. The channel filling here is composed of a lower sand with freshwater shells, from which rootlets penetrate the underlying sediment, and an upper silty mud with leaf impressions. This silty mud was seen to contain sand-tubes, probably formed by marine fauna associated with the transgression gravel overlying it. The shelly sand is probably to be correlated with bed c or e further east, and pollen analysis shows the silty mud is to be correlated with bed f of the main exposure of the West Runton Freshwater Bed.

(ii) *Sedimentary history of the central channel filling*

The first sediment of the channel filling is the marl and mud of bed a. At WRAQ the surface of this horizon reaches 2·90 m OD. A similar height is reached by the younger sediments of bed b at WRAR. Subsequent to the deposition of these two horizons there was evidently a period of erosion and reworking of the basal sediments, marked most obviously by the unconformity at the top of the marl. This erosion is probably associated with the spread of the fluviatile shelly sands and gravels of beds c and e, and the reworking of the marl to form the breccia of bed d. Thus it appears that the erosion was caused by a shifting fluviatile channel. The shelly sands and breccia aggrade to a height varying from 3·60 to 2·60 m OD in the centre of the section, and to 4·50 m OD at the west margin, and their deposition is succeeded conformably by much more organic quiet-water sediments of bed f. At this time the channel currents ceased to flow rapidly and aggradation of detritus muds took place to a level of 3·95 m OD at least in the central part of the channel and to 4·70 m OD at least on the western margin. The present height difference may be caused by compaction. The top of the detritus mud is remarkably level along the section, apart from a drop at the west end (BQ) to 2·0 m OD. This may be a result of later erosion. Where the contact with an overlying sediment is seen, it is a sharp junction, most often with the basal gravel of the Cromerian marine transgression, indicating that the mud was truncated (as at the west margin), by marine transgression. But at the eastern end of the channel (WRCT, Fig. 5) the laminated silt of bed g overlies the mud, indicating a resumption of fluviatile conditions before the transgression.

(iii) *The two swale fillings*

The two swale fillings, one west of Woman Hithe (Fig. 4) and one east (Figs. 10, 11; Plate 2), show a black, very humified sandy organic sediment, with scattered flint pebbles, up to 50 cm thick, and designated Z in the figures of the sections. The base of the organic horizon is at 2·60 m OD in the eastern swale and at 3·60 m OD in the western swale. No variation of the sediment fillings occurs, such as is seen in the central channel, and organic deposition probably resulted from mud or peat sedimentation under very wet soil conditions, governed by the local water table. The organic sediment appears to have been truncated by the transgression gravels as in the central channel. The occurrence of iron-stained sands and gravels at the margins of the swales, possibly parts of soil profiles associated with the Cromerian land surface, is seen well at the eastern margins of both swales (Fig. 4, 525–555 m; Fig. 10, 221–244 m). Valentine & Dalrymple (1975) have investigated possible soil profiles associated with the western swale, and concluded that only one, 228 m west of Woman Hithe, contained evidence that a palaeosol was present.

(b) Post-West Runton Freshwater Bed succession

(i) *Lithostratigraphy*

The sediments above the West Runton Freshwater Bed, reaching a maximum thickness of about 5 m, can be divided into beds as follows.

a. The basal sediment, immediately overlying the West Runton Freshwater Bed, is a flint pebble gravel up to 30 cm thick. West of Woman Hithe (Fig. 7) it contains casts of marine shells, principally *Mya truncata* Linn., found in the position of life, as described by Trimmer (1845). East of Woman Hithe shell fragments were found in the gravel at WRBQ (127 m, Fig. 6), and in a similar gravel 55 cm higher. The basal gravel is clearly distinguishable where it overlies organic sediments, as over the organic swale fillings and central channel filling of the West Runton Freshwater Bed (e.g. east of Woman Hithe at WRBT (519 m, Fig. 4), and WRAJS (278 m, Fig. 6) and west of Woman Hithe at WRBXb (16 m, Fig. 7) and WRBE (272 m, Fig. 11)). It is also distinguishable where it overlies pre-Cromerian sands, as 20–50 m west of Woman Hithe (Fig. 7). Where it overlies pre-West Runton Freshwater Bed gravels it can hardly be separated in the absence of fossils.

b. This bed (Plates 4 and 5) is of grey laminated silts, silty clay and sand, up to 2 m thick. The bedding indicates tidal deposition. Two sediment units are in places apparent within the horizon: a lower pale, more sandy, unit, with thin grey silt streaks, and an upper unit richer in laminated silt (e.g. at 40 m west of Woman Hithe, Fig. 7). These sediments overlie the basal gravel, both east of Woman Hithe (e.g. at WRCT, 355–368 m, Fig. 5; at WRBQ, 127 m, Fig. 6, where a thin shelly gravel is intercalated within them, and west of Woman Hithe along much of the section, except where they directly overlie the West Runton Freshwater Bed at 235–250 m west of Woman Hithe (Figs. 10, 11). Further west, at WRBCa and WRBCb (406–408 m, Fig. 11), a laminated silty clay overlies a weathered horizon with rootlets (Plate 3). The laminated sediments have a flora showing a marine influence and they are placed in this bed. The rootlet horizon is correlated with pre-West Runton Freshwater Bed sediments seen further east.

c. This horizon is of more massive brown-grey, clayey silt, up to 2 m thick, laminated, with thin sand horizons, some current-bedded. At WRAY, WRAYa and WRAYb, 159–163 m west of Woman Hithe (Fig. 10), the silt was divisible into an upper and a lower part by a thin sand with freshwater shell fragments.

A sample from the sand with shells (s, site AYa, Fig. 10), between the sediments of Cr IVa and Cr IVc, was analysed by B. W. Sparks. He recorded a very poor and fragmentary fauna, including a few broken opercula, probably *Bithynia inflata*, some very weathered fragments of *Pisidium amnicum* and possibly also of *Sphaerium* sp., and also of a larger completely unidentifiable bivalve. The fauna, though sparse, is important in indicating a regression between Cr IVa and Cr IVc.

The silt above this sand contained very friable freshwater shells. In places, this silt horizon thickened into channels which evidently cut down well into the tidal sediments below, e.g. at 20 m west of Woman Hithe (Fig. 7) and at 216 m west (Fig. 10).

d. Between the till and the massive silts of bed c west of Woman Hithe are waterlaid current-bedded sands (Fig. 7). They are not much more than a metre and a half thick over 20 m west of Woman Hithe, but at 15 m west they cut down through a channel filling of bed c until just east of Woman Hithe (WRBP, Fig. 7), they rest on the tidal sediments of bed b, and at WRBQ (Fig. 6) they rest on gravel with marine shells. Further east, at WRCT (360 m, Fig. 5) they rest on a metre of tidal sediments, and at 500–555 m (Fig. 4) directly on the gravel of bed a. Thus east of Woman Hithe these sands thicken up to 4 m, with their upper limit, a sharp junction with till, remaining fairly level.

e. The water-laid sediments described above are sealed by till, included by Reid (1882) in his Contorted Drift (see also Banham, 1975). The till usually rests on the sands of bed d, as it does east of Woman Hithe. West of Woman Hithe, the till may also rest on the massive silt of bed c or the tidal silts of bed b, or on pre-West Runton Freshwater Bed gravels. West of 420 m west of Woman Hithe, the till cuts down to the stone bed and the Chalk, which is here seen at the base of the cliff.

(ii) *Periglacial stratigraphy*

Along the West Runton sections both ice-wedge casts and involutions are seen in the sections, as follows.

Ice-wedge casts

1. Penetrating bed b, sealed by bed c:
   272 m west of Woman Hithe (WRBE, Fig. 11).
   216 m west of Woman Hithe (Fig. 10).
2. Penetrating bed b, top not seen:
   16 m west of Woman Hithe (Fig. 7).
3. Penetrating bed b, sealed by bed d:
   368 m east of Woman Hithe (Fig. 5).
4. Penetrating bed a, sealed by bed d:
   485 m east of Woman Hithe (Fig. 4), if the thin gravel above is a wash from the slope of the underlying gravels.
   525 m east of Woman Hithe (Fig. 4).
5. Penetrating horizon d, near junction with till:
   550 m east of Woman Hithe (Fig. 4, Plate 7).

The stratigraphical relations of these casts indicates ice-wedge systems developed at the following times:

a. In an interval after the deposition of sediment of bed b, but before deposition of part of the massive silt of bed c type (1 above).

b. In an interval prior to the deposition of sediments of bed d (3, 4 above). This interval may be the same as interval a above, but is likely to be later since it must post-date erosion of the higher level silts west of Woman Hithe.

c. In an interval towards or at the end of the period of deposition of bed d, probably just prior to the glaciation which deposited the till of bed e (5 above).

Involutions

Disturbed sediments occur at the base of bed c (217 m, 240 m and 256 m west of Woman Hithe, Fig. 10) and at the top of bed c (48 m west of Woman Hithe, Fig. 7). The lower horizon is closely related to the level of ice-wedge casts a above.

(iii) *Sedimentary history*

Later than the deposition of freshwater sediments in the swales and channel of the West Runton Freshwater Bed, marine transgression first resulted in the formation of the thin gravel of bed a, containing a marine fauna. A marine fauna of this and perhaps closely related horizons is listed by Reid (*see* Appendix 2). The presence of *Mya truncata* in the position of life in the gravel indicates a later intertidal situation when *Mya* was able to live in the gravel. This intertidal situation is related to the presence of tidal sediments, bed b, above the gravel, at first formed in a higher energy environment with much sand (lower intertidal?), later with an increase of silty laminations. The tidal sediments reach a maximum height of 7·5 m OD, and are well seen in Plate 6.

A regression followed, with the deposition of the massive silts of bed c, with freshwater shells, indicating a period of alluviation. This was accompanied at first by channelling of the older sediments, and ice-wedge formation occurred before alluviation was complete (group a of ice-wedge casts above). Involutions indicate disturbance of the upper and lower contacts by load effects or freeze-thaw. Further down-cutting occurred before the deposition of the water-lain sands of bed d. No clearly autochthonous marine fossils were seen in this horizon, and a resumption of fluviatile conditions is indicated, with current-bedded sands filling a wide channel, the western margin of which is seen a few metres west of Woman Hithe. The ice-wedge casts below the sand indicate a further permafrost period before its deposition, and after the down-cutting which left beds a and b at the land surface east of Woman Hithe at 368–525 m (Figs. 4 and 5). This is not likely to be the same permafrost horizon as that demonstrated before the deposition of the massive silts of horizon c, although it is possible that the casts seen were indeed from that period and subsequently truncated by the down-cutting. The final period of ice-wedge formation is towards or at the end of the deposition of the sands of bed d, and this period may be closely related to the climatic conditions which subsequently resulted in till formation.

(c) Biostratigraphy of the West Runton Freshwater Bed and post-West Runton Freshwater Bed succession

Table 4 lists the biozones distinguished in the pollen diagrams from West Runton (Figs. 12, 13 and 14). Table 5 lists the macroscopic plant remains. We shall here consider the West Runton Freshwater Bed and post-West Runton Freshwater Bed succession, and later the pre-West Runton Freshwater Bed succession.

The biozone sequence (Table 4) is divisible into two main parts, a lower one with high NAP values, here named late Beestonian (l Be), as it represents the final part of a cold stage not separable from the Beestonian, and an upper one with higher AP values, constituting the Cromerian (Cr) temperate stage. The following p.a.b.s are distinguished:

Cr IV	c	*Betula* p.a.b.
	a	*Pinus–Alnus–Picea–Betula* p.a.b.
Cr III	b	*Abies–Carpinus* p.a.b.
	a	*Carpinus* p.a.b.
Cr II	b	*Quercus–Ulmus–Tilia* p.a.b.
	a	*Pinus–Quercus–Ulmus* p.a.b.
Cr I	b	*Pinus–Ulmus* p.a.b.
	a	*Betula–Pinus* p.a.b.
l Be	b	Gramineae–Cyperaceae–*Betula* p.a.b.
	a	Gramineae–Cyperaceae–*Artemisia* p.a.b.

Before considering the vegetational history, the relation of the late Beestonian and Cromerian biozones to the stratigraphy may be summarised as follows. In the West Runton Freshwater Bed, the marl of bed a at WRAQ contains in its lower part the late Beestonian biozones, and in its upper part Cr Ia and Ib. Cr Ia is also seen spanning bed a at WRCGB and WRCGC. Cr Ib is also seen in stony mud (bed b) at WRAR, in the brecciated sediments (bed d) at the base of WRU and in WRCGB, and in the detritus mud (bed f) at WRT. The first mixed oak forest biozone, Cr IIa, occurs in the brecciated sediments (bed d) at WRU, WRAQ, and WRCGB and in the overlying detritus mud (bed f) at WRU and at WRT. The later mixed oak forest biozone, Cr IIb, occurs in the upper part of the brecciated sediments at WRCGB, and in the overlying detritus mud (bed f) at WRU and WRAR.

The biozones of Cr III and Cr IV occur in the post-West Runton Freshwater Bed sediments. An assemblage (Cr IIIa) with *Carpinus*, *Quercus* and *Ulmus* is seen in transgression sediments at WRBCa (Fig. 11; Plate 3). The *Abies–Carpinus* p.a.b. (Cr IIIb) is seen in laminated silts and sands (bed b) at WRAW, WRAYa, WRAYb, WRAZ and WRBZ. The biozones of Cr IV occur in the laminated silts of bed c at WRAYa and WRAY, with a *Pinus–Alnus–Picea–Betula* p.a.b. (Cr IVa) and a *Betula* p.a.b. This is named Cr IVc, since good evidence of a *Pinus–Betula* zone (Cr IVb) occurs elsewhere on the coast.

The correlation of biozones and stratigraphy shows that freshwater sedimentation started in the central channel in late Beestonian times and continued through to Cr II times. The marine transgression of the post-West Runton Freshwater Bed succession had occurred in or by Cr III times, with the regression taking place in Cr IV times.

The complexity of the fill of the central channel of the

West Runton Freshwater Bed is shown by the variation of sediment in each biozone (see Table 4). The same can be seen by applying the zonation to the seven pollen diagrams through the West Runton Freshwater Bed made by Duigan (1963), with Duigan's West Runton pollen zones equated with the proposed biozones as follows: a1, l Be; a2, Cr Ia, b; b1, Cr IIa, b.

A consideration of Table 4 and Figs. 5–7 shows uniform sedimentation in the central channel in l Be times and in Cr Ia times (bed a). During Cr Ib and IIa sediments included in beds a, b, d and f were deposited, with the breccia containing sediments of bed d being associated principally with Cr Ib and IIa. The final episode of infilling, the detritus mud of bed f, was deposited mainly in Cr IIa (Duigan's E, F and G in the western part of the channel) and in Cr IIb (in WRU and Duigan's C and D).

(d) Vegetational history of the West Runton Freshwater Bed and the post-West Runton Freshwater Bed succession

(i) *Late Beestonian a*

The Gramineae–Cyperaceae–*Artemisia* p.a.b. of this division of the substage l Be is recorded in the marl of bed a at WRAQ (286 m, Fig. 6). The AP frequency is low, principally *Betula* (*B. nana* type also present), with *Salix*, and low frequencies of *Juniperus* and *Pinus*. The NAP is largely composed of Gramineae, Cyperaceae, *Artemisia* and Compositae Liguliflorae. Pollen is also present of Chenopodiaceae, *Plantago maritima*, *P. major/media*, Compositae Tubuliflorae, *Filipendula*, Rosaceae, *Ranunculus*, *Rumex*, *Thalictrum* and Umbelliferae. Aquatics are represented by *Myriophyllum verticillatum*, *Sparganium* and *Typha latifolia*. The pollen assemblage indicates the presence of herb-rich grassland with minor *Betula*, *Salix* and *Juniperus* communities, surrounding a eutrophic pool with marginal reedswamp and aquatics.

Analyses of macroscopic plant remains were made from WRAQ 80–90 cm in this biozone and from 60–80 cm, partly in this biozone. The ecological groups of the floras are shown in Fig. 15. Both samples are dominated by *Eleocharis*, the former with *E. palustris* subsp. *palustris*, the latter by *E. palustris* + *E. uniglumis*, indicating extensive spreads of *Eleocharis* at the margins of the pond, probably relating to the high frequencies of Cyperaceae pollen in the assemblage. Shrubs are represented by fruits of *Betula nana* and budscales of *Salix*. Dry-ground herbs include *Atriplex*, *Chenopodium murale* and *Sonchus oleraceus* (pollen of Compositae Liguliflorae). Damp-ground herbs are represented by *Selaginella selaginoides*, muddy ground by *Alisma plantago-aquatica* as well as the *Eleocharis*, fen and reedswamp by *Carex* cf. *riparia*, *Cicuta virosa* and *Ranunculus sceleratus* and aquatics by *Chara*, *Hippuris vulgaris*, *Potamogeton friesii*, *P. natans*, *Ranunculus* subg. *Batrachium*, *Scirpus lacustris* and *Typha latifolia*. The list bears out and gives specific details to the reconstruction based on pollen.

(ii) *Late Beestonian b*

The Gramineae–Cyperaceae–*Betula* p.a.b. of the later division of the substage l Be is recorded in the marl bed a at WRAQ 35–65 cm. The AP frequency remains low, with its constitution much the same as in l Be a except for rising frequencies of *Pinus* and *Betula* (including *B. nana* type) and decreasing frequencies of *Salix* in the later part of the biozone. The NAP content is also similar, except for far lower frequencies of *Artemisia* pollen. Evidently a change in the constitution of the grassland and a slight expansion of woodland or scrub occurred in this biozone. The following also appear for the first time in the WRAQ sequence: Caryophyllaceae, *Epilobium*, Rubiaceae, *Valeriana officinalis* and *Nuphar*.

The macroscopic analysis from WRAQ 50–60 cm falls into this biozone, and that from WRAQ 35–40 is shared with the succeeding zone (Cr Ia). In both these samples shrubs are represented by *Betula* cf. *nana*; the montane element by *Ranunculus hyperboreus*; dry-ground herbs by *Atriplex* and *Atriplex hastata*, and *Thalictrum minus*; muddy ground by *Alisma plantago-aquatica*, *Eleocharis* spp. including *Eleocharis uniglumis*, and *Ranunculus* cf. *flammula*; fen and reedswamp by *Carex riparia*, *Epilobium parviflorum*, *Lycopus europaeus*, *Ranunculus sceleratus* and *Urtica dioica*; aquatics by *Chara*, *Hydrocharis morsus-ranae*, *Lemna* sp., *Nuphar lutea*, *Ranunculus* subg. *Batrachium*, *Sagittaria sagittifolia*, *Scirpus lacustris*, *Typha angustifolia* and *latifolia* and *Zannichellia palustris*. The aquatic flora is evidently richer (the thermophilous *Hydrocharis*, *Nuphar*) than in the earlier biozone, and the fen and reedswamp element is also more strongly represented. This enrichment and the increasing AP frequencies in the later part of the biozone indicate the effects of climatic amelioration and foreshadow the spread of forest which marks the end of substage l Be and the beginning of the Cromerian temperate stage.

(iii) *Cromerian Ia*

The *Betula–Pinus* p.a.b. of this first subdivision of the Cromerian temperate stage is recorded in the marl of the upper part of bed a at WRAQ and, to the east, in two sections of calcareous mud at WRCGB and WRCGC, resting on sands of pre-West Runton Freshwater Bed age. The AP frequency rises to above 50%. *Betula* and *Pinus* are the predominant tree pollen taxa, with *Betula* reaching its highest frequencies at this time. *Salix* frequencies decline, and *Ulmus* appears in WRCGB, and substantial frequencies of Filicales spores begin to appear. The fall in NAP results from declining frequencies of Gramineae and Cyperaceae pollen. Compositae Liguliflorae are still well represented and a number of other NAP taxa are represented. The aquatics remain much the same as in the previous biozone with increased frequency of *Typha latifolia* pollen.

The assemblage indicates the spread of *Betula–Pinus* woodland, with *Ulmus* beginning to occur. The local pool vegetation pollen appears to have remained similar to that in the previous biozone.

Three macroanalyses come entirely from sediments of this biozone: WRAQ 10–25, WRCGB 25–40, WRCGB 0–20. One, WRAQ 30–40, overlaps with the previous l Be b biozone and has already been considered, and another WRAQ 5–20 overlaps with Cr Ib and will be considered with that subdivision. The WRAQ sample shows aquatics to be most strongly represented and WRCGB both aquatics and fen/reedswamp helophytes. Presumably the WRCGB site was nearer the margin of the water body, also indicated by the section in Fig. 6. WRCGB also shows a higher representation of trees, containing numerous cone scales of *Betula*, *B. pendula* and *B. pubescens*, with many fruits and cone scales of *B. pendula/pubescens*. WRAQ contains a single fruit ascribed to the last taxon, and also a single seed of the now-maritime shrub *Frankenia laevis*. At both WRAQ and WRCGB the dry-ground herbs are represented only by *Atriplex hastata/patula*, *Atriplex patula*, *Chenopodium glaucum*, *Hypericum humifusum*, *Sagina* and *Sonchus oleraceus*; damp-ground herbs are sparsely represented by *Scirpus setaceus*, *Selaginella selaginoides* and *Senecio aquaticus*, and muddy ground by *Alisma plantago-aquatica* and *Eleocharis*, including *E. palustris* subsp. *palustris* and *E. uniglumis*. Fen and reedswamp helophytes include *Cyperus longus*, *Hypericum tetrapterum*, *Lycopus europaeus*, *Ranunculus lingua*, *R. sceleratus*, *Urtica dioica* and *Viola palustris*. Aquatics include *Chara*, very abundant in the marl of WRAQ, *Hippuris vulgaris*, *Myriophyllum spicatum/verticillatum* (*M. verticillatum* pollen recorded), *Nuphar lutea*, *Potamogeton friesii*, *P. natans*, *P. pectinatus*, *Ranunculus* subg. *Batrachium*, *Sagittaria sagittifolia*, *Scirpus palustris*, *Typha angustifolia* and *T. latifolia*, and *Zannichellia palustris*. The unclassified group includes *Cerastium*, *Cirsium* spp., Gramineae, *Mentha arvensis/aquatica*, *Rumex*, *Saxifraga*, *Solanum* and *Valeriana*.

The macroanalyses indicate the development of an extensive pool and marginal flora, with the reduction of the muddy substrate flora in WRCGB and the local appearance of *Betula pendula* and *B. pubescens*. Dry-ground herbs are sparsely represented, indicating pool conditions with slow water flow and marginal reedswamp and fen reducing inwash of terrestrial plant remains, apart from *Betula*.

(iv) *Cromerian Ib*

The *Pinus–Ulmus* p.a.b. of this subdivision is recorded in a variety of sediments: in marl of bed a at WRAQ, in stony mud of bed b at WRAR, in the breccia-containing mud of bed d at WRU and WRCGB, and in the detritus mud of bed f at WRT. In this biozone *Pinus* reaches its highest values, with *Betula* frequencies falling. This change is accompanied by a continuous pollen curve for *Ulmus*, shown with high values in WRAR, WRT and WRAQ, but with lower values in WRCGB. *Picea* and *Alnus* are both represented by low frequencies. The NAP is again predominantly Gramineae and Cyperaceae, with much lower values of the latter taxon. Compositae Liguliflorae continues frequent, and WRT shows a high frequency of Caryophyllaceae pollen. Aquatic pollen content in WRAQ, WRU and WRCGB is similar to the preceding biozone. *Azolla filiculoides* occurs for the first time in WRU. The analyses from WRT and WRAR show a reduced variety of aquatic pollen, possibly related to their topographically higher situation. Filicales spores reach their peak in this biozone.

The assemblage indicates the development to *Pinus–Ulmus–Betula* woodland, with the beginning of spread of *Picea* and *Alnus*. The Filicales may be associated with the spread of carr communities. *Alnus* fruits are present, though the pollen frequency has yet to rise substantially.

The macroanalysis from WRU 130–140 cm falls in this biozone, with that from WRAQ 5–20 cm overlapping with the previous biozone and that from WRU 120–130 cm overlapping with the succeeding biozone of Cr IIb. The WRAQ 5–20 list is very short, showing only single examples of cf. *Alisma* and Gramineae, several achenes of *Ranunculus sceleratus* and abundant nuts of *Urtica dioica*, giving a predominant representation of the fen/reedswamp helophyte group. This may signify increasing closeness of the pool margin and shallowing water. The WRU 130–140 cm and WRU 120–130 samples also show a predominance of the same group but they have a strong representation of dry-ground herbs and aquatics. Of the trees *Alnus glutinosa* fruits are present (though the pollen percentages are not high), and *Betula pendula/pubescens* fruits are also recorded. Shrubs are represented by *Rubus* cf. *idaeus* (from fen carr?) and *Salix*, maritime herbs by *Euphorbia paralias*, dry-ground herbs by *Atriplex*, *A. hastata* and *A. hastata/patula*, *A. patula*, *Cerastium holosteoides*, cf. *Chenopodium album*, *C. glaucum*, *Pastinaca sativa*, *Plantago major*, *Polygonum lapathifolium* ssp. *nodosum*, *Rumex acetosella*, *Sagina*, *S. ciliata*, *S.* cf. *procumbens*, and *Viola arvensis*. This extensive list indicates inwash from areas of open ground or herb vegetation in the neighbourhood. *Barbarea vulgaris/stricta*, *Conium maculatum* and *Ranunculus repens* represent a damp-ground element and *Alisma plantago-aquatica*, *Eleocharis*, *Rorippa nasturtium-aquaticum* and *Rumex maritimus* the muddy substrates element. The fen and reedswamp flora is restricted, with *Lythrum salicaria* and very abundant *Ranunculus sceleratus* and *Urtica dioica*, as in WRAQ 5–20 cm. The aquatic flora now includes *Callitriche*, *Ceratophyllum demersum*, *Potamogeton lucens*, *Sparganium angustifolia* as well as many taxa already present in the previous biozone.

(v) *Cromerian IIa*

The *Pinus–Quercus–Ulmus* p.a.b. of this subdivision is recorded in the breccia-containing mud of bed d at WRU, WRAQ and WRCGB, and in the detritus mud of bed f at WRU and WRT. In this biozone the *Quercus* pollen curve rises to levels similar to the *Ulmus* curve. The *Pinus* frequencies are falling, *Picea* forms a continuous curve and the *Alnus* curve rises. The NAP falls to lower levels, with Gramineae predominant and low levels of Cyperaceae. Compositae Liguliflorae are still frequent and there are records for Chenopodiaceae, Caryophyllaceae, Compositae Tubuliflorae, *Bidens*, *Centaurea nigra*, *Heracleum*, *Matricaria*-type, *Filipendula*, *Geranium*, *Plantago lanceolata*, *P. major/media*, *Polygonum aviculare*, *P. persicaria*-type, *Ranunculus*, Rubiaceae, *Rumex*, *Thalictrum*, Umbelliferae and *Valeriana officinalis*. The aquatic pollen is principally *Sparganium*. Frequencies of Filicales spores fall after their maximum in the previous biozone.

The assemblage indicates the diversification of the woodland with thermophilous genera becoming more widespread, with *Picea* and *Alnus* also becoming more important. The NAP variety indicates the persistence of local open habitats.

A macroanalysis from WRAQ 0–+15 cm falls in this biozone. The WRAQ 5–20 cm analysis overlaps with the previous biozone and has already been discussed. The analysis from WRU 110–120 cm overlaps with the biozone of Cr IIb. The WRAQ 5–20 cm analysis shows 34 taxa, with aquatics and fen/reedswamp helophytes predominant. There is a smaller representation of trees, dry-ground herbs and herbs of muddy substrates. The trees are *Alnus glutinosa* and *Betula pendula/pubescens*. One fruit of *B.* cf. *nana* is recorded, though *B. nana*-type pollen is not recorded later than l Be b. The dry-ground herbs include *Atriplex hastata/patula*, *Chenopodium rubrum/botryodes* and *Polygonum lapathifolium* ssp. *nodosum*. The representation of dry-ground herbs, herbs of muddy substrates, fen and reedswamp helophytes, and aquatics is much as in the preceding biozone at WRU, but very much contrasting with the very restricted flora from the underlying sediment at WRAQ, indicating the change in conditions with coarse organic sedimentation (bed f) from the earlier marl deposition. At WRU deposition of breccia (bed d) ceased and mud was deposited in this biozone. As this occurred, representation of dry-ground herbs and fen/reedswamp helophytes fell while aquatics increased. A slowly rising water level following the erosion of the marl of bed a may have been the cause of these changes.

(vi) *Cromerian IIb*

The *Quercus–Ulmus–Tilia* p.a.b. of Cr IIb is recorded in the breccia-containing mud of bed d at WRCGB 140–110 cm, and in the detritus mud of bed f at WRU, WRAR and WRBXb. In this biozone occurred the deposition of the detritus mud which forms the bulk of the West Runton Freshwater Bed. The AP reaches high frequencies. The representation of the thermophilous trees *Ulmus*, *Quercus* and *Tilia* reaches a maximum in this biozone, together with *Corylus*, and there are lower frequencies of *Pinus*. *Alnus* remains very strongly represented throughout, and *Fraxinus* is sporadically present. At WRU, WRCGB and WRAR there are low frequencies of Gramineae and Cyperaceae with the scattered occurrence of a variety of taxa, including herb vegetation indicators Chenopodiaceae, *Plantago maritima*, *P. major/media*, *Artemisia*, and *Helianthemum*. *Sparganium* pollen and Filicales spores form continuous curves. The NAP at WRBXb, at the western margin of the central channel, is rather different. There is a higher frequency of Chenopodiaceae pollen and of Ericales types pollen, changes probably associated with the beginning of the marine transgression which occurs near the beginning of Cr III. A macro analysis from the same horizon gave a short list including *Azolla filiculoides*, a seed of *Erica* cf. *tetralix*, an embryo of cf. *Chenopodium* and seeds of *Typha latifolia*.

The pollen assemblage of Cr IIb marks a period of mixed oak forest dominance, with *Alnus* and *Pinus*. Open herbaceous vegetation still persisted in the area as shown by the NAP content.

Twelve macroanalyses were made from Cr IIb sediments at WRU, together with the one from WRBXb considered above. A few isolated records have been made from finds in bed f. They include *Ilex aquifolium* and *Prunus domestica*. Fig. 15 shows the frequencies of the ecological groups at the section WRU. The major groups throughout are the fen and reedswamp helophytes and aquatics. Trees (*Alnus glutinosa*, *Betula pendula*, *Betula pendula/pubescens*) are best represented near the base of the biozone. The high frequencies of *Alnus* probably derive from local fen carr, as may the remains of *Betula pendula/pubescens*, and substantiate the pollen record. They are accompanied by records of *Rosa*, *Rubus idaeus* and *Salix*. Dry-ground herbs remain poorly represented throughout the section, except for a rise at 30–50 cm, at the time of minimum macro totals recorded, indicating less of local macro productivity. Possibly the *Atriplex* remains (*A. patula* recorded) constitute a type of 'long-distance' component. Guppy (1906) recorded good buoyancy of *A. patula*. Other dry-ground taxa recorded are *Ballota nigra*, *Calamintha*, Chenopodiaceae, *Chenopodium album*, *C. glaucum*, *C. rubrum/botryodes*, *Hypericum humifusum*, *Peplis portula*, *Polygonum lapathifolium*, *Rumex acetosella*, *Silene dioica*, *S. nigrum* and *Urtica urens*. This list indicates herb communities on both base-rich and base-poor soils in the area.

The damp-ground herbs include *Ajuga reptans*, *Barbarea vulgaris*, *Carduus* cf. *acanthoides*, *Juncus compressus*, *Ranunculus acris*, *R. repens*, *Solanum dulcamara* and *Stachys sylvatica*, associated with either woodland or herb vegetation. In the muddy substrate group *Alisma plantago-aquatica* is present throughout the WRU section, with *Rumex maritimus*, and *Eleocharis* is less frequent. Of the fen and reedswamp helophytes, *Lycopus europaeus*, *Ranunculus sceleratus* and *Urtica dioica* are the most abundant, with less frequent *Apium*, *Bidens*, *Carex* cf. *aquatilis*, *C. dioica*, *C. riparia*, *Cicuta virosa*, *Cladium mariscus*, *Cyperus longus* and *Eupatorium cannabinum*.

Of the aquatics the most abundantly represented or persistent are *Chara*, *Hydrocharis morsus-ranae*, *Nuphar lutea*, *Potamogeton friesii foliosus*, *P. natans*, *Ranunculus* subg. *Batrachium*, *Sagittaria sagittifolia*, *Scirpus lacustris*, *Stratiotes aloides*, *Typha latifolia*, *Zannichellia palustris*. *Azolla filiculoides* and *Equisetum* should also be alloted to this category. Also present are *Callitriche platycarpa*, *C. stagnalis*, *Ceratophyllum demersum*, *Lemna trisulca*, *Najas minor*, *Nymphaea*, *Potamogeton alpinum*, *P. berchtoldii*, *P. densus*, *P. friesii*, *P. lucens*, *P.* cf. *panormitanus*, *P. pectinatus*, *P.* cf. *perfoliatus*, *P. trichoides*, *Scirpus lacustris/tabernaemontani*, *Sparganium angustifolia*, *S. emersum* and *Typha angustifolia*. A rich aquatic and reedswamp flora evidently persisted throughout Cr IIb times at the WRU section, with minor fluctuations in the relation between the two elements, such as that at 30–40 cm, when the aquatic element is reduced. The aquatics and *Azolla filiculoides* reach their maximum at the top of the section, perhaps related to rising water levels as marine transgression approached. The Ericales types pollen and *Erica* remains from WRBXb indicate presence of heath, perhaps associated with acidic communities developing in the hydrosere from fen, as in the Norfolk Broads in the Flandrian.

(vii) *Cromerian IIIa*

The *Carpinus* p.a.b. of Cr IIIa is recorded in a single sample from WRBCa, in the laminated silt (bed b) of the Cromerian transgression. The pollen spectrum shows 60% AP, with the *Ulmus*, *Quercus*, *Carpinus*, *Corylus* and *Alnus*. The NAP is principally Gramineae and Chenopodiaceae (12%); *Armeria* and *Plantago maritima* are also present. The assemblage indicates the prevalence of mixed oak forest with *Carpinus*, but *Tilia* is not recorded, as it is further south-east (Corton, Pakefield) in Cr IIIa, where biozones recording this substage are more extensive.

(viii) *Cromerian IIIb*

The *Abies–Carpinus* p.a.b. is recorded in the tidal sediments of bed b of the post-West Runton Freshwater Bed succession, at WRAW, WRAYa, WRAYb, WRAZ and WRBZb. At WRBQ a sample from the base of bed b shows only low frequencies of *Abies* and *Carpinus*, with much higher frequencies of *Quercus* and *Ulmus*, but is tentatively included in Cr IIIb. The other spectra in this biozone show higher frequencies of *Abies* and *Carpinus* pollen, and lower frequencies of pollen of *Ulmus*, *Quercus* and *Tilia* than are seen in Cr IIb. This change is of the same kind that is seen in the later part of the Hoxnian temperate stage (Ho III). Mixed oak forest is reduced, with the expansion of *Abies*.

The NAP shows very low values throughout Cr IIIb. There is an increase in pollen of *Calluna*, *Empetrum* and Ericales, with Gramineae and Chenopodiaceae pollen and Filicales spores well represented. There is little other NAP.

The biozone is divisible into two parts. WRAYb, in tidal silts just above the beach gravel of bed a, shows higher frequencies of *Carpinus* than Cr IIIb sediments at WRAW, WRAYa, WRBZb and WRAZ. At WRAW and WRAYa the sediments are well above the beach gravel, but at WRBZb and WRAZ they lie close to it. There are clearly considerable unconformities in the post-beach gravel tidal sediments.

(ix) *Cromerian IVa*

A *Pinus–Alnus–Picea–Betula* p.a.b. is recorded in analyses from sediments of bed c in WRAYa and WRAY. *Abies*, *Carpinus* and pollen of mixed oak forest trees fall to low frequencies. The NAP remains similar to the previous biozone, but Ericales taxa are not represented. The spectrum from WRAY 95 cm is present within a sequence ascribed to Cr IVc. Whether this abrupt change is caused by facies change (freshwater to tidal?) or not is impossible to say.

The Cr IVa assemblage indicates that the trend of revertence to coniferous forest, started in Cr IIIb, continued in this substage, with the woodland characterised by *Pinus*, *Picea*, *Alnus* and *Betula*.

(x) *Cromerian IVc*

The *Betula* p.a.b. is recorded in the upper part of the massive silts with freshwater shells of bed c at WRAY and WRAYa. AP frequencies remain high with *Betula* pollen frequencies reaching 60%, and corresponding reductions of pollen of *Pinus*, *Picea* and *Alnus*. The NAP is principally Gramineae, with rather higher frequencies of Cyperaceae pollen than in the previous biozone. *Ranunculus* pollen is present consistently, probably derived from the aquatic flora. By this time, *Betula* woodland had evidently replaced coniferous forest of the earlier part of Cr IV, a change which accompanied regression of the Cromerian sea.

(xi) *General*

The floras of the West Runton Freshwater Bed and the later tidal sediments give a clear idea of the principal vegetation types which followed the amelioration of the climate at the beginning of the Cromerian temperate stage. The floras are principally derived from woodland, fen and fen carr, reedswamp and aquatic communities. The presence of floating leaf aquatics (*Nuphar lutea*, *Potamogeton natans*), with *Scirpus lacustris*, and the presence of an inorganic fraction in the sediment indicate a rather slowly flowing, silting, fluviatile environment, with a meadering mainstream responsible for the irregular deposition of fluviatile shelly sands. This river may have been quite close to a coastal environment, since there are records of *Plantago maritima* pollen in Cr Ib, IIa and IIb, with seeds of *Frankenia laevis* and *Euphorbia paralias* in Cr Ia and IIa/b. There is also some evidence for the local development of acidic communities in the presence of *Calluna*, *Empetrum* and Ericales pollen and *Erica* macroscopic remains (*E.* cf. *tetralix*) at site WRBXb in Cr IIb, with *Viola* cf. *lactea* in Cr Ia and *Carex pauciflora* in Cr IIa. As mentioned above acidification would be a natural development in the local development of the hydrosere, as in present-day Norfolk. The Ericales pollen taxa increase their representation in the tidal sediments of Cr III, a change which may be associated with taphonomic differences between the freshwater and tidal sediments. This type of change is observed at many other sites along the coast and is discussed on p. 5.

(e) Land and freshwater Mollusca of the West Runton Freshwater Bed, by B. W. Sparks

The Freshwater Bed at West Runton has been well known for the quantity of its non-marine Mollusca for a century or more. Indeed, apart from the problematic deposit at Little Oakley, Essex, which has never been properly dated (Warren, 1940), it is the only mollusc-rich British Cromerian deposit. As might be expected from its age, it contains several species which either seem to be confined to this horizon or which are now extinct though occurring in other interglacials: examples of the two types are *Valvata goldfussiana* and *Pisidium clessini* respectively. A number of species were described by Sandberger (1880) as new, though not all of them have stood the test of time.

During the present work non-marine Mollusca were extracted from the same samples as the plant macrofossils at WRU and WRBZ in Fig. 6 and at WRBZb in Fig. 7. The WRU series includes Late Beestonian to Cr II biozones and the WRBZ and WRBZb samples are in the sandy facies of the channel filling, Cromerian II in age. The analyses of non-marine Mollusca are listed in Table 6 (freshwater Mollusca) and Table 7 (land Mollusca): nomenclature is according to Waldén (1976) and Kerney (1976).

Unfortunately they do not range as widely through the bed as plant macrofossils which occur virtually throughout section WRU. Mollusca are absent from the top 70 cm of WRU. This is much more likely to be a weathering phenomenon than an original absence of shells. The upper 70 cm of WRU are characterised also by an absence of small vertebrate debris. In place of the Mollusca and vertebrate remains is an abundance of tiny gypsum crystals so perfect that they must have been formed *in situ* and cannot be detrital. Many occur as beautiful rosettes of aggregated crystals. It seems likely that these owe their origin to chemotrophic bacteria, some of which derive their energy from the oxidation of sulphur and produce sulphuric acid. This acid may well have attacked the original calcium carbonate of the mollusc shells and so caused the formation of gypsum.

The preservation of the shells is also far from perfect in the lower horizons and chemical corrosion and mechanical shatter make it very difficult to be sure of some of the identifications. This affects primarily the large shells especially *Unio*, *Anodonta* and *Viviparus*, but it also means that some juvenile *Lymnaea palustris* may be listed as *L. truncatula* and vice versa. It makes the differentiation of young *Gyraulus laevis* and *albus* difficult when the diagnostic spiral striae may have been corroded from the latter, though the young of *albus* do seem to have slightly more laterally compressed whorls than the young of *laevis*. The Succineidae are extremely difficult and one can only say that they are *putris/pfeifferi* in form and not *oblonga/arenaria*. Again, all the *Vallonia* may not be *pulchella* but there is no positive evidence for any more species than those indicated. The division between *Limax* and *Deroceras* is based on size and, if one believes that the shells of the latter are specifically identifiable, they are mostly *agrestis* type with some *reticulatus* and a few *laevis*. Finally, all the *Bithynia* shells are *inflata*, though some of the opercula are bluntly angulated above and not as oval as *leachi*. Earlier (Duigan, 1963) I had recorded *B. tentaculata* based on opercula alone, but I now believe that only *inflata* is present.

Reference to the pollen diagram (Fig. 13) indicates that the levels containing Mollusca in WRU range from zone Cr Ib to Cr IIb, while in WRAQ the only appreciable number of Mollusca is at the top in Cr IIa, though small numbers are found in the upper part of the Beestonian late glacial. However, apart from the lower levels of WRAQ, where the smaller numbers of Mollusca might be due to less favourable conditions as none of the species is inconsistent with such conditions, the general fauna must be described as fully interglacial. Indeed, on the evidence of the Mollusca alone, one could not assert with any confidence that the small numbers in the lower levels of WRAQ were due to less genial climatic conditions.

This general uniformity also prevails to a considerable degree when one analyses the relative percentages of land and freshwater Mollusca and the balance of the crude freshwater ecological groups I have used before (Sparks, 1964). In addition, for section WRU I have put the ratio of the totals of two groups of *Pisidia*: the total of the moving water group, *amnicum*, *clessini*, *henslowanum*, *supinum* and *moitessierianum* is divided by the total of the catholic group, *subtruncatum*, *milium* and *nitidum*.

	Freshwater ecological groups (%)				*Pisidium* ratio	Land Mollusca (%)
	1	2	3	4		
WRU						
140–130	3	25	4	65	2·0	31·6
130–120	2½	25	4	66	3·3	23·8
120–110	4	37	5	53	1·9	11·1
110–100	6½	42	8½	42	1·3	1·0
100–90	3½	14½	13	69	7·3	4·3
90–80	0	12	20	68	6·0	0·9
WRAQ 0– +15	3	36	6	45	—	7·9
WRBZ	2	16	2	79	—	5·6
WRBZb	6	2	0	92	—	0

The combination of groups 2 (catholic species tolerant of a variety of habitats except the worst) and 4 (moving water species), which together form always over 80% and often over 90% of the freshwater total, clearly shows that this was a good molluscan environment. The presence of large shells, *Viviparus*, *Unio* and *Anodonta* confirms this. In this connection the low total of the slum group 1 is also significant. These freshwater Mollusca clearly point to a moving body of water. This might have been either a permanent pond or mere or a slow well-vegetated stream. The marked decline of land species in WRU above 130–120 probably indicates an increase in size of the body of water rather than a change in the nature of the body of water as the general composition of the freshwater groups remains the same. Again in the lower levels of WRU the land Mollusca are all those that one would expect in somewhat marshy conditions at the edge of the water, and, as far as one can tell, their general nature does not change upwards where the proportion of the total decreases. The same comments apply to the top of WRAQ and WRBZ.

Although numbers are small the Mollusca in WRAQ seem to indicate a general expansion of the freshwater environment at about the same stage as in WRU, i.e. about the junction of Cr Ib and Cr IIa.

The odd section out is WRBZb. This obviously represents much more active stream conditions. A number of facts point to this: the sandy nature of the sediment (which is also probably responsible for the Mollusca being some of the worst preserved I can remember seeing); the absolute dominance of group 4 of the freshwater Mollusca; the presence of a number of heavy thickened species, in particular *Valvata goldfussiana* and *Sphaerium bulleni*. Ellis (1962) regards the latter as a thickened form of *S. corneum*, but it is a striking and very heavy form worth recording separately. *V. goldfussiana* was first described as *woodwardi* by Kennard from a bed just east of West Runton gap – the present bed is to the west of the gap.

Generally speaking the non-marine Mollusca point to a reasonable sized and clean body of water which expanded in the milder parts of the interglacial, Cr IIa. Unfortunately the record at higher levels has been destroyed by weathering in the main section, WRU.

(f) Pre-West Runton Freshwater Bed succession: west of Woman Hithe

(i) *Lithostratigraphy*

The pre-West Runton Freshwater Bed stratigraphy is far more complex and less well exposed than that of the younger deposits. The lower part of the sequence is below HWM and usually hidden by beach. It is also much more variable laterally than the younger sequences. The better exposures have been to the west of Woman Hithe, where the Chalk slowly rises towards HWM as we get nearer Sheringham, so that at the west end of the section (Fig. 11) Chalk is seen at the base of the cliff. This part of the section will be described first, then the part to the east of Woman Hithe. Plate 4 shows a characteristic section. The following beds can be distinguished west of Woman Hithe:

a. The surface Chalk at the west end of the section is disturbed and brecciated (WRBN, WRBM; 705, 1066 m, Fig. 11), so that it strongly resembles the coombe rock of southern England (Plate 11).

b. At the west end of the section and at low tides further east, a stone bed up to 30 cm thick is seen lying on the Chalk (see Reid, 1890, p. 139–140). It is composed mainly of flint pebbles and nodules up to 20 cm across, embedded in an iron-cemented sandy matrix, which contains marine shells and shell fragments, and it may be sealed by grey silty clay. An elephant tooth was found cemented on the surface of the bed at WRBN (705 m, Fig. 11). At one site, WRBM (1066 m, Fig. 11), 20 cm of sand and grey silty clay intervened between the stone bed and the Chalk (Plate 10).

c. A sand, with abundant marine shells (principally *Macoma balthica* (Linné)) in its lower part overlies the stone bed at WRBG (37 m, Fig. 7) and WRV (417 m, Fig. 11). A similar sand with shells occurs at the base of the cliff at 80–100 m west of Woman Hithe (Fig. 9), WRCQb (148 m, Fig. 10), WRAS and WRAT (190–205 m, Fig. 10) and WRBCa and b (405–409 m, Fig. 11). In all these exposures the sediment is a pale or red sand with grey or brown silt laminations and lenses, reaching a maximum thickness of 70 cm. It is the Weybourne Crag of Reid (1890, p. 140).

di. A blue-grey silty clay with pale, fine, sand laminations rests unconformably on the sand of bed c, as seen at WRCU at Woman Hithe (fig. 8) and at 60–80 m west of Woman Hithe (Fig. 9). At WRCU (Fig. 8) the silty clay interdigitates with the shelly sand. At 76 m this silty clay is cut out by the overlying sands of bed f. The sediments of this horizon show a tidal structure, with thin wavy laminations and the contained fossils support this interpretation. They reach a maximum observed thickness of 1·70 m at AX, 36 m west of Woman Hithe (Fig. 7). Their high silt content and thin intercalations of cross-bedded sand indicate a high tidal regime. The lack of bioturbation may indicate low salinity and/or rapid accumulation. At Woman Hithe (Fig. 8), the clays are interrupted by shelly sands.

e. At the west end of the section an iron-cemented gravel overlies the sand of bed c, as at WRBCa and b (405–409 m, Fig. 11). It reaches a maximum thickness of *c.* 1 m in this part of the section. Further east (215–245 m, Fig. 10) this indurated gravel has an uneven surface, showing an unconformity with overlying current-bedded sands. At 147 m

(Fig. 10) the gravel may be represented at WRCQb, and at 100 m (WRBF) the gravel in a narrow channel overlying bed c may be of the same age. At WRBG (37 m, Fig. 7) an iron-cemented gravel lies on the silty clay of bed d i. R. W. Hey reports that the constitution of the gravels of WRCQb, WRBF and WRBG is similar. If these all belong to the same gravel horizon, the presence of clay pebbles in the channels in bed c at 95 and 100 m (Fig. 9) will have resulted from the reworking of the silty clay of d i. Fig. 9 shows that by 1973 the 95 m channel section with clay conglomerate had receded to show grey-brown laminated silt in the same stratigraphic position as the silty clay of bed d i. The correlation of the gravels suggested above is supported by these observations, with deposition of reworked pebbles of the silty clay of bed d i together with the gravels of bed e.

It is possible that the gravels at WRBG, WRBF and WRCQb and those at the west end of the section (215–245 m, 272–417 m (WRV)) are of the same age since at WRBF, WRCQb, WRBCa and b, and WRV they overlie shelly sands of bed c. But the indurated gravels at the west end of the section may well be older.

f. This horizon of red cross-bedded sands with strings of pebbles is clearly distinguishable for extensive lengths of the section both east and west of Woman Hithe. The sets have a consistent apparent dip towards the west, closely parallel to the present coastline (R. W. Hey, personal communication). Between them and at the base of the horizon, there may be thin gravels. Fragments of marine shells were found in the sands at 40 m west of Woman Hithe (Fig. 7). The maximum observed thickness is about 2 m.

West of the ice-wedge cast, WRBD, 31 m west of the datum, these sands overlie unconformably the sands of bed c, the silty clay of bed d i and the gravels of bed e. East of the ice-wedge cast, towards the datum point, there are again sands and gravels but much less well exposed. The cross-bedding, seen further west, was not apparent. The sediments are at a lower level; they are more gravelly and they contain worn fragments of *Mya* and *Arctica*. As they lie in exactly the same stratigraphical position as the cross-bedded sands further west, they are taken to be a lateral variation of this horizon.

At the west end of the section pale and red sands with occasional silt partings occur at the same level as the cross-bedded sands, and above the cemented gravel of bed e. They may also be a lateral variation of this horizon, but in the absence of continuous sections, this correlation remains uncertain.

g. Near or at the top of the cross-bedded sands of bed f occur thin silt and clay horizons, often disturbed, together with small shallow basins of silty clay (e.g. 40 m, 54 m, Fig. 7; 56 m, 74 m, 94 m, 107 m, Fig. 9; 134 m, 148 m, 216 m, 222 m, Fig. 10).

h. Above the cross-bedded sands of bed f at 32–40 m west of the datum (Fig. 7) a pale, level-bedded, silty clay with sandy partings occurs in a channel cut in the underlying sediments, retaining a very characteristic lithology, and extending to the datum. It reaches a maximum observed thickness of a metre (WRCO, 35 m, Fig. 7). Between 40 and 43 m west of the datum the pale silty clay rises westwards and tapers to a thin silty sand. A thin gravel and disturbed silt occurs below it in this region (40 m). The sand partings contain fragments of marine shells, but the silty clay contains freshwater molluscs, including *Pisidium* and *Sphaerium* with valves united, and gastropods; larval cases of a caddis fly constructed of sand grains were also found in the silty clay (Plate 8). The horizon can then be considered to be a freshwater sediment with derived marine shells in the coarser laminations.

i. This is the topmost bed of the pre-West Runton Freshwater Bed succession and consists of gravels, sands and silts, with marine shells. It reaches a maximum observed thickness of 2 m (at 180 m west of datum, Fig. 10, where they reach 6·90 m OD). Silt basins of the underlying horizon have been truncated, but there is no very marked unconformity at the base of this horizon. From the datum to 44 m west (Fig. 7), the sediments are mainly sandy, with silty laminations (tidal bedding); *Macoma balthica* was common at 8 m west. West of 44 m, the horizon rises above 4 m OD, so that its base is considerably higher than at the east end of the section. Concomitantly, the shelly sands and silts are replaced by rounded pebble gravels with a beach structure and occasional marine shell fragments. The beach facies is well represented from 52 m west to 230 m west, where the gravels are cut out by the swale filled with the West Runton Freshwater Bed. At 405 m (WRBCa, Fig. 11), gravels at the same level reappear. They have not been seen further west in the West Runton section.

(ii) *Periglacial stratigraphy*

Ice-wedge casts

32 m (Fig. 7; Plate 9), penetrating bed f, sealed by bed h (see West, 1968).
105 m (Fig. 9), penetrating bed f, sealed by bed i.
131 m (Fig. 10), penetrating bed f, sealed by bed i.
191 m (Fig. 10), penetrating bed c, sealed by red sand, bed f?

Involutions

56 m, 94 m (Fig. 9), 149 m, 215 m, 222 m (Fig. 10): near top of bed f.
65 m (Fig. 9), 123 m (Fig. 10), in lower part of bed i.

(iii) *Sedimentary history*

The Chalk at its surface is strongly brecciated and appears as a coombe rock (bed a), implying an old land surface affected by solifluction. Subsequent marine transgression resulted in the formation of a stone bed (bed b) with marine molluscs, some in the position of life (see Reid, 1890, p. 139), and also with remains of terrestrial vertebrates, presumably derived from the old land surface. Marine conditions continued with the deposition of the shelly sands of bed c (Weybourne Crag). The position of the ice-wedge cast which penetrates this bed (191 m, Fig. 10) is problematical. It appeared to be sealed by red sand, possibly of bed f. If so, a time of permafrost is indicated between this marine horizon and the deposition of the cross-bedded sands of bed f. This point is discussed further in the light of pollen-analytical evidence in (h)(iv) below.

Further east, the marine sands are cut into by, and interdigitate with, the tidal silty clays of bed d i, indicating a low energy marine environment. Subsequently thin gravels of bed e, probably also marine, were deposited over a wider area than that covered by the tidal silty clays. Reworking of the clays is indicated by clay conglomerates associated with this horizon (95–100 m, Fig. 9).

Marine conditions continued with the deposition of the shelly sands and cross-bedded sands of bed f, which transgress all the older sediments are and continuously exposed for much of the section west of Woman Hithe, with their consistent apparent dip of the sets westwards. The structure of the sands indicates shallow conditions with strong tidal currents moving west parallel to the coast.

The finer sediments near the top of this horizon (bed g), often disturbed and forming involutions, indicate a slackening of flow conditions, perhaps indicating regression. Certainly regression and permafrost are indicated by the ice-wedge cast at WRBD (32 m, Fig. 7), part of a polygon system (Plate 9) described previously (West, 1968).

Subsequently freshwater deposition occurred with the formation of the silty clay of bed h. The ice-wedge cast at 131 m (Fig. 10) may also date from the same permafrost period as the one at 32 m; its cast does not affect the deposition of the overlying beach gravel of bed i. But the cast filling at 105 m (Fig. 9) does affect the bedding of the beach gravels, which are down-faulted above the cast. It is possible that this wedge was formed immediately before the deposition of the beach gravel, implying permafrost during or after the deposition of the pale silty clay of bed h. In any event, after the deposition of this freshwater silty clay, marine conditions again obtained, and the transgression resulted in the formation of the shelly sands, tidal sediments and beach gravels of bed i, up to 6·90 m OD. Disturbances of silt horizons in the beach may be the result of freeze-thaw conditions or loading.

Regression followed, with the erosion of the swales and channel in which the West Runton Freshwater Bed was deposited.

(g) Pre-West Runton Freshwater Bed succession: east of Woman Hithe

(i) *Lithostratigraphy*

East of Woman Hithe the pre-West Runton Freshwater Bed sediments are badly exposed, being normally covered by beach. They do not appear in the cliff base except towards Wood Hill, for the channel in which the West Runton Freshwater Bed lies east of Woman Hithe cuts down to near or below HWM. After a scour, the silty clay of bed d i outcrops for a few metres east of the Woman Hithe concrete gangway at the level of the upper part of the beach. Further seawards, at a lower level, iron-cemented gravels with marine shell fragments and containing clay conglomerate stretch some 50 m east. A horizon of similar lithology occurs, with many small basin structures a few metres across, from 500 m east of Woman Hithe all the way to East Runton. Above this horizon sections are visible in the cliff from 444 m (Fig. 4) eastwards, and other sections have been seen from time to time near LWM. But east of 543 m correlation of the sediments is made difficult by the paucity of exposures. The succession can be divided into the following horizons, using the same lettering as for the beds west of Woman Hithe.

b. A stone bed up to 30 cm, of similar lithology to that west of Woman Hithe, lies directly on the Chalk, which was not seen to be brecciated at its surface in this area.

c. A grey or red shelly sand, up to 40 cm thick, was seen resting on the stone bed at WRQ and WRE (inset, Fig. 4). At WRE, it contained a thin lamination of silt.

d ii A sequence of grey silty clays up to 30 cm thick associated with detritus mud horizons up to 10 cm thick was seen in the lower part of the beach opposite WRBV, about 625 m east of Woman Hithe (WRF, WRAC, inset, Fig. 4). These clays rested on shelly sands of bed c.

e. Higher in the beach, in the same area, occur the ferruginous gravels with marine shells and clay conglomerates of the horizon already mentioned, and which stretch east to East Runton and west for some distance. They are similar to the sediments filling the channel at WRBF, 100 m west of Woman Hithe. The inset of WRCM 15 m north of BV, 628 m, Fig. 4, shows a characteristic section. This horizon of gravel, with redeposition of clay pebbles and a marine fauna, is estimated to be some 2–3 m thick. At 486 m east of Woman Hithe (Fig. 4), laminated blue silt 50 cm thick occurs above the ferruginous gravel and clay conglomerate of bed e, the sediment resembling that of bed d i west of Woman Hithe.

f. From 443–486 m (Fig. 4) the cross-bedded sands of bed f west of Woman Hithe are seen. They again occur in WRBS, 514 m, resting on ferruginous gravel. A thin seam of silt occurs in this bed at 470 m and 514 m.

g. Silty clay, partly laminated and containing an organic fraction, occurs above the cross-bedded sands in WRCNA, 443–454 m (Fig. 4). At 464–486 m thin silts occur, often disturbed, at the top of the underlying sands, as they do west of Woman Hithe.

h. Underneath the West Runton Freshwater Bed at WRAR (344 m) and WRCI (340 m) in Fig. 6, occur pale, laminated silts with freshwater shells, larval cases of caddis flies constructed of sand, and ferruginous impressions of small leaves. The sediment is identical to that of bed h west of Woman Hithe. It again appears to be in a small channel. At WRBT, 519 m east of Woman Hithe, a grey clay immediately below the organic swale filling of Cromerian age contains friable freshwater molluscs with valves united. It is penetrated by rootlets, presumably from vegetation associated with the swale filling. This horizon is also placed in bed h.

i. Gravels of exactly the same type as bed i west of Woman Hithe are seen above the laminated silty clays of bed g (443–487 m, Fig. 4). Gravels with marine shells which may belong to this horizon occur above silty clay of bed h in WRAR, 345 m (Fig. 6), and below the West Runton Freshwater Bed at other sites further west (WRAQ, 285 m; WRBR, 150 m). Two ice-wedge casts penetrate the gravels at 467m and 485 m (Fig. 4).

(ii) *Sedimentary history and correlation with sediments west of Woman Hithe*

The first difference from the succession west of Woman Hithe is the appearance of the substantial thickness of ferruginous shelly gravel and clay conglomerate (bed e) at the same level as the silty clays of bed d i west of Woman Hithe. The redeposition of the clay pebbles and gravel will have occurred at a time of marine sedimentation under conditions of changing flow regimes. The second difference is the occurrence of freshwater detritus muds (bed d ii) below the ferruginous gravels and clay conglomerates, indicating a period of regression after the deposition of the shelly sands and before the deposition of the ferruginous gravels and clay conglomerate. A hiatus between beds c and d ii is indicated by the pollen analytical evidence from Beeston.

The ice-wedge cast observed in the upper part of the beach gravel (bed i) at 485 m east of Woman Hithe probably is of the same age as those which are sealed by the sands, bed d of the post-West Runton Freshwater Bed succession, with reworking of the surface gravels to seal them. But the wedge at 467 m may possibly be associated with the final stage of beach deposition, as it appears to be sealed within the gravel itself.

Finally the pre-West Runton Freshwater Bed succession suffered erosion, with swale and channel formation in which the West Runton Freshwater Bed was deposited. A significant hiatus is therefore present between the Beestonian beach gravels of bed i and the deposition of the late Beestonian marl at the base of the West Runton Freshwater Bed in the central channel.

(h) Biostratigraphy of the pre-West Runton Freshwater Bed succession

Three pollen assemblage biozones have been distinguished in this sequence (Table 4), all from sediments associated with marine conditions. In bed di, west of Woman Hithe, at WRAX, WRBD and WRM, laminated silty clays contain a *Pinus–Quercus–Carpinus–Ulmus* p.a.b. indicating regional temperate forest. This is correlated with the mixed oak forest substage of the Pastonian (temperate) stage (Pa II). A similar forest assemblage is seen east of Woman Hithe, at WRF and WRAC (0–5 cm), in bed dii (silty clays and associated detritus muds) and at WRCMa, in bed e (ferruginous gravels, silty clays, clay conglomerates). A local p.a.b. with high Gramineae and Chenopodiaceae frequencies is separated at WRAC 15–10 cm.

At WRAA, a single pollen sample in silty clay with marine shells of bed e, east of Woman Hithe, contained high frequencies of *Carpinus* pollen, with *Tsuga* present. This bed is separated from the previous p.a.b. It may be included in a later stage of the Pastonian (Pa III). At WRCNA, in bed g, an *Alnus–Betula–Pinus–Quercus* p.a.b. is also included in Pa III.

Pollen analyses of silts associated with shelly sand of bed c, east of Woman Hithe, at WRQ and WRE, and at WRBN with silt associated with the stone bed west of Woman Hithe (bed b) have already been described in Chapter 2 and are included in substage Pre-Pa a.

Though no pollen was obtained from the sediments between the Pastonian and late Beestonian at West Runton, ferruginous impressions of leaves and freshwater molluscs were found in the pale laminated silts of bed h. These are described in the last section of this chapter.

(i) Vegetational history of the pre-West Runton Freshwater Bed succession

(i) *Pre-Pastonian a*

The *Pinus–Betula*–Gramineae–Ericales p.a.b. of the substage has already been described and discussed in Chapter 2.

(ii) *Pastonian II*

The *Pinus–Quercus–Carpinus–Ulmus* p.a.b. correlated with this substage is recorded in the tidal silty clay of bed di at WRAX, WRM and WRBD, in the silty clay and detritus mud of bed dii at WRF and WRAC, and in silty clay laminae in the conglomerate of bed e at WRCMa. Taking first the spectra from beds di and e (WRAX, WRM, WRBD, WRCMa), these show very high AP frequencies, with *Pinus*, *Picea*, *Alnus* and *Quercus* most frequent, with lesser frequencies of *Betula*, *Ulmus*, *Quercus*, *Carpinus* and *Corylus*. There are low frequencies of *Calluna*, *Empetrum* and Ericales. Of the other NAP, Gramineae is the most abundant, with curves for Chenopodiaceae and Compositae Tubuliflorae. There are high frequencies of Filicales spores and curves for *Osmunda* and *Sphagnum*. The assemblage indicates regional forest, but deposition under tidal conditions may have enriched the conifer representation. There is a more local pollen component of Gramineae and Chenopodiaceae from maritime communities, and spores derived from fen communities (Filicales, *Osmunda*).

The spectra from the silty clays and detritus mud of bed dii (WRF, WRAC) show variations associated with sediment type. They show higher NAP frequencies, particularly WRAC 10–15 cm, which level is separated to form a Gramineae–*Pinus–Quercus–Ulmus–Carpinus* p.a.b. The sediments have a freshwater aspect, as shown by the increased pollen frequencies of *Sparganium* and *Typha latifolia* compared with the spectra from the tidal sediments of bed di. There is a greater representation of Umbelliferae, possibly associated, with Filicales and *Osmunda*, with a fen component. Chenopodiaceae and Composiate Tubuliflorae frequencies rise and fall together, both probably derived from local salt marsh. There are higher frequencies of dinoflagellates, particularly *Operculodinium*. The highest Gramineae, Chenopodiaceae and Compositae Tubuliflorae frequencies are associated with the detritus mud levels of WRF and WRAC, indicating that these are local components of the pollen assemblage.

More information on local vegetation comes from macro analyses at WRF. The frequency diagram of the ecological groups (Fig. 15) shows aquatics to be most abundant in the detritus mud and underlying silty clay (13–25 cm), with an increase of the fen reedswamp helophyte and muddy substrate/shallow-water groups in the upper silty clay (0–13 cm). Taking the flora as a whole, there are no macroscopic remains of trees, shrubs or the heath or maritime element, though the foraminifer *Jadammina macrescens* occurred in the middle mud, indicating brackish conditions on site or nearby. In the dry-ground (or maritime) group *Atriplex*, Chenopodiaceae and *Chenopodium album* are present. *Alisma plantago-aquatica* is abundant in the upper silty clay, the only representative of the muddy substrate/shallow water group. It is accompanied by *Carex riparia* and *Urtica dioica* in the fen/reedswamp helophyte group. The latter occurs with *Ranunculus sceleratus* in the lower silty clay. In the aquatic group, *Hippuris vulgaris* is very abundant in the lower silty clay, with *Lemna trisulca*, *Ranunculus* subg. *Batrachium*, *Typha angustifolia* + *latifolia* and *Azolla filiculoides*. In the middle mud the same taxa are present, but *Potamogeton pectinatus* is more abundant, and *Ceratophyllum* (including *C. demersum*) is present. In the upper silt *Hydrocharis morsus-ranae*, *Lemna trisulca*, *Ranunculus* subg. *Batrachium*, *Typha* and *Azolla* (very abundant) are present.

Most of the aquatics recorded are found in brackish conditions as well as freshwater: *Ceratophyllum demersum*, *Hippuris vulgaris*, *Lemna trisulca*, *Potamogeton pectinatus*, *Ranunculus* subg. *Batrachium*, *Typha angustifolia*. The assemblage indicates brackish (lagoonal?) conditions with organic or silt deposition, probably antedating the deposition of the tidal silty clays of bed di, but following the Pre-Pa a substage biozone found in the underlying shelly sands. There must therefore be a hiatus in deposition corresponding to the regression noted at Beeston and East Runton between Pre-Pa a and the Pastonian.

(iii) *Pastonian III*

Three pollen spectra are tentatively ascribed to a substage later than Pa II. At WRAA, a high AP assemblage similar to the Pa II biozone is seen, but with higher frequencies of *Carpinus* and with *Tsuga*. Both *T. canadensis* and *T. diversifolia* types were recorded. Since a biozone with similar characteristics occurs above Pa II biozones at Happisburgh, a separation between biozones of this type and the Pa II type seems reasonable, though the evidence is scanty. The separation of Pa II and Pa III is further discussed on p. 100.

At WRCNA, in silty clay with organic laminations of bed g, an *Alnus–Betula–Pinus–Quercus* p.a.b. was recorded. The NAP is mostly Gramineae, with low frequencies of *Calluna*, *Empetrum*, Ericales and Compositae pollen, and high frequencies of Filicales spores. The assemblage indicates regional woodland, with low representation of thermophilous trees. The macroscopic assemblage from WRCNA is one of the richest discovered. Fifty-seven taxa are recorded, with a preponderance of aquatics, but containing representatives of all the other groups except the montane element (Fig. 15). Trees are represented by abundant fruits of *Alnus glutinosa*, fruits of *Betula* tree ssp. × *B. nana*, and a leaf of cf. Coniferae, shrubs by budscales of *Salix*, the maritime element by seeds of *Salicornia*, and the heath element by seeds of *Empetrum*, and spindles of *Eriophorum vaginatum*. The dry-ground herbs include cf. *Atriplex*, *Campanula rotundifolia*, *Chaenorhinum minus*, *Chenopodium*, *Peplis portula*, cf. *Picris*, *Plantago major*, *Polygonum lapathifolium*, *Ranunculus* cf. *bulbosus* and *Thalictrum minus*. The damp-ground element only contains *Ajuga reptans*, and the muddy substrates/shallow water element *Alisma*, *Limosella*, *Ranunculus flammula/reptans* and *Rumex maritimus*. The fen/reedswamp helophytes are *Bidens*, *Eupatorium cannabinum*, *Lythrum salicaria*, *Osmunda*, *Ranunculus sceleratus* and *Urtica dioica*. The aquatics include four species of *Callitriche*, *C. hermaphroditica*, *C. intermedia*, *C. platycarpa* and *C. stagnalis*, *Chara*, *Elatine hydropiper*, *Hippuris vulgaris*, *Hydrocharis morsus-ranae*, *Menyanthes trifoliata*, *Nitella/Tolypella*, *Potamogeton*, *Ranunculus* subg. *Batrachium*, *Sagittaria sagittifolia*, *Stratiotes aloides*, *Typha* cf. *latifolia*, *Zannichellia palustris*, *Azolla filiculoides* and *Salvinia natans*. The unclassified group includes *Carex*, Cruciferae (very abundant seeds), *Dryopteris*-type (sporangia), *Juncus*, *Potentilla*, *Saxifraga*, Umbelliferae and *Viola*.

The mixed nature of the assemblage is related to the origin of the sediment. The organic seams in the silt are composed of drift mud, derived from a variety of sources, maritime, freshwater, fen, fen carr (*Alnus* pollen and fruits), and terrestrial communities. Similar mixed assemblages have been recorded from drift mud associated with recent Flandrian tidal sediments on the Essex coast. The assemblage was formed by alluviation in a hollow on the surface of the cross-bedded marine sands of bed f, indicating it is a sediment formed in regression.

(iv) *Other pre-Cromerian pollen spectra*

Pollen spectra from thin silt seams in shelly sand (Weybourne Crag) were obtained at WRAS and WRAT (190 m and 204 m respectively, Fig. 10). The WRAS sample was very poor in pollen, and only a very low total could be counted. *Pinus* is very abundant, with *Picea*, *Betula* and Filicales spores, and abundant dinoflagellates, especially *Operculodinium*. The pollen content may have been altered by weathering. This spectrum bears some similarity to those from crag at Sidestrand

(SSV 10, 12) and Trimingham (BHT 16 4), placed in *Pinus–Picea* p.a.b.s and allocated to Pa Ib, and the WRAS spectrum is tentatively correlated to the same substage.

The WRAT sample gave a better spectrum, with *Pinus*, *Picea*, *Betula* and *Alnus*, with a good representation of *Empetrum* and Ericales. This assemblage may be Pre-Pa a. It is unfortunate that pollen preservation was not better in this stratigraphical context, since the samples came from *Macoma balthica* crag on the stone bed, and are earlier than a permafrost period indicated by an ice-wedge cast which penetrates the crag. It is possible that the hiatus between Pre-Pa a and Pre-Pa b at Beeston and between Pre-Pa a and Pa II at West Runton may have been a time of weathering of Crag and ice-wedge formation.

(j) Beestonian flora and fauna

The laminated silty clays of bed h on either side of Woman Hithe contained no identifiable pollen, but at WRAV, 25 m west of Woman Hithe, WRCO, 35 m west of Woman Hithe (Fig. 7), and at WRAR, 344 m east of Woman Hithe (Fig. 6), freshwater shells (see below) and caddis fly larval cases of coarse sand grains were found in the clays. At the latter site WRAR, a teleost fish scale (? *Perca fluviatilis*) and abundant ferrugineous impressions of small leaves were found in the bedding planes of the clay. Leaves of *Betula nana* L. were identified, but most of the leaves were of the small *Salix* type such as are found in *S. phylicifolia*, and preservation was not complete enough for identification.

As described in section (f)(iii) above, the laminated clays of bed h were deposited in channels after the Pastonian regression, and after a period of permafrost. The flora is suggestive of cold stage conditions, as is the lithology of the sediments, and bed h is placed in the Beestonian cold stage, together with the beds showing evidence of permafrost, and the marine sands and gravels above, since these are overlain by the pre-Cromerian late-glacial (late Beestonian) initial filling of the central channel of the West Runton Freshwater Bed.

We may here note that a leaf flora and caddis fly larval cases of sand were also found in the Beeston section BAM (Fig. 16), in laminated muddy silt of bed r, placed in the Beestonian cold stage. At this site the flora was much better preserved (*see* Chapter 4, *p.* 44).

Three samples of laminated clays of bed h were analysed for molluscs by B. W. Sparks. All the faunas found were very poor and fragmentary, but freshwater conditions were indicated. 25 m west of Woman Hithe at site WRAV he found three bivalve fragments, all very badly worn; one is *Pisidium amnicum*, one a *Sphaerium* sp., and the other a *Pisidium* sp., possibly *P. casertanum*. At site WRCO, 35 m west of Woman Hithe, he recorded *Sphaerium* sp. with both valves united and unidentifiable but freshwater gastropods. East of Woman Hithe (519 m), at site WRBT he recorded a few broken fragments, including one or two of the size and thickness of *Sphaerium* or a comparable *Pisidium*, and several others which could be of a larger gastropod or bivalve.

4: Beeston

(a) Lithostratigraphy

The sections at Beeston, Figs. 16–18, extend from the 1964 eastern limit of the coast protection revetment, the datum point for distances eastward, for about 350 m east towards the stream at Old Kiln. The till cuts down to the level of the Chalk some 50 m west of this stream, but between this point and the revetment, good sections showing up to 7 m of sediment below the till were seen in the period 1964–74. The exposures for the first 200 m of the section were reasonably continuous, but were more scattered for the rest of the section, to 343 m.

The sections are more complex than those at West Runton. As at West Runton, however, a primary threefold division of the succession may be made. No correlation with the threefold at West Runton is thereby implied. The middle member of the three is a sediment series which shows marine transgression sediments, grey-blue, laminated silty clays (bed k), overlying freshwater muds (bed i), sometimes with a thin pebble gravel (bed j), interpreted as beach, between the two (e.g. at 28 m, Fig. 18; BAX, 127 m, Fig. 17). This sequence extends along much of the Beeston sections. Below it (beds a to h) and above it (beds l to w) are sequences of organic sediments, gravels, sands and silts, sequences which vary laterally very considerably. The sequences of the middle member are quite clear along the section and so are the sequences older than this. The post-transgression sequences at the west end of the section are also clear, but with those at the east end, where exposures were short and infrequent, the interpretation cannot be so satisfactory. Nevertheless correlations from exposure to exposure are suggested. In fact, the biostratigraphy of the post-transgression sequence shows little variation, so errors in correlation will not necessarily lead to major errors in the interpretation of environmental history.

The Beeston sediments can be grouped into a number of distinct beds which will now be described. Table 11 summarises this succession, and Plate 12 shows part of the sequence.

a. The Chalk is exposed at low water along this part of the coast, and after a scour may also be seen up to HWM near the cliff base. Much of the exposed surface shows brecciation, as at the west end of the West Runton sections, giving the appearance of coombe rock (Plate 11). Brecciated Chalk is shown in the sections at BTA (137 m, Fig. 17), and BKF, BKC and BKA (65–72 m, Fig. 18). In other places the surface Chalk appeared undisturbed.

b. A stone bed, up to 30 cm thick, rests on the Chalk. The lithology is the same as already described from West Runton. At BE (300 m, Fig. 16) and BKC (68 m, Fig. 18) up to 15 cm of pale calcareous silt with shell fragments was seen between the Chalk and ferruginous stone bed.

c. At BA–B (43 m, Fig. 18) 30 cm of current-bedded sand lies over gravel, with a similar thickness over the stone bed at BR–S (23 m, Fig. 18).

d. At BR–S (23 m, Fig. 18), 45 cm of red and grey laminated silt and sand overlies the sand of bed c. A similar thickness occurs to the east at BA–B (43 m, Fig. 18). Sediments at both sites have a similar pollen flora.

e. At BR–S (23 m, Fig. 18) 25 cm of dark grey silty mud was seen resting on the laminated silt and sand of bed d. At BA–B, 20 m to the east, with the top of the laminated silt a metre higher in level, the mud is absent. It evidently deposited in a shallow basin.

f. At BKF–BKA (63–75 m, Fig. 18) a stratified gravel rests directly on the marine gravel of the stone bed. It is up to 1·5 m thick at BKF, where it contains remains of freshwater plants in lenses of mud. The pollen flora of these lenses is similar to that of the silts of bed g at BR–S. Eastward from BKA, there is stratified gravel over the stone bed at BTA (137 m, Fig. 17), at BTB (184 m, Fig. 16), which probably also belongs to this horizon.

g. At BR–S (23 m, Fig. 18) 45 cm of grey sand with thin laminae of silt was seen above bed e. A similar sediment, but oxidised, was found at a similar level at BA–B (43 m, Fig. 18), with a thickness of 35 cm, and at 63–76 m (Fig. 18), with a thickness of 50 cm, and overlying gravel of bed f. Further east, 87–118 m (Fig. 17), a sand with silt laminae up to 60 cm thick, is found in the same stratigraphical position. Included in this bed are the cross-bedded sands with lenses of silt and organic mud seen at the base of the section at 87–102 m, Fig. 17 (up to 30 cm coarse red sand), at BTA (75 cm of coarse current-bedded sand, 137 m, Fig. 17), where it lies on gravel (bed f), at BTB (2 m thick, 184 m, Fig. 16), at BU (2·5 m thick, 287 m, Fig. 16), at BW (1·5 m thick, 302 m, Fig. 16), and at BV (1·5 m thick, 325 m, Fig. 16). At BW and BU there are thin strings of gravel in the sand. The pollen floras from silts and organic muds in this bed indicate a general contemporaneity of deposition under freshwater conditions along the length of the Beeston section.

h. At 103–120 m (Fig. 17) a pale gravel, up to 20 cm thick, overlies pale sand and underlies the mud of bed i. It appears to occupy the base of a channel which later filled with the mud. Further east a gravel also occurs below the mud, but in the absence of a lower boundary it is not possible to say whether this is bed f or h.

i. At BR–S (23 m, Fig. 18; Plate 12) 60 cm of dark grey, freshwater, silty mud with occasional stones occurs above the grey sand with silt laminae of g. This horizon is very extensive in the Beeston sections. It is the peaty loam of the Beeston cliffs described by Reid (1882*a*, *p.* 26). It was correlated by Reid with the Cromerian West Runton Freshwater Bed, though it in fact belongs to a temperate stage older than the Cromerian, as will be shown later. This horizon of mud extends east to 36 m where it is cut out by gravel, appears again at 63 m and 65 m, and as very thin lenses at 73 m and 76 m. In Fig. 17 it is seen extending from 87 to 101 m, where it is cut out by a channel of later age, from 107 to 112 m, in the base of the channel, from 115 to 130 m, and from 137 to 156 m. In Fig. 16, it extends from 158 to 170 m, from 178 to 184 m, from 304 to 301 m, and from 324 to 327 m.

In certain parts of the section, this horizon is subdivided by lenses of current-bedded sand and of muddy silt. Thus from 116 to 126 m (Figs. 17, 18) a pale sand (i–ii) 20 cm thick separates a lower (i–i) from an upper (i–iii) mud. At 141–156 m, a much thicker (to 60 cm) sand, current-bedded in places, subdivides the mud. At BC, 180 m (Fig. 16), the upper division of the mud is divided by a muddy silt.

j. A thin pebble gravel, or pebbly sand, up to 60 cm thick (at BKA, 65 m, Fig. 18), succeeds the mud. It extends from 27 to 39 m, and from 63 to 77 m. It is probably a beach gravel.

k. At BR–S (23 m, Fig. 18), 23 cm of grey-blue laminated silty clay with sandy partings and with a tidal structure, rest on the mud, with a narrow transition between them. This horizon is very extensive along the Beeston sections. West of BR–S, it is seen at BQ (21 m) and BG (17 m). Eastward (27 m) the silty clay overlaps the beach gravel. It is seen in Fig. 18 to 39 m, from 41 to 48 m, at 63 m and 65 m, and from 68 to 77 m. In the part of the section covered by Fig. 17, the silty clay loses its laminated structure, and becomes thicker (90 cm) and more massive. The absence of pollen in the sediment in this part of the section, compared with its abundance in the silts further west, suggests a heavy weathering. The grey-blue silty clay extends from 117 to 132 m, from 136 to 143 m, and from 149 to 156 m. In Fig. 16, it becomes up to 1·3 m thick and extends from 158 to 170 m, at 200 m and 222 m, from 286 to 290 m, from 310 to 311 m, from 324 to 327 m and is seen at 342 m. In most of its exposure, the silty clay rests on the beach gravel or the mud, but it overlaps older sediments, such as beds g (43 m, Fig. 18), or d (290 m, 301 m, Fig. 16). At BKA, marine shells were found in the silt at 2·12 m from the Chalk surface at the base. At BZ (150 m, Fig. 17) the contact between the silty clay and the underlying sand horizon (i–ii) within the mud is disturbed. This is where the upper mud (i–iii) is missing and the disturbance appears to be a result of diapir formation at a point of weakness, taking place to some extent even after the deposition of part of the overlying gravel.

l. At BR–S (23 m, Fig. 18) 15 cm of compressed brown peat with wood lies conformably on the tidal silts. To the west this bed is broken up into discrete lumps (Plate 12); to the east it thins rapidly and disappears. This wood peat is seen in no other part of the Beeston sections.

m. At BR–S (23 m, Fig. 18), 90 cm of red sand with many silt laminae occurs above the wood peat. To the east this horizon thins rapidly, and appears to be cut out by the overlying sand.

n. In a channel in the freshwater mud and transgression sediments, with its eastern margin at 115–120 m (Fig. 17) occur laminated sandy silts, current-bedded sands and gravel strings, up to 2 m thick. The margin of the channel is lined with thin gravel. To the west, this horizon is cut off by later channelling at 95–99 m. On the upper surface at 113 m occurs a small channel with more massive silt, cut by erosion on the west side. The laminated silts are otherwise concentrated in the lower half of the channel filling. This horizon may belong to the same period of deposition as bed m. Both are incised by the sands of bed q.

o. Stratified gravel covers this channel filling, and extends eastward over the blue-grey silt of bed k. Thence it can be traced, becoming more sandy in places, along the section above bed k from 121 to 130 m, at 136 m, from 137 to 156 m (Fig. 17), in Fig. 16 from 159 to 170 m, from 178 to 181 m, where it rests directly on the mud of bed i, at 184 m, thickening to 3·4 m at BAV (200 m), at 209 m, 248 m, 275 m, 284 m, 287–290 m (where it is underlain in part by a current-bedded sand), from 300 to 311 m, and from 324 to 327 m.

p. A brown, faintly laminated, muddy silt occurs in a basin on top of the gravel of bed o at 113–118 m (Fig. 17). This horizon may also be represented by the small silt lenses and the silty mud horizon at or near the upper boundary of the gravel (bed o) at BV (324–327 m and 341–343 m, Fig. 16), BW (301 m, Fig. 16) and BU (BU 3, 289 m, Fig. 16).

q. Stratified sand, with some silt laminae, a gravelly base and gravel lenses occupies a channel cut into the sands and gravels of beds n and o. The eastern margin of the channel is at 95–99 m (Fig. 17). This horizon reaches a thickness of 1·6 m in the channel at 91 m, and 2·1 m at 60 m.

Westward this sand rests on the mud of bed i at 94–87 m, on the silt of bed k at 77–68 m, at 65 m and 63–59 m, and 48–32 m, on the beach gravel of bed j at 33 m, and on the red sand of bed m at 32–23 m. The basal gravel of this bed at 76–43 m may partly be the same age as the gravel of bed o.

Eastward, the sand extends over the silt of bed p and the gravel of bed o, with much cross-bedding, drift mud streaks, and occasional gravel lenses. The boundary between beds o and q is not always clear. A similar sand or gravel becomes distinct at BL (209 m, Fig. 16), BD (275 m), BU (290 m), BW (306–311 m), BX (316 m) and BV (324–327 m), resting on sandy gravel.

r. Red and brown laminated silts lie in channels cut in the sands of bed q at 23–49 m (Fig. 18) and from 77–114 m (Fig. 17). The channel fillings reach a thickness of about 1·5 m. The base of the filling in places is grey and much more organic (e.g. 44 m), but as a whole the silt is much oxidised. The appearance is of a strong weathering at this level. The surface is level from 23 m, right across the sand interfluve at 49–77 m, and over the second channel filling as far as about 105 m (Fig. 17), where the surface of the silt starts to rise. At 40–44 m there appears to be a lower channel in the sand of bed q, containing a sand rich in silt laminae. The gravel below this probably belongs to bed q, which does contain lenses of the same kind elsewhere.

The east end of the section also shows a strong weathering horizon, with rootlets penetrating oxidised silt in places. We correlate this weathering horizon with that so clearly seen at 35–115 m. We can then trace the silt of bed r along the section as follows. It is seen at 120 m at the eastern end of the channel, and runs east along at the same level in the cliff section, containing in places freshwater shells, to 161 m, where weathered silty clay with rootlets marks the western edge of a second channel, and to a 'rootlet bed' at BD (209 m), near the eastern end of this second channel. At BAM (281 m) a further rootlet horizon occurs. At both BAM and BD the silts contain freshwater shells and muddy horizons. At BAM caddis larval cases (of sand grains) occur in the sediment. The silt of bed r is then traceable along to 342 m, with a rootlet bed at BV (325 m).

s. Overlying the brown and grey silt of bed r is a thin gravel, up to 30 cm thick, seen at 114 m, 158–173 m, 178–181 m, 199 m, 280 m, 288 m and 316 m. It is at its thickest in the two channels, at the eastern end of the western channel (105–115 m, Fig. 17) where it interdigitates with the sand of bed t, and in the eastern channel (BP, 161–180 m, Fig. 16), where it again interdigitates with the sand of bed t. There is a red ferrugineous sand above the gravel in this channel.

t. Cross-bedded sands with thin gravel strings and streaks of drift mud lie on the silts of bed r at the east and west ends of the section. They are associated with the channels west of 115 m (Fig. 17) and at BP (161–180 m, Fig. 16). In the western channel they reach a thickness of 1·8 m at 100 m.

u. Associated with the two channels is a further horizon of brown and grey silt, usually laminated. It reaches a maximum thickness of 70 cm at BR–S (22 m, Fig. 18), thinning eastward in this channel. At the BP channel (161–173 m, Fig. 16) it is much thinner.

v. Overlying this silt is a further cross-bedded sand, with a maximum thickness of about 1·6 m. This extends from one end of the Beeston section to the other, and separates the silts of beds r and u from the overlying till.

w. The whole succession is sealed by the till, already described in the West Runton section.

(b) Periglacial stratigraphy

Ice-wedge casts

These are all above the transgression and regression sequence of beds i to l. From west to east they occur as follows:

32 m (Fig. 18), penetrating bed q, sealed by bed r, 4
35 m (Fig. 18), penetrating bed q, sealed by bed r, 4
41 m (Fig. 18), penetrating bed t, sealed by bed u, 6
57 m (Fig. 18), penetrating bed q, sealed by bed t, 5
62 m (Fig. 18), penetrating bed q, sealed by bed t, 5
76 m (Fig. 18), penetrating bed q, sealed by bed t, 5
89 m (Fig. 17), penetrating bed r, sealed by bed t, 5
92 m (Fig. 17), penetrating bed r, sealed by bed t, 5
114 m (Fig. 17), penetrating bed n, sealed by bed n, 1
114 m (Fig. 17), penetrating bed o, sealed by bed p, 3
117 m (Fig. 17), penetrating bed o, sealed by bed p, 3
184 m (Fig. 16), penetrating bed i, sealed by bed o, 2
307 m (Fig. 16; Plate 13), penetrating bed k, sealed by bed o, 2

These ice-wedge casts can be grouped into 6 groups, numbered as given above, in order of age, 1 the oldest. These groups can be related to the sequence (see Table 11), showing six periods of permafrost, as follows:

1. During the deposition of bed n.
2. Before the deposition of bed o, after the deposition of beds i and k. This may be the same period as 1.
3. Before the deposition of bed p, after the deposition of bed o.
4. Before the deposition of bed r, after the deposition of bed q.
5. Before the deposition of bed t, after the deposition of beds r and q.
6. Before the deposition of bed u, after the deposition of bed t.

Involutions

These occur at the following points in the section:

30 m (Fig. 18), base of bed r 2
45 m (Fig. 18), base of bed u 3
115 m (Fig. 17), within bed n 1
128 m (Fig. 17), base of bed r 3
150 m (Fig. 17), base of bed r 3
155 m (Fig. 17), base of bed r 3
180 m (Fig. 16), base of bed r 3
275 m (Fig. 16), base of bed r 3
310 m (Fig. 16), base of bed r 3
316 m (Fig. 16), in bed r 3

These are grouped into three periods of involution as numbered, 1 the oldest. Table 11 shows their relation to the sequence.

(c) Sedimentary history

The surface of the Chalk is brecciated, appearing as a coombe rock (bed a) and implying an old land surface subjected to solifluction. A subsequent marine transgression deposited the stone bed (bed b), containing marine molluscs. Pale calcareous silts associated with this horizon indicate deposition of a chalk wash before the deposition of the stone bed. Lying on the stone bed are coarse sands (fluviatile or marine?) of bed c, and over these, at BR–S and BA–B, are the laminated silts and sands of bed d, with a pollen flora indicating freshwater conditions, which must have been deposited following a regression.

The silty mud of the succeeding bed e is confined to the depression at the west end of the section (Fig. 18), and there must have been open water here with sediment-productive vegetation at the time. The succeeding grey sand with silt, bed g, overlies the mud of bed e at BR–S (23 m, Fig. 18), and overlaps it to the east on to bed d at BA–B (43 m), and onto the fluviatile gravel of bed f at 67–76 m. The gravel of bed f and the sand of bed g are part of the same fluviatile aggradation, which evidently occurred along the length of the Beeston section. Lenses of silt (BTA, etc.) and organic mud (BU 1, etc.) in the sand and in the gravel (BKF) contain pollen floras, so allowing the lateral correlation of the aggradation.

The mud of bed i accumulated on the surface (and within the upper part) of the sand of bed g. Channelling of these sands resulted in a wide shallow depression to the east of 87 m in which the thin gravel, bed h, was first deposited. Then the silty mud of bed i was formed in this depression, overlapping towards the west onto the sand and silt of bed g. This mud thins out where the underlying deposits rise in the section (68–71 m) and thickens again to the west, where the level of the underlying sediments fall, at 63–65 m and 36–17 m (Fig. 18).

Eastward from 87 m this organic horizon marks the existence of a large pool extending to 184 m, where the underlying sands (bed g) rise and the mud becomes thinner and is finally cut out by the gravels of bed o. Further east, 304–327 m, Fig. 16, a further occurrence of the mud indicates another pool of the same age, formed on the surface of the sand of bed g. The sand (bed i–ii) and muddy silt levels within bed i show the renewal of fluviatile conditions in the pool from time to time.

The mud of bed i is succeeded in places by the gravels of bed j. Elsewhere, there is a transition to the grey-blue silt and silty clay of bed k, which contains marine molluscs, foraminifera and pollen of maritime plants. A change in land/sea-level relations allowed a transgression, with the formation of thin beach gravels in some places and tidal facies fine sediments in others. The extensive exposures of this silt and silty clay show the transgression sediments overlapping the organic mud of bed i, resting on older sediments elsewhere (e.g. 280–290 m, Fig. 16 on sands of bed d), and channelling these older sediments in places (301 m, Fig. 16). They reach a height of *c.* 6 m OD at 140 m (Fig. 17).

At the west end of the section (BR–S) thin wood peat (bed l) directly overlies the transgression sediments, indicating regression and the growth of forest vegetation on the old regression surface. This horizon is broken up at 22 m and then covered by sand with many silt laminae (bed m) probably of fluviatile origin, indicating a renewal of fluviatile conditions. It is possible that this period of deposition is represented by the sands and laminated silts deposited in the channel incised in the mud eastward from 87 m (Fig. 17), but the conformity of the sediment of bed m to those underlying it at 21–30 m (Fig. 18) suggests a renewal of fluviatile conditions, which, at a later date, was also associated with dissection and channel formation and the subsequent aggradation of the channel sediments.

This channel filling (bed n, 95–119 m, Fig. 17) starts with laminated silts and continues with sands with gravel strings. Involutions and an ice-wedge cast occur within it, indicating permafrost conditions obtained at an interval during the filling.

The succeeding gravel, bed o, truncates the channel filling and overlaps eastward onto the surface of the grey-blue transgression sediment, which is here heavily weathered. It can be traced eastward along a considerable part of the section, sometimes being replaced by sand with gravel at the base (140 m, Fig. 17), continuing to 342 m as an easily recognisable horizon. At 184 m and 307 m ice-wedge casts predate the gravel, indicating a period of permafrost before the gravel deposition. It is likely and possible that this is the same period that produced the cast in the channel filling (bed n), but in the absence of closer stratigraphical control it is not possible to confirm this. No fossils were found in this gravel but its post-Pastonian stratigraphical position and its composition, analysed by R. W. Hey (see Hey, 1976, Beestonian gravel of sites 1 and 2), is very similar to the beach gravel, bed h, at West Runton, which does contain marine molluscs. If this correlation is correct a transgression must have occurred at this stage.

The muddy silt, bed p, occurs in a small basin above the gravel at 113–118 m (Fig. 17), being separated from it by the permafrost period indicated by the two ice-wedge casts in group 3. Organic sediments at BU (290 m), BW (301 m) and BV (342 m) may also belong to this episode. Regression must have occurred after the gravel deposition, then permafrost development, then deposition of muddy silt.

The stratified sands of bed q, at 87–99 m (Fig. 17) fill a channel cut into the gravel (bed o) and sands (bed n) of the earlier channel filling. This later channel filling shows a wide aggradation to the west end of the section later than the incision, with the filling finally overlapping the channel edge at 99 m and spreading eastward over the gravel of bed o. The sand of this aggradation eastwards is difficult to separate from the sandy facies of bed o, but at the east end of the section (BW) it becomes clear again.

Subsequent to the deposition of the sands and thin gravels of bed q, permafrost developed (group 4 of the ice-wedge casts), and the laminated silts of bed r, with organic horizons and freshwater shells, were deposited along much of the section, with involutions occurring at the base. At the west end of the section the silt occurs in wide channels, and at the east end laminated silts occur widely over sands ascribed to bed q. After this period of alluviation, permafrost again developed (group 5 of the ice-wedge casts). After sedimentation ceased much weathering evidently took place, with the oxidation of sediments and the formation of rootlet beds.

A further fluviatile stage deposited the cross-bedded sand of bed t, seen overlying the channel silt, bed r, in the channels at the west end (22–115 m) and east end (BD, 161–173 m) of the section, with gravels of bed s associated with the margins of the channels and with the channel fillings. At BU, fluviatile activity cut the silts of bed r and again deposited gravel of bed s. A further period of permafrost followed this fluviatile aggradation (group 6 of the ice-wedge casts), then alluviation of the laminated silts of bed u took place in the channels at the west and east ends of the section above the sands of bed t. Involutions (group 3) were frequently observed at the base and within this horizon.

Finally fluviatile conditions again obtained along the length of the section, with the deposition of the sand, bed v. Ice then deposited the till, bed w, which sealed the pre-glacial succession.

Table 11 summarises the sequence described above.

(d) Biostratigraphy

Table 9 lists the biozones distinguished in the pollen diagrams from Beeston (Figs. 19–21). Table 10 gives the plant macroscopic analyses and Table 11 gives a summary of the biozone sequence and vegetational history.

The most complete sequence of pollen analyses comes from section BR–S (23 m, Fig. 18), with the pollen diagram in Fig. 20. There are four high NAP biozones at the base, then six biozones indicative of forest conditions, a sequence of biozones with high frequencies of NAP, and a final biozone with higher AP frequencies. The forest stage is placed in the Pastonian temperate stage, the upper high NAP biozones in the Beestonian cold stage, and the lower high NAP biozones in the Pre-Pastonian cold stage.

(i) *Pollen assemblage biozones: Pre-Pastonian*

The sequence of Pre-Pa a, b, c and d is erected on the basis of analyses through 135 cm sediment at BR–S.

Pre-Pa a. This includes the *Pinus–Betula*–Gramineae–Ericales p.a.b. associated with the stone bed (bed b).

Pre-Pa b. Gramineae–*Artemisia* p.a.b. Beds d and e.

Pre-Pa c. Gramineae–*Betula–Pinus–Alnus* p.a.b. This is seen in the mud within gravel of bed f (BKF) and in the silt of bed g (BR–S).

Pre-Pa d. Gramineae–Cyperaceae–*Betula–Pinus* p.a.b., in the mud of bed i at BR–S and BAX.

(ii) *Pollen assemblage biozones: Pastonian*

The next sequence of biozones shows higher frequencies of AP, with the development and retrogression of temperate forest. All are placed within the Pastonian temperate stage.

Pa Ia. *Betula–Pinus*–Gramineae p.a.b. In the silts of bed g at BTA and BTB and in the mud of bed i at BKA.

Pa Ib. *Pinus* p.a.b. In the mud of bed i at BR–S, and at BW and BV in muddy silt horizons within bed g.

Pa II. *Pinus–Quercus–Carpinus–Ulmus* p.a.b. In the tidal silty clays of bed k. The transition from the underlying biozone is well seen in the pollen diagram from BQ (Fig. 20). This biozone shows the maximum development of temperate forest.

Pa III. *Pinus–Betula–Quercus* p.a.b. A few analyses from the upper part of bed k (BR–S 40 cm; BKA 60–78 cm) show a reduced representation of temperate forest, and an increased Ericales frequency. These are tentatively placed in a separate biozone.

Pa IVa. *Pinus–Alnus–Picea–Betula*–Ericales p.a.b. In the brown peat with wood of bed l at BR–S, and in silty clay of bed k in BG and BKA.

Pa IVc. *Betula–Pinus* p.a.b. In silty laminae of bed m. This is referred to Pa IVc, since there is evidence elsewhere on the coast of a *Pinus–Picea* p.a.b. (Pa IVb) antedating it.

(iii) *Pollen assemblage biozones: Beestonian*

The remainder of the biozones at Beeston, younger than the Pastonian, show high frequencies of NAP and a very restricted representation of thermophilous trees. They are named and related to their stratigraphic position in Table 9. The spectra from beds n, o, p, q and r belong to the periods of channel-cutting and aggradation which occurred after the end of the Pastonian. They show high frequencies of Gramineae pollen, with *Betula*, *Pinus* and *Alnus* being dominant in the AP figures, and a variable, sometimes strong, representation of Cyperaceae pollen. The assemblage from bed u, BP (166 m, Fig. 11) and BA–B4 (45 m, Fig. 18), the final silt horizon in the sequence, shows rather lower NAP frequencies. It is possible that this is related to a post-Beestonian time of more temperate conditions. The possibility that these horizons may be Cromerian is discussed later (*p.* 44).

In Table 42, the Beestonian biozones are not separately named. Uncertainties about exact stratigraphic relationships and the general similarity of the high NAP biozones make a grouping of biozones of no additional use.

(e) Vegetational history: Pre-Pastonian

The vegetational history of substage Pre-Pa a at Beeston has already been described in Chapter 2. The pollen assemblages were derived from sediments associated with the stone bed (bed b) on the Chalk, and indicated cold conditions. Here we are concerned with the remainder of the substages Pre-Pa b, c and d. It may be noted that other pollen assemblage biozones in this group of substages are found at East Runton, Ostend and Happisburgh.

(i) *Pre-Pastonian b*

The Gramineae–*Artemisia* biozone of this substage is recorded in the silts and sands of bed d and the succeeding silty mud of bed e. The record is based on pollen, since the single macrosample of mud (BR–S 180–200 cm) contained only fragmentary wood and organic remains with one nut of *Urtica dioica*. At this time of fluviatile and open-water conditions, there are high frequencies of NAP. The principal trees and shrubs are *Betula* and *Pinus*, with lower frequencies of *Alnus* and *Salix*. The herbs are mainly Gramineae and Cyperaceae, but marked frequencies of Compositae Tubuliflorae and *Artemisia* pollen are present, together with records of *Plantago maritima*, *Linum catharticum*, *Polygonum bistorta/viviparum*-type, *Ranunculus*, *Rumex* and *Thalictrum*. Pollen of aquatics is scarce, only including records of *Sparganium* and *Typha latifolia*. The assemblage must derive from mainly herbaceous communities adjacent to a river, with scattered but restricted woodland of *Betula*, *Salix* and *Alnus* with some *Pinus*.

(ii) *Pre-Pastonian c*

The Gramineae–*Pinus–Betula–Alnus* biozone of this substage is recorded in the laminated silt of bed g and in the organic lenses in the gravel of bed f. Fluviatile conditions continued in this substage. The pollen spectra show higher AP frequencies, with increases in *Pinus*, *Alnus* and *Ulmus* over the preceding biozone, and reductions in *Salix*, *Artemisia* and Compositae Tubuliflorae. The variety of herb taxa remains, with Chenopodiaceae, *Plantago maritima*, *Filipendula*, *Polygonum bistorta/viviparum*. The aquatic pollen shows an increased representation, with *Sparganium*, *Nuphar* and *Alisma*, and there are increased frequencies of Filicales spores. The assemblages indicate a slight expansion of woodland in this substage, with herbaceous communities still very well represented. A slight amelioration of climate is probably indicated.

Macroscopic plant remains were obtained from both

sites yielding pollen evidence. At BKF, washings from the pollen samples gave *Potamogeton* endocarps and achenes of *Ranunculus* subg. *Batrachium*. A longer list was obtained from BR–S 140–155 cm. Fig. 15 gives frequencies of ecological groups in this sample. The remains are mostly those of aquatics, herbs of shallow-water/muddy substrates and damp-ground herbs. The list includes *Thalictrum alpinum*, *Selaginella selaginoides*, *Alisma*, *Eleocharis*, *Chara*, *Potamogeton*, *Typha* and *Azolla*, and indicates a fairly rich local herb flora. No tree or shrub remains were found.

(iii) *Pre-Pastonian d*

The Gramineae–Cyperaceae–*Betula–Pinus* biozone of this substage is recorded in the mud of bed i at BR–S and BAX. The AP frequencies are lower than in the previous biozone, with *Betula* and *Pinus* the principal trees. *Picea*, *Alnus* and *Salix* are represented by low frequencies. Gramineae and Cyperaceae pollen form the bulk of the herbaceous pollen, but a variety of herb taxa is also recorded, including Chenopodiaceae, Compositae, *Artemisia*, *Plantago media/major*, *Polygonum bistorta/viviparum*, *Thalictrum* and Umbelliferae. *Sparganium* pollen frequencies remain as in the previous biozone. The spectra indicate a recession of woodland, with pollen frequencies rather similar to those of Pre-Pa b.

A single macroscopic sample was analysed from the muddy sediment (bed i) included in this biozone at BAX 170–190. The sediment also includes part of the subsequent biozone of Pa I. A rich aquatic flora is predominant, with *Callitriche*, *Hippuris*, *Nuphar*, *Potamogeton* spp., *Ranunculus* subg. *Batrachium*, *Sagittaria*, *Stratiotes* and *Typha*. The *Potamogeton* spp. (cf. *epihydrus*, *gramineus*, *lucens*, cf. *natans* and *trichoides*) indicate base-rich conditions on an organic substrate with rather shallow water. Few taxa of other ecological groups are present. A single fruit of *Betula* and a *Salix* bud scale are recorded; few remains of dry-ground taxa are represented. The flora is thus predominantly aquatic and local, indicating a rather large and productive pool, as indicated by the sediment and its wide occurrence in the Beeston sections. This period also appears to be represented by the organic silt at East Runton (Chapter 5).

(f) Vegetational history: Pastonian

(i) *Pastonian Ia*

The *Betula–Pinus*–Gramineae biozone of this division of substage Pa I is recorded in silt in the sand of bed g (BTB 1, BTA 9, 8) and in the mud of bed i. The AP frequencies rise, with *Betula* the most abundant tree. The spectra indicate that fluviatile and pool conditions continued as climatic amelioration allowed the spread of birch forest. *Betula* is accompanied by *Pinus* and *Juniperus* with low frequencies of *Picea*, *Alnus* and *Salix*. Gramineae is the most abundant herb pollen, accompanied by Cyperaceae, Chenopodiaceae, Compositae, *Artemisia*, *Filipendula*, Ranunculaceae, *Thalictrum* and Umbelliferae. Aquatics include *Myriophyllum spicatum* and *verticillatum*, *Sagittaria*, *Sparganium* and *Typha latifolia*.

Analysis of macroscopic remains was made at BKA. The sample at BAX 170–190, already considered, possibly also covers part of this biozone, as may the 'Stratiotes' sample, extracted from a fallen mass of mud at site BKA. The BKA sample shows again the preponderance of aquatics, and a paucity of dry-ground taxa. *Cladium mariscus* is the only fen/reedswamp representative. The aquatics include *Hippuris*, *Nuphar*, *Potamogeton natans*, *P.* cf. *perfoliatus*, *P. trichoides*, *Ranunculus* subg. *Batrachium*, *Sagittaria sagittifolia*, *Stratiotes* and *Typha*. The flora indicates the continuance of base-rich conditions.

(ii) *Pastonian Ib*

The *Pinus* biozone of this division of substage Pa Ib is recorded at many sites, all in the mud of bed i, except in the lower part of BW where it is present in the silt of the sand of bed g. Fluviatile and pool conditions continued in this biozone, in which *Pinus* becomes very much the dominant tree pollen taxon. *Picea*, *Betula* and *Salix* are present in lower frequencies. Of the herb flora Gramineae is the most abundant taxon, accompanied by lower frequencies of Cyperaceae, with a continued variety of herb taxa, including Chenopodiaceae, Compositae in a variety of types (*Arctium*, *Bidens*, *Centaurea nigra*, *Cirsium*, *Matricaria*), *Epilobium*, *Filipendula* and others. The aquatic pollen flora is well represented, with *Alisma*, *Myriophyllum spicatum* and *verticillatum*, *Sagittaria*, *Sparganium*, *Stratiotes* and *Typha latifolia*.

The spectra represent the replacement of *Betula* woodland by *Pinus* woodland, with the continuation of a herb element related probably in part to the pool conditions of sedimentation.

An analysis of macroscopic remains in this biozone comes from BR–S 80–90 cm. The flora is again largely aquatic, but with an increased representation of the tree, shrub, maritime and dry-ground herb elements, including cf. *Alnus*, *Betula pendula/pubescens*, *Salix*, *Aster tripolium*, *Salicornia*, *Corispermum* (*see* Wilson, 1973) and *Rumex acetosella*. The aquatics include *Callitriche*, *Chara*, *Potamogeton pectinatus* and *Zannichellia*. The presence of *Aster* and *Salicornia* indicate neighbouring maritime conditions, and the water in the pool may have become brackish, as is also perhaps indicated by the presence of *Potamogeton pectinatus* and *Zannichellia*, previously unrecorded in the BR–S section. This is a likely change, since marine transgression sediments are present in the subsequent biozone.

(iii) *Pastonian II*

The *Pinus–Quercus–Carpinus–Ulmus* biozone of this substage is recorded in the grey-blue silts and clays with tidal structure of bed k. The pollen spectra, with high AP frequencies, show the maximum representation of thermophilous tree and shrub genera in the Pastonian, with *Ulmus*, *Quercus*, *Carpinus* and *Corylus* present. The transition from the previous biozone to this one is best seen in BQ (Fig. 20), where the gradual change from the mud i to the tidal silt and clay of bed k marks the change in pollen content. The thermophilous trees are accompanied by *Pinus*, *Picea*, *Alnus* and *Betula*. The herb pollen flora is dominated by Gramineae and Chenopodiaceae, but the variety is less than in the earlier biozones. *Armeria* is present, with Compositae, *Thalictrum* and Umbelliferae. The aquatic pollen flora is also more restricted. The pollen spectra indicate the spread of thermophilous forest into the region, with pollen deposition under tidal conditions.

An analysis of macroscopic remains was made from the tidal silts and clays at BR 40–50 cm. The aquatic element is reduced, with the tree, maritime and heath elements greatly increased. The tree remains include fruits and cone scales of *Betula* cf. *pubescens*, conifer leaves, and a leaf of *Picea* cf. *omorika* (*see* Wilson, 1973). The maritime element includes *Aster tripolium*, *Ruppia* sp., *Salicornia* sp. and Gramineae cf. *Spartina* sp. The heath element includes *Andromeda polifolia*, *Chamaedaphne calyculata* (*see* Wilson, 1973), *Erica* spp., *Erica* cf. *tetralix* and *Eriophorum vaginatum*. The shallow-water, muddy substrate and fen elements include *Alisma plantago-aquatica*, *Ranunculus sceleratus* and *Urtica dioica*. The aquatic flora is much reduced in numbers of remains; it includes *Elatine*, *Hippuris*, *Menyanthes*, *Potamogeton* spp. and *Typha angustifolia*. This macroflora is obviously derived from a variety of nearby environments – which must include maritime communities (salt marsh), heath, fen and freshwater. In view of the Chalk basement rock of the area, it seems likely that the heath must have been ombrogenous in origin, and perhaps bore the same topographic relation to saltmarsh as the coastal ombrogenous bogs of the North-Sea coast do at the present day.

(iv) *Pastonian III*

The *Pinus–Betula–Quercus* biozone ascribed to this substage is based on few pollen analyses from the upper part of the tidal silty clays of bed k. The spectra have similarity with those of Pa II, but show a reduction in the pollen of thermophilous trees (*Ulmus*, *Quercus*, *Carpinus*, *Corylus*) and an increase of Ericales pollen and *Sphagnum* spores. The macroscopic sample from BR–S 40–50 cm, already discussed, is partly included in this biozone. The spectra indicate a regression of the temperate forest and the beginning of the expansion of heath.

(v) *Pastonian IVa*

The *Pinus–Alnus–Picea–Betula*–Ericales biozone of this first division of substage Pa IV is present in the uppermost part of the tidal silty clays of bed k at BG and BKA and in the wood peat, bed l, overlying this clay at BR–S. Regression therefore appears to have taken place during this biozone. The tree-pollen frequencies remain high in the biozone. A great reduction in pollen of thermophilous trees is accomapnied by rises in pollen frequency of *Pinus* and *Picea*, and by rise in the frequencies of *Calluna*, *Empetrum* (including *E. hermaphroditum*-type) and Ericales tetrads. The latter change is most marked in the analyses from the wood peat, the macroscopic remains from which show the local development of heath vegetation. But it is also reflected to a lesser degree in the coeval tidal sediments. *Bruckenthalia* pollen is also present in the wood peat of BR–S. The NAP component of the wood peat, apart from Ericales taxa, is low, with few taxa represented. On the other hand, *Sphagnum* frequencies are very high. In the spectra from the tidal sediments, the NAP is evidently derived from wider sources. Gramineae, Cyperaceae, Chenopodiaceae and *Sphagnum* frequencies are considerable and a variety of other NAP taxa is present. The regional forest was evidently dominated by *Pinus*, *Picea*, *Alnus* and *Betula*, with associated heath, with the resultant pollen rain sorted into two assemblages associated with local heath and tidal deposition.

The contribution of the heath is more accurately seen in the macroanalysis from the wood peat BR–S 20–30. The predominant groups are the heath (91 %) and trees. Tree remains include fruits of *Betula pendula/pubescens*, cone scales of *Betula*, Coniferae leaves, bud and budscale of *Picea*, and wood and seeds of *Pinus*. The heath group includes fruits with seeds of *Calluna vulgaris*, seeds of *Chamaedaphne calyculata*, seeds of *Erica* (very abundant), seeds of *Erica lusitanica*, *E.* cf. *tetralix*, *E. umbellata*, sclerenchyma spindles of *Eriophorum vaginatum*, and fruits of *Vaccinium myrtillus*. *Phragmites* rhizomes were also present, and *Menyanthes*, together with leaves and capsules of *Sphagnum*. The identification of the *Erica* spp., on the basis of scanning electron microscopy, is described by Wilson, Marchant & Oldfield (1973).

This assemblage, both pollen and macroscopic plant remains, is phytogeographically one of the most interesting of the pre-glacial floras. It combines a number of oceanic west European species (*E. lusitanica*, *E. tetralix*, *E. umbellata*) with the boreal circumpolar *Chamaedaphne calyculata*, and the Balkan taxon *Bruckenthalia spiculifolia* (Andersen, 1973). These taxa must all have been present in ombrogenous heath, developed on the surface of tidal silts, after the regression of the Pastonian sea.

(vi) *Pastonian IVc*

The *Betula–Pinus* biozone ascribed to this last division of substage Pa IV is recorded from a silty lamination near the base of the sand of bed m at BR–S. *Betula* is the predominant tree, with reduced pollen frequencies of *Pinus*, *Picea* and *Alnus*. The Ericales pollen taxa and *Sphagnum* are also greatly reduced from the frequencies seen in the previous biozone. The NAP is predominantly Gramineae, with low frequencies of Cyperaceae, Chenopodiaceae, *Plantago maritima*, Compositae Liguliflorae, and *Ranunculus*-type. *Sparganium* is the only aquatic pollen type.

The spectrum indicates a further change of the vegetation to birch woodland, with the apparent reduction of heath. This was accompanied by a change to fluviatile deposition.

(g) Vegetational history: Beestonian and post-Beestonian

The post-Pastonian sequence of biozones in the complex succession of the Beeston cliffs is best dealt with horizon by horizon, the oldest first.

(i) *Bed n*

This is the channel filling of laminated silt, sand and gravel, with involutions and an ice-wedge cast, at 104–120 m (fig. 17). Pollen spectra BF 1 (Gramineae–*Betula–Pinus* p.a.b.) and BF 2 (Gramineae–*Betula* p.a.b.) show high NAP frequencies. The principal tree-pollen types are *Betula*, *Pinus* and *Alnus*. There are high frequencies of Gramineae and Cyperaceae pollen, with a variety of other NAP taxa recorded: Chenopodiaceae, Caryophyllaceae, Compositae Liguliflorae and Tubuliflorae, *Artemisia*, Cruciferae, Ranunculaceae, *Thalictrum* and Umbelliferae.

There are low frequencies of other trees: *Picea*, *Ulmus*, *Quercus*, *Carpinus* and *Corylus*. The presence of thermophilous taxa in these and younger high NAP pollen spectra poses a problem. The sedimentary evidence and the high NAP frequency indicate a cold-stage type of environment. The origin of the low frequencies of thermophilous trees may then perhaps be sought in the likelihood of their derivation from the temperate stage sediments cut through by the channel formation. This interpretation, though impossible to prove, is adopted as the most likely cause of the presence of the thermophilous tree pollen taxa. The spectra then indicate predominantly herb vegetation, probably grassland, with scattered *Betula*, *Alnus* and *Pinus*.

(ii) *Bed o*

Two spectra are associated with the lower part of the stratified gravel of this horizon, BAV (Gramineae–*Betula–Pinus* p.a.b.) and BM 1 (Gramineae–*Betula* p.a.b.). There are the same problems of interpretation as with the previous spectra. NAP frequencies are 60–70%, and include marked frequencies of *Artemisia*. The gravels cut off and overlie earlier temperate stage sediments, so opportunity for derivation is clearly present. The spectra indicate that herb vegetation persisted during the time of deposition of the gravel.

(iii) *Bed p*

Two shallow basins of muddy silt BF 3 (Gramineae–Cyperaceae–*Betula–Pinus–Alnus* p.a.b.) and BV 12, 13 (Gramineae–Cyperaceae p.a.b.) gave pollen spectra with high NAP frequencies. Sample BU 3 (Gramineae–Cyperaceae–*Betula–Alnus* p.a.b.) is probably also associated with this time. The BF 3 spectrum is similar to those of BF 1 and 2 spectra, except that a higher Cyperaceae value is shown. The silt fills a basin showing ice-wedge casts at its base, so again there is a relation between permafrost conditions and a high NAP spectrum, as with BF 1. The BU spectrum is similar to the BF spectrum. The BV spectra from the more organic sediment show greatly increased values of Cyperaceae and very depressed tree values. *Betula* and *Alnus* are below 5% and *Pinus* below 10%, but frequencies of *Ulmus*, *Quercus* and *Carpinus* are still shown. The NAP includes Chenopodiaceae, Caryophyllaceae, Compositae Tubuliflorae, *Artemisia*, *Epilobium*, *Potentilla*, Ranunculaceae and *Thalictrum*. Aquatic pollen includes *Myriophyllum*, *Sparganium* and *Caltha*.

Two macroscopic samples were analysed from BV. The ecological grouping (Fig. 15) shows a predominance of aquatics, with the lower sample (BV 13), from more organic sediment, showing a bigger component of shallow-water, damp-ground taxa. Trees or shrubs include cone scales of *Betula* sp. and *Betula pendula/pubescens*, and buds and budscales of *Salix*. Amongst the herbs occur seeds of the alpine species *Cherleria sedoides*, achenes of *Potentilla nivea/tabernaemontani* or *Sibbaldia procumbens*, seeds of *Juncus* cf. *effusus* and *J. subnodulosus*, and *Eleocharis* nuts. *Ranunculus flammula*, *R. sceleratus*, *Rorippa microphylla*, *Chara*, *Hippuris vulgaris*, *Nitella/Tolypella*, *Potamogeton* cf. *nodosus*, *Ranunculus* subg. *Batrachium*, *Sagittaria* and *Typha* are also present, with *Carex*, Gramineae, Cruciferae, *Saxifraga* and *Taraxacum*. The flora indicates low shrub with *Betula* and *Salix*, and an abundant herb-rich grassland and local damp-ground/aquatic flora. The flora may be considered a cold stage flora, again indicating that the low frequencies of thermophilous tree pollen are derived. The same may apply to the megaspores of *Azolla filiculoides* found in the macroscopic samples.

(iv) *Bed q*

Pollen spectra were obtained from two samples of silt in this stratified sand, at BF 4 (Gramineae–*Betula–Pinus–Alnus* p.a.b.) and at BA–B 2 (Gramineae–*Betula–Pinus* p.a.b.). Both show very high NAP. *Betula* is the best represented tree in both spectra. The BF 4 spectrum has a remarkably high frequency of Caryophyllaceae pollen.

(v) *Bed r*

Pollen spectra were obtained first from the more organic base of the laminated silts of bed r at BA–B 3 (44 m, Gramineae–*Betula–Pinus–Alnus* p.a.b.). The spectrum shows high NAP, with *Betula* the most abundant tree, high Gramineae and *Artemisia* frequencies, with pollen of *Plantago maritima*, *P. major/media* and *Polygonum*. Spectra were also obtained from silts at BK (136 m, Gramineae–Cyperaceae–*Pinus–Betula–Alnus* p.a.b.), BM 2 (170 m, Gramineae–Cyperaceae–*Pinus* p.a.b.), and from BAM (281 m, 130–220 cm, Gramineae–Cyperaceae–*Pinus–Alnus–Betula* p.a.b.; 45 cm, Cyperaceae–Gramineae–*Pinus* p.a.b.). The basal spectrum of BAM shows rather higher AP frequencies (*Pinus*) than the other spectra. The uppermost BAM spectrum shows very high Cyperaceae values, probably a result of local pollen production from this organic horizon with freshwater shells. In this respect it is similar to the BV spectra. Otherwise the spectra are typical of the high NAP Beestonian spectra, with *Betula*, *Pinus* and *Alnus* being the principal AP taxa, and Gramineae and Cyperaceae the principal NAP taxa.

Two macroscopic samples were analysed from this horizon, one from BA–B 3, and one from BAM 50–60 cm. The analysis from BA–B 3 shows a variety of ecological groups (Fig. 15) represented, including a substantial dry-ground group. Trees or shrubs are represented by fruits of cf. *Betula* and *Betula* cf. *nana*, herbs by *Armeria maritima*, *Atriplex patula*, *Cerastium arvense*, *Chenopodium rubrum/botryodes*, *Potentilla anserina*, *Thalictrum minus*, fen and reedswamp by *Carex aquatilis/bigelowii* and *Ranunculus sceleratus*, aquatics by *Callitriche hermaphroditica*, *Hippuris vulgaris* (very abundant), *Myriophyllum*, *Ranunculus* subg. *Batrachium* and *Zannichellia palustris*. All of these taxa are characteristic of cold stage floras of the Upper Pleistocene, and it is clear that this assemblage has the same environmental interpretation in the Beestonian, indicating open conditions of vegetation with disturbed ground under a cold climate. The flora is associated with permafrost in the sense that ice-wedge casts immediately predate the silt deposition of bed r.

The macroscopic analysis from BAM 50–60 cm is from laminated muddy silt. It is the best example of a cold stage 'leaf flora' which has yet been analysed from the coastal pre-glacial sediments. The leaves included cf. *Betula nana*, *Salix* cf. *polaris*, *Salix* cf. *repens*, *Salix* cf. *reticulata* and *Saxifraga oppositifolia*. A calyx of *Armeria maritima* was recorded and an achene of *Thalictrum alpinum*. A number of nuts of *Carex aquatilis/bigelowii* type were recovered, perhaps relating to the high frequencies of Cyperaceae pollen in the pollen spectrum from the sediment (BAM 55 cm). The flora indicates the presence of montane dwarf shrub communities at this time, and has taxa in common with the flora from BA–B 3.

(vi) *Bed u*

Pollen analyses were obtained from the silt of this bed at BA–B 4, in the western channel, and at BP in the eastern channel. A *Pinus–Betula–Alnus*–Gramineae–Cyperaceae p.a.b. is present at the former site, and a *Pinus–Betula–Alnus*–Gramineae p.a.b. at the latter site. The AP at both sites totals about 60% of the total pollen. The principal taxa are *Pinus*, *Alnus*, *Betula* and *Picea*, but there is a minor representation of *Ulmus*, *Quercus*, *Carpinus* and *Corylus*. The principal NAP taxa are Gramineae, Cyperaceae and Chenopodiaceae. In showing relatively high AP frequencies these spectra differ from the other post-Pastonian samples, except for the spectrum at BAM 220 cm. This spectrum is older than the cold-stage flora of BAM 50–60 cm and therefore older than the silts of bed u, from which they are both separated by the strong period of weathering. It is possible that the hiatus represented is pre-Cromerian, and may be the same as that which resulted in weathering below the Cromerian transgression sediments at WRBCa (406 m, Fig. 11) and further east. If this correlation is right, then the assemblages of bed u may be late Cromerian, Cr IVa, represented at West Runton by a *Pinus–Alnus–Picea–Betula* p.a.b. in silts just below the till in the upper part of the cliff section. The macroscopic analysis from BA–B 4 is rather distinct from the earlier macroscopic plant remains assemblages from Beeston. It contains *Eupatorium cannabinum*, *Isoetes histrix*, many megaspores of *Azolla filiculoides* and a few of *Salvinia natans*. These give the flora a more temperate aspect than the earlier post-Pastonian floras. These floras of bed u occur in laminated silt closely below the till in the upper part of the pre-glacial sections at Beeston in a stratigraphical position comparable to Cr IVa sediments further east and provide support for the tentative correlation of the assemblages to Cr IVa, made above on the basis of the floras. It is not clear whether the silts are tidal in origin, as are the Cr IVa silts at West Runton. At BA–B 4 they contain 8% pollen of Chenopodiaceae, a figure usually signifying marine influence and a seed of *Juncus gerardii*, as well as remains of freshwater plants.

5: East Runton

(a) Lithostratigraphy

Site ERA (location in Fig. 2; section in Fig. 22) was briefly exposed in 1966, very near low tide mark, 400 m east of East Runton Gap. It showed 135 cm dark grey, freshwater, muddy silt. The Chalk surface with the ferrugineous stone bed on it was exposed at about the same level nearby. C. Reid (1882, *p.* 28) noted that there was 'exposed at low water two or three feet of black mud' three-quarters of a mile north-west of Cromer, the same location as ERA. The relationship of the muddy silt to under- and overlying sediments could not be seen in 1966. But Reid (*loc. cit.*) reported that 'This mud can be traced for about 100 yards, cutting through the Weybourne Crag, and in one place for a few feet into the Chalk, so that it extends to extreme low water mark. Resting on and overlapping it there is greenish laminated sandy clay full of wood and containing occasional marine shells and fir-cones'.

Thus this freshwater organic horizon overlies crag (≡ bed c at West Runton, east of Woman Hithe) and underlies laminated sediments which are almost certainly equivalent to those of beds di and dii at West Runton. It occupies a similar position, near low water, as the thin freshwater muds east of Woman Hithe (Fig. 4), and indicates regression in a period between the crag and the silty clays of beds di and dii at West Runton. The biostratigraphical evidence indicates that the freshwater silt was deposited during substage Pre-Pa d, bearing out evidence for regression in Pre-Pa times after substage Pre-Pa a.

(b) Biostratigraphy

The muddy silt shows a single biozone (Table 12, Fig. 25), a Gramineae–Cyperaceae p.a.b., correlated with the Gramineae–Cyperaceae–*Betula*–*Pinus* p.a.b. of Pre-Pa d at Beeston.

(c) Vegetational history

The pollen diagram shows low frequencies of AP, mainly *Pinus* and *Betula*, with high NAP (65–80%) composed mainly of Gramineae and Cyperaceae, but also with pollen of Chenopodiaceae, Caryophyllaceae, Compositae Tubuliflorae, *Artemisia*, Cruciferae, *Ranunculus*, Rubiaceae and *Thalictrum*. The regional vegetation was evidently herbaceous – a typical cold stage regional assemblage. Table 13 gives a list of macroscopic plant remains from this site. Fig. 15 gives ecological groupings of the flora. A wide variety of habitats is represented, the species being all herbaceous except *Salix* (*Salix* pollen present). Aquatics form the largest group, but many terrestrial taxa are present, including *Ranunculus hyperboreus*, a characteristic member of late Pleistocene cold stage floras. The pollen frequencies of Gramineae and *Ranunculus* are paralleled by the abundant macroscopic plant remains of Gramineae and *Ranunculus*, including *R. hyperboreus*, *R. sceleratus* and *R.* subg. *Batrachium*.

6: Cromer and Overstrand

(a) Lithostratigraphy

The sections of the Cromer Forest-bed Formation along this part of the coast have been obscure for many years. One section, CA, was recorded east of Cromer and one at Overstrand, OB. The locations of these two sections are shown in Fig. 2. In addition, samples from boreholes at the base of the Overstrand cliffs, made in connection with land-slide studies (Hutchinson, 1976), were provided by J. N. Hutchinson. The sections observed and the stratigraphy of the boreholes are shown in Fig. 22.

The sediments can be grouped into beds as follows. The succession is summarised in Table 14.

a. Stone bed. The boreholes show up to 15 cm of sand with flint cobbles resting on the Chalk at *c.* −4·25 m OD. In borehole 1B grey clay and marine shells were present.

b. In boreholes 2B and 3B grey sand with marine shells, with a maximum thickness of 4·3 m, succeeds the stone bed. A mollusc fauna from this horizon is recorded by Markham (1973).

c. In borehole 4S a pebbly sand a metre thick occurs on grey sand, base at −3·1 m OD.

d. A thin seam of grey silty clay is found in grey sand at −2 m OD in borehole 4S.

e. 45 cm of grey pebbly sand overlies bed d in borehole 4S. A similar sediment is seen at the same level in borehole 1B.

f. 1·5 m of grey sand overlies the upper pebbly sand in borehole 4S.

g. In boreholes 2B and 3B a grey, laminated, silty clay, with seams of grey sand, overlies grey sand at 0·2 m to 1·7 m OD. In borehole 3B it contains gravel and clay conglomerate at around 90 cm OD. In section CA, 800 m to the west and at about the same level, grey laminated silty clays were seen above a horizon with clay conglomerate 50 cm thick. A branch of wood a metre long was seen in the upper part of the section. At OB, 1800 m to the east and at +2 m OD, grey laminated silty clay is again seen. This clay, with clay conglomerate, was seen to outcrop at the base of the cliff for about 80 m west of OB. Its surface is penetrated by rootlets at OB.

h. This and the succeeding horizons have been observed only at site OB. 30 cm of ferruginous gravel and sand, with clay conglomerate, lie on the surface of the laminated clays.

i. 60 cm of pale sand with oblique bedding, apparent dip west, and gravel strings, was seen on the ferruginous gravel. This horizon is similar to the cross-bedded sand of bed f at Paston–Mundesley, and occupies a similar stratigraphical position.

j. A further horizon of ferruginous gravel, 25 cm, with an irregular base cutting into the bedded sand, follows.

k. 30 cm of grey laminated silty clay with tidal structure rests on the ferruginous gravel.

l. 95 cm of sand with a brown silty sand in the middle and more gravelly to the base rests on the laminated silts.

m. 25 cm of brown detritus mud, more silty at the top, and with large freshwater bivalves in the basal part succeeds the sand.

n. The succession is sealed by till.

(b) Sedimentary history

The stone bed (bed a) with marine molluscs, was deposited on the Chalk after marine transgression. A series of sands with marine shells (bed b) were then laid down. The gravel of bed c may be later than the shelly sand. It was only found in borehole 4S, perhaps implying a regression and channelling of the shelly sand, with the deposition of gravel, then grey silty clay (bed d) and the further gravel (bed e) of boreholes 4S and 1B, both seaward of the other boreholes. The sand of bed f and then the laminated clays of bed g were deposited. This latter horizon, observed along a long stretch of coast in the Cromer–Overstrand area, indicates tidal conditions of sedimentation. Regression followed with the growth of vegetation on the surface of the clay (OB – 'rootlet bed'). Deposition of gravels (beds h and j) and an intervening cross-bedded sand followed, and then a further bed (k) of tidal sediments was formed, indicating transgression. A series of sands, with silt horizons and gravel near the base, possibly beach and tidal sediments, was deposited, followed by the formation of freshwater muds (bed m), indicating another regression.

(c) Biostratigraphy

The biozones identified in pollen diagrams from Cromer and Overstrand (Fig. 25) are listed in Table 12. Table 14 gives a summary of the biozone sequence and vegetational history.

In the marine laminated silty clay (bed g) of section CA at Cromer, two biozones are present, a lower one (a), *Pinus–Quercus–Carpinus–Ulmus* p.a.b., correlated with Pa II, and an upper one (b), *Pinus–Picea* p.a.b., correlated with Pa IVb.

At Overstrand, clay in the basal stone bed (bed a) shows a *Pinus*–Gramineae–Ericales p.a.b., correlated with Pre-Pa a at Beeston. A *Pinus–Quercus–Ulmus–Carpinus* p.a.b. (biozone b) is found in clay at a higher level (bed d), and is correlated with Pa II. In the upper part of the borehole 3B, a *Pinus–Picea–Alnus–Betula* p.a.b. (biozone c) is correlated with Pa IVa, and is succeeded by a *Pinus–Picea* p.a.b. (biozone d), correlated with Pa IVb. This latter biozone is similar to those found in the clay (bed g) at section OB and in the upper part of the section CA at Cromer.

In the upper part of the section at OB an *Abies–Carpinus* p.a.b. (biozone e) is found in the tidal laminated clay of bed k, and a *Pinus–Betula* p.a.b. (biozone f) in the freshwater mud of bed m. These two biozones are correlated with Cr IIIb and Cr IVb respectively.

(d) Vegetational history

(i) *Pre-Pastonian*

A sample from the stone bed (bed a) in Overstrand borehole 1B gave a *Pinus*–Gramineae–Ericales p.a.b. ascribed to this substage (see Chapter 2).

(ii) *Pastonian II*

A *Pinus–Quercus–Ulmus–Carpinus* p.a.b. is recorded in grey silty clay of bed d in Overstrand borehole 4S and in tidal silty clays (bed g) at the base of the Cromer section CA. *Ulmus*, *Quercus*, *Carpinus* and *Corylus* are well represented in these spectra, though *Pinus* and *Alnus* pollen is more frequent. In the NAP, Ericales taxa have a low representation as usual in the Pa II tidal sediments and Gramineae and Chenopodiaceae pollen frequencies are considerable. Regional deciduous temperate forest with *Pinus* is indicated, with local maritime communities.

(iii) *Pastonian IVa*

A *Pinus–Picea–Alnus–Betula* p.a.b. is recorded in the silty clay of bed g in the Overstrand borehole 3B. With these genera, pollen of thermophilous trees is poorly represented. There is a low frequency of Ericales taxa, with *Sphagnum*, and Gramineae pollen is well represented. Regional boreal forest is indicated.

(iv) *Pastonian IVb*

A *Pinus–Picea* p.a.b. is recorded in silty clay of bed g at a higher level in Overstrand borehole 3B, in the basal laminated silty clay (bed g) of the Overstrand section OB and in the upper part of the laminated silty clay of the Cromer section CA. These spectra show high AP frequencies, with higher *Pinus* and *Picea* frequencies than in the spectra of Pa IVa. Otherwise they have a similar pollen content, with low frequencies of pollen of Ericales taxa, Gramineae well represented, and low frequencies of Chenopodiaceae pollen. In the Overstrand spectra of OB there are higher frequencies of pollen of aquatics than elsewhere, with *Nymphaea*, *Sparganium* and *Typha latifolia*, perhaps indicating regression, known to occur at Beeston in the early part of Pa IV. This regression is also indicated by the presence of rootlets penetrating the silty clay at the OB section.

(v) *Cromerian IIIb*

The characteristic *Abies–Carpinus* p.a.b. of substage Cr III is recorded in the tidal laminated silty clays of bed k at the Overstrand section OB. As at West Runton, there are low frequencies of *Ulmus*, *Quercus* and

Tilia. Empetrum, Gramineae and Chenopodiaceae pollen form the bulk of the low NAP total and Filicales spores are also frequent.

(vi) *Cromerian IVb*

A *Pinus–Betula* p.a.b. is recorded in freshwater detritus mud, bed m, at the top of the OB section. The AP frequency is high, and the only other tree taxa represented are *Alnus* and *Ulmus*. No Ericales taxa pollen is present and the low NAP total is mostly Gramineae. There is a greater variety of NAP taxa than in the earlier marine sediment; Cyperaceae, Chenopodiaceae, Caryophyllaceae, Compositae Liguliflorae, *Matricaria*-type, Cruciferae, *Polygonum aviculare*, *Ranunculus* and Umbelliferae are present, with *Sparganium*, *Typha latifolia* and *Salvinia natans* in the aquatic group. The assemblage indicates a regional boreal forest dominated by *Pinus* and *Betula*, with local herb communities. A degree of thermophily is suggested by the presence of massulae of *Salvinia natans*.

7: Sidestrand

(a) Lithostratigraphy

Along the Sidestrand stretch of coast there has been much disturbance of the Pleistocene pre-glacial deposits by the thrusting-up of large masses or rafts of Chalk together with the superincumbent Pleistocene deposits (Plates 14, 15, 16 and 19) (Peake & Hancock, 1970; Banham, 1975). As a result it is impossible to determine the original levels of the sediments, but this is more than counterbalanced by the opportunity so given to examine in open section sediments which would otherwise be hidden below sea level. The sections are especially of interest in showing clearly the relations between Pre-Pa, Pa and Cr sediments and biozones.

The datum point for locating the sections is shown in Fig. 2. Some of the sections are on the foreshore (SSN, SSP) and others in the cliff. Certain of the cliff sections show Chalk at the base (SSB, SSV, SSF, SSE). The other cliff sections (SSR, SSJ, SSD, SST, SSU) may also be resting on displaced Chalk, but talus and beach concealed their relation to the Chalk. The sketch inset in Fig. 22 shows the relation of certain sections to the displaced Chalk masses. This figure also shows all the sections, except SSV, which is given in Fig. 23.

The stratigraphy of this stretch of the coast is at first sight difficult to interpret, even though the sections may be good, since there are rapid lateral changes of details of lithology imposed on the predominantly sandy nature of the deposits. The most extensive section, SSV (Fig. 23; Plate 16) is taken as a standard and the other sections are related to it as far as possible. Some (SSP, SSN) are not possible to place on stratigraphic evidence alone. The origin and structure of the marine sediments at Sidestrand are further discussed in Chapter 16. Plates 14, 15, 16 and 19 show the arrangement of the strata.

The following beds are recognised. Table 17 summarises this succession.

a. Directly on the Chalk, as seen in the displaced masses, is a stone bed of massive flints, up to 60 cm thick (SSV, Fig. 23). It contains marine molluscs and lenses of grey silty clay. At SSE the massive stone bed is absent.

b. On the stone bed at SSB there is 2·8 m of sand with marine shells and occasional thin or massive grey silty clay laminae, and seams of clay conglomerate and gravel. At the south-east end of SSV there is up to 1·5 m of sand with rare marine shells on the stone bed. Funnell (1970) has described the foraminiferal assemblage from this shelly sand at SSB as being impoverished and suggesting cold conditions. Norton (1967) has described the mollusca fauna as being recent in type and predominantly intertidal.

c. At the south-east end of SSV up to 1·2 m of gravel with marine shell fragments rests on the shelly sand.

d. 1 m of coarse stratified sand with thin grey silt laminae and marine mollusc fragments overlies the gravel of bed d at the south-east end of SSV.

e. 20 cm of ferruginous gravel overlies the coarse sand at the south-east end of SSV. It thins out to the north-west.

f. At the south-east end of SSV up to 30 cm of grey sand with grey silty clay laminae, showing a tidal structure, overlie the ferruginous gravel and, where that is cut out, the coarse sands of bed d. This horizon thins out to the south-east. A similar sediment lies on the Chalk at SSE

(660 m, Fig. 22), and is here 1·3 m thick. The stone bed here consists only of a few small pebbles at the base of this sediment (*see* West & Banham, 1968, p. 504). At SSU and SST similar laminated sediments are associated with pebble gravel interpreted as beach. At both of these sites there are seams of silty drift mud within the laminated sequence.

g. In SSV up to 5 m of red and yellow sands, with cross-bedded and horizontally bedded sets, overlies bed f. This sand contains many small basins of coarse drift mud, with rolled mud pebbles and wood, as well as horizontal laminations of coarse drift mud (Plate 18). It also contains levels with clay conglomerate. At SSE, a similar sand, with drift mud streaks and 14·7 m thick, overlay sediments of bed f type. Beds f and g differ in their structure, the former showing more regular lamination, with clay drapes. When weathered, however, as beneath the unconformity at SSU, the sand becomes red and the silty horizons ferruginous, giving a resemblance to bed g. On the whole, bed g is more sandy, and shows more marked cross-bedding, and the two horizons are therefore separated. They may, however, be contemporaneous with lateral variation of facies and weathering. Chapter 16 further discusses sediments of beds f and g.

h. In SSV, towards the top of bed g, there are gravels lying in three channels cut in the sand. The channels are filled with cross-bedded sand with strings of gravel and thin blue silty clays. The channel at 31 m shows several centimetres of dark blue silty mud/clay sealing in its filling, immediately beneath the gravel of the overlying horizon. A similar channel, cut into bed f, is seen in SST (Plate 20). Here it is filled with 1·5 m of sediment, mainly dark silty mud with stones (bed i), becoming grey, more silty and laminated in the upper part. The filling of this SST channel may be of the same age as the channel fillings of stratified sand of SSV. Gravels at the top of sections SSB and SSF and base of section SSR may belong to this same horizon of gravel.

i. This bed is the freshwater silty mud filling the channel of SST, described above.

j. Right across section SSV a thin ferruginous gravel overlies beds g and h (Plate 17). It truncates the channels associated with bed h. In section SST a similar thin gravel occurs on the truncated cross-bedded set at the top of bed g, but does not extend across the channel filling.

k. A further series of tidal sediments, pale sands with grey silty clay laminae, occurs conformably on the freshwater mud of bed i at SST and on the gravel of bed j at SST and SSV. At the north-west end of SSV, these sediments contained pebbly gravel, interpreted as beach. At SSV this bed reaches a thickness of 2 m, and at SST 40 cm. At this latter site the silty clay is associated with a channel and the channel contains erratics of mud and silty clay. At SSR a similar sediment, but disturbed by burrows, reaches a thickness of 2·4 m.

l. At SSR and SST (Plate 20) a more massive brown and grey-blue silty clay, stony in places, up to 80 cm thick, occurs above bed k. There are sand partings in the upper part in SST, giving a laminated structure, and shell fragments are also present in the sediment at this site.

m. At SST 3 m of coarse-bedded sand overlies the laminated clay of bed l. At SSR 20 cm of similar sediment occurred in the same position.

n. At SSR and SST till overlies the sediments of bed m.

(b) **Sedimentary history**

Marine transgression over the Chalk resulted in the formation of a stone bed (bed a), with marine molluscs. Shallow marine conditions continued with the deposition of shelly sands (bed b), gravel (bed c) and a further sand with marine shells (bed d). Surface texture of sand grains from bed b show the effects of glacial action followed by attrition in either a moderately turbulent aqueous environment or on a beach (Krinsley & Funnell, 1965, sample D 25). Ferruginous gravel (bed e) caps this part of the sequence. The tidal sediments of bed f were then deposited, but there appears to be an unconformity between these and the stone bed, as at SSE they are found lying directly on the Chalk. Unconformity is clearly demonstrated by the biostratigraphy of the Sidestrand sections between the basal stone bed and crag (as at SSB) and the overlying marine sediments, as seen in Fig. 24. Marine deposition continued with the formation of bed g. At SSTB and SSU this sequence is split by beach gravel, which separates Pa II sediments from Pa IV sediments. Much drift mud and wood is present in this horizon, indicating a nearby land source of such debris.

Surface textures of sand grains in the gravels at SSR (150–190 cm), probably belonging to bed g or bed h, again show evidence of glacial action followed by beach action (Krinsley & Funnell, 1965, sample D 15).

Regression followed and these sediments were channelled, the channels first filled with gravel (bed h), then sand and also freshwater mud (bed i). A later transgression formed silty clays above the muds of bed i, and then a gravel horizon marking an erosional surface (bed j), truncating the channels at SSV. The conformable deposition of tidal sediments followed, with laminated sand and silty clay and beach gravels (bed k).

Surface textures of sand grains from the tidal sediments of bed k at SSR again show evidence of glacial action followed by beach action (Krinsley & Funnell, 1965, sample D 20). A more massive silty clay with sand partings was then deposited, probably under marine conditions, then finally water-lain coarse-bedded sands with drift mud streaks (bed m), which may be fluviatile.

The relation of these sediments to substages and the Chalk surface is drawn diagrammatically in Fig. 24. The successions at each site vary considerably and the relationships indicate periods of erosion, with resultant unconformities after the deposition of the Pre-Pa a sediments and between the deposition of the Pa and Cr sediments. There is no evidence for the presence of Beestonian sediments.

Table 17 summarises the sequence described above.

(c) Biostratigraphy

Table 15 lists the biozones identified in the pollen diagrams from Sidestrand (Figs. 25, 26, 27, 38). Table 16 gives the macroscopic plant analyses and Table 17 gives a summary of the biozone sequence and vegetational history.

Biozone a. *Pinus*–Gramineae–Ericales p.a.b. This biozone is based on analyses from the stone bed (bed a) and the overlying sands with their contained silty clays and clay conglomerates (bed b). The spectra show *Pinus* to be dominant with varying but low frequencies of *Picea*, *Betula* and *Alnus*, and with varying, usually considerable, frequency of pollen of Gramineae and Ericales. In spite of these variations the spectra are grouped together in one biozone, since subdivision is difficult. The biozone is correlated with Pre-Pa a. There are two other assemblages associated with this biozone, as follows.

Biozone b. Gramineae–*Alnus*–*Pinus* l. p.a.b. This local biozone is based on a single sample (SSH 20) from sand with streaks of drift mud, near the base of bed b.

Biozone c. Gramineae p.a.b. Two erratics of mud within gravel (SSF 600) and clay conglomerate (SSJ) showed pollen spectra with high frequencies of Gramineae pollen.

Biozone d. *Alnus*–*Pinus*–Gramineae–*Artemisia* p.a.b. This is represented by a single sample, at a higher level in section SSB than biozone a. The correlation is uncertain but may be with a Pre-Pastonian zone (Pre-Pa c ?).

Biozone e. *Pinus*–Gramineae p.a.b. This is again identified by a single sample. It occurs below the next biozone in section SSN and may perhaps be correlated with Pa I.

Biozone f. *Pinus*–*Quercus*–*Ulmus*–*Carpinus* p.a.b. This biozone is represented in several sections and is correlated with Pa II. There is a variation in representation of the various constituents, notably *Alnus* (correlated with horizons of drift mud), but only the following local zone, with high frequencies of *Alnus* and absence of *Carpinus*, is separated.

Biozone g. *Alnus*–*Pinus* l. p.a.b. This is identified in a drift-mud seam in the sand of bed g (SSF 950) and from balls of mud in drift-mud-rich seams in bed g (SSV 11, 2).

Biozone h. *Pinus*–*Picea*–Cupressaceae–*Alnus*–*Betula* p.a.b. This is present in SSP where Cupressaceae pollen frequencies remain high over a depth of three samples. Apart from Cupressaceae the spectra are similar to those of Pa IVa.

Biozone i. *Pinus*–*Picea*–*Alnus*–*Betula* p.a.b. This biozone is identified at several sites from sediments of beds f or g. It is correlated with Pa IVa. A local p.a.b. with much *Alnus* is separated off as follows.

Biozone j. *Alnus*–*Pinus* l.p.a.b. This is identified by a single sample from section SSE (1500).

Biozone k. *Pinus*–*Ulmus*–*Quercus* p.a.b. This biozone is from the freshwater mud filling the channel in section SST. It is correlated with Cr IIa.

Biozone l. *Abies*–*Carpinus* p.a.b. This characteristic assemblage is identified from several sites. It is correlated with Cr IIIb.

Biozone m. *Pinus*–*Alnus*–*Picea*–*Betula* p.a.b. A single sample from SSV, bed k, shows this assemblage, which is correlated with Cr IVa.

Biozone n. *Pinus–Betula* p.a.b. This biozone is seen at the top of section SST, in bed l. It is correlated with Cr IVb.

(d) Vegetational history

(i) *Pre-Pastonian a*

Pollen assemblage biozones ascribed to this substage are described and discussed in Chapter 2, and listed in Table 15. It will be noted that there is considerable variation within the high NAP pollen spectra, probably related to the taphonomy of the assemblages, with a *Pinus*–Gramineae–Ericales p.a.b. characteristic of marine shelly sediments (Weybourne Crag) and a Gramineae p.a.b. in erratics of detritus mud.

(ii) *Pre-Pastonian*

A single spectrum from a thin seam of organic silty clay at the top of the SSB section (SSB 0 cm) shows an *Alnus–Pinus*–Gramineae–*Artemisia* p.a.b. The relation of this biozone to the divisions recognised in the Pre-Pa stage is uncertain. The *Alnus* and *Artemisia* frequencies are high, and in this the spectrum resembles that from the drift mud at SSH 20 cm, which is certainly a variant within Pre-Pa a. High *Artemisia* frequencies are characteristic of Pre-Pa b freshwater sediments at Beeston, and it may be that these variations relate to the taphonomy of the assemblages, with a contrast between *Alnus–Artemisia* in assemblages principally derived with terrestrial sediments and *Pinus*–Ericales derived with marine sediments.

(iii) *Pastonian Ib*

A *Pinus–Picea* p.a.b. is recorded near the base of the SSV section in sands with silt seams (bed d) above shelly sand and gravel (SSV 12, 10). AP frequency is high, composed predominantly of *Pinus* (to 70%) and *Picea*, with lower frequencies of *Betula* and *Alnus*. *Empetrum*, Gramineae, Filicales, *Osmunda* and *Sphagnum* comprise most of the NAP and spores. The spectra bear resemblance to those of the *Pinus* p.a.b. of freshwater Pa Ib sediments at Beeston, but *Picea* frequencies are higher, and *Empetrum* pollen is present in substantial frequencies. Both these differences may be a result of marine deposition as opposed to freshwater deposition at Beeston. *Picea*, usually considered under-represented in pollen assemblages (e.g. Phillips, 1976, *p.* 241), was evidently present with *Pinus* in the forest of this substage.

(iv) *Pastonian II*

The *Pinus–Quercus–Ulmus–Carpinus* p.a.b. of this substage is recorded in a number of sections: in silts and drift muds within the cross-bedded sands of bed g at SSF, SSR, SSV, SSTB and SSU, and in tidal silty clays of bed f at SSE, SSTB, SSU and SSN. There are also spectra showing an *Alnus–Pinus* p.a.b. from drift-mud seams (SSF 4) and erratics of mud in bed g (SSV 11, 2).

It will be most convenient to deal separately with each section showing Pa II assemblages.

SSV. Samples 13 and 4 at the base and top respectively of the Pa II sequence are from silty laminae in the sands of bed g. The AP is high, with the characteristic assemblage of thermophilous trees, *Ulmus*, *Quercus* and *Carpinus*, and low frequencies of Cupressaceae b, *Empetrum*, Gramineae and Chenopodiaceae. Pollen of *Armeria*, Compositae, *Artemisia*, *Plantago media/major*, *Ranunculus*, *Thalictrum* and Umbelliferae is present. Aquatics include *Azolla*, *Potamogeton*, *Sparganium* and *Typha latifolia*. Filicales and *Osmunda* (both *claytoniana*-type and *regalis*-type) are frequent, especially in the upper spectrum SSV 4, and will be derived from local fen communities. The identity and origin of the Cupressaceae b pollen is problematical. The matter is discussed in Chapter 18, *p.* 109, where the possibility is raised that it may derive from local swamp communities with *Chamaecyparis*.

Pollen spectra SSV 11 and 2 are from erratics of mud within the sands of bed g. *Pinus* wood was identified from the SSV 2 horizon. Both pollen spectra show higher frequencies of *Alnus*, Filicales spores in SSV 2, Cyperaceae and Umbelliferae and an absence of *Carpinus*. The mud erratics evidently derive from fen carr (SSV 11) or fen (SSV 2).

Pollen spectra from 3a and 3b are of interest in showing variation in content with lithology (Plate 18). 3a is from a silt lamina in cross-bedded sand, 3b from drift mud lying unconformably and horizontally on the set containing 3a. 3a is similar to the spectra from the silts of SSV 13 and 4. 3b has higher frequencies of *Alnus*, Cupressaceae b, Filicales spores and *Osmunda claytoniana* spores, and lower frequencies of *Pinus*, *Carpinus*, *Empetrum* and Chenopodiaceae. The pollen assemblage is evidently derived by inwash with drift mud from local communities, principally fen carr and fen, with a lower representation of more distantly derived taxa.

SSF. The three spectra show high *Alnus*, Filicales and *Osmunda* spores in the lower two samples (SSF 4, 3), which are from seams of detritus mud. The uppermost sample (SSF 5), from a seam of grey silt, shows the more usual assemblage with *Quercus*, *Carpinus* and Chenopodiaceae better represented.

SSE. The two basal samples (SSE 6, 7) from silty clay overlying the Chalk show a usual Pa II assemblage. The next sample above (SSE 5) is from drift mud in sand and shows a spectrum with higher *Alnus*, Filicales and *Osmunda* frequencies and no *Carpinus*. It is a variant similar to the spectra from the drift mud horizons analysed from other sites, but Ericales frequencies are higher.

SSTB. The two basal samples (SSTB 265, 400 cm) are from drift mud (265 cm) and a silty clay lamina in marine sediments of bed f or g. The drift mud again shows high frequencies of *Alnus*, Filicales spores and *Osmunda claytoniana*-type and *regalis*. A macroscopic analysis was made of this drift mud.

SSR, *SSN*. The analyses from these sites, from samples in silty clay laminae, show characteristic Pa II assemblages.

SSU. The analyses are from drift-mud laminae in bed f or g and show high *Alnus*, Cupressaceae b, Gramineae and *Osmunda claytoniana*-type and *O. regalis*.

A macroscopic analysis was made from the drift mud at SSTB 260–70 cm. The maritime group is the best represented in the assemblage, together with aquatics and water ferns, with low frequencies of the other groups. Trees include only *Quercus*, shrubs *Rubus idaeus*, and the heath group *Erica* cf. *tetralix*. Seeds of *Salicornia* were very abundant. Dry-ground herbs include cf. *Atriplex*, *Chenopodium album* and *Diplotaxis tenuifolia*, damp-ground herbs *Selaginella selaginoides* and *Solanum dulcamara*, the muddy substrates/shallow water group *Alisma plantago-aquatica* and *Eleocharis*, and the fen/reedswamp helophytes *Ranunculus lingua* and *R. sceleratus*. The aquatic group is the richest in taxa: *Elatine hydropiper*, *Hippuris vulgaris*, *Lemna*, *Menyanthes trifoliata*, *Najas marina*, *Nuphar lutea*, *Potamogeton berchtholdii*, *P. filiformis/vaginatus*, *P.* cf. *gramineus*, *P. pectinatus*, *Ranunculus* subg. *Batrachium*, *Sagittaria sagittifolia*, *Scirpus lacustris*, *Typha angustifolia*, *Zannichellia palustris* and abundant *Azolla filiculoides*.

The mixed assemblage is characteristic of tidally deposited drift muds. The rarity of tree remains and prevalence of aquatics indicates the area of deposition received detritus principally from nearby areas of freshwater and brackish lakes and fens.

(v) *Pastonian IVa*

The *Pinus–Picea–Alnus–Betula* p.a.b. is recorded in marine sediments of beds f and g at SSU 0 cm, SSTB 0–225 cm, SSTA 40 cm, SSE 1, 3 and 4, and SSP 50–0 cm. The spectra show high NAP, with *Pinus* dominant, and with lesser frequencies of *Picea*, *Alnus* and *Betula*. The frequencies of thermophilous trees are low. Ericales frequencies are higher than in Pa II, and *Osmunda* frequencies lower. Very low frequencies of *Abies* are recorded, SSTB 0–225 cm and SSE 1, as they are at Beeston in Pa IVa (BR/S).

Two local p.a.b.s are distinguished. At SSP 30–0 cm Cupressaceae b pollen is very frequent and a separate p.a.b. is distinguished. The possible role of the Cupressaceae b taxon in the local vegetation is discussed above. The second local p.a.b. is a *Pinus–Alnus* p.a.b. of sample SSE 2, from a horizon of drift mud in sand of bed g. There is no enrichment of Filicales or *Osmunda* concomitant with the *Alnus* rise as in earlier biozones, and presumably the origin is very local.

(vi) *Cromerian Ib*

A *Pinus–Ulmus–Betula* p.a.b. is recorded in the freshwater mud and silty mud of section SSTC, bed i, extending into laminated silty clay (bed k) on the uppermost part of the channel filling (samples C 6 and C 70 cm). The assemblage resembles the *Pinus–Ulmus* p.a.b. at West Runton in its low *Alnus* and *Quercus* pollen frequencies and it is therefore correlated with Cr Ib. The AP frequency is high. In addition to the trees already mentioned, *Salix*, *Corylus* and *Fraxinus* are also present. Ericales taxa are absent except in the uppermost silty samples, where they are accompanied by higher frequencies of Chenopodiaceae pollen. Marine transgression is probably indicated by these changes. Of the NAP, Gramineae, Cyperaceae and Chenopodiaceae show sustained percentages. Also present is a variety of NAP taxa: *Plantago maritima*, Caryophyllaceae, Compositae Liguliflorae and Tubuliflorae, *Artemisia*, *Bidens*, Cruciferae, *Plantago media/major*, *Ranunculus*, Rubiaceae, *Rumex*, *Thalictrum* and Umbelliferae. Pollen of aquatics includes *Myriophyllum verticillatum*, *Nuphar*, *Sparganium* and *Typha latifolia*, with *Salvinia* massulae. Filicales spores are frequent throughout. The assemblage indicates the presence of *Pinus–Ulmus–Betula* woodland, with deposition in a freshwater pool, subject later to marine influence and with local herb communities.

Two macroscopic samples were analysed from the muds of the channel filling, at SSTC 180–200 cm (60 taxa) and 130–150 cm (39 taxa), with identification of further plant remains at 90 cm. Fig. 28 shows that fen and reedswamp helophytes and aquatics form the majority of remains represented, with a lesser representation of dry-ground herbs and the muddy substrates/shallow-water group. Since there are no great differences between the analyses, we may consider the macroscopic flora as a whole. In spite of the high AP, trees are only recorded by a few fruits of *Betula pendula/pubescens*, with shrubs represented by thorns of *Rosa* and *Rubus*. A heath element includes *Eriophorum vaginatum*; *Myrica gale*, placed in this group, is more likely to have been present in local fens. A large number of taxa of dry ground indicate wide opportunities for colonisation of local habitats presumably on the sandy banks and surrounds of the channel, which must have been formed largely by Pastonian and earlier sands and gravels. The taxa include: *Arenaria serpyllifolia*, *Atriplex* sp., *Cerastium holosteoides*, *Chenopodium* sp., *C.* section *Pseudoblitum*, *C. album*, *Cirsium vulgare*, *Galeopsis angustifolia*, *Lamium purpureum*, *Origanum* sp., *Plantago major*, *Polygonum aviculare*, *Rumex* cf. *acetosa*, *Stellaria media*, cf. *Thlaspi* sp., *Thymus* sp., and *Urtica urens*. The damp-ground herbs are *Cirsium palustre*, *Conium maculatum*, *Solanum dulcamara* and *Stellaria neglecta*. The muddy substrate/shallow-water group is a strong one, with *Alisma plantago-aquatica*, *Eleocharis carniolica*, *Rorippa* sp., *R.* cf. *sylvestris*, *Rumex maritimus* and *R. palustris*.

The fen and reedswamp helophytes also form a substantial group, with cf. *Angelica sylvestris*, *Bidens* sp., *B. tripartita*, *Carex riparia*, *Cicuta virosa*, *Cyperus longus*, *Eriophorum latifolium*, *Hypericum tetrapterum*, *Lycopus europaeus*, *Oenanthe aquatica*, *Ranunculus sceleratus* and *Urtica dioica*. The aquatic flora is not so rich with *Ceratophyllum demersum*, *Chara* sp., cf. *Menyanthes trifoliata*, *Potamogeton distinctus*, *Potamogeton natans*, *Ranunculus* subg. *Batrachium*, *Sagittaria sagittifolia*, *Scirpus lacustris*, *Sparganium erectum* ssp. *erectum*, *Stratiotes aloides*, *Typha angustifolia* and *T. latifolia*, with the water fern *Salvinia natans*. Many of these taxa can withstand slight salinity and it is possible the regime is brackish rather than freshwater. The evidence of the sediments points to transgression in the uppermost part of the pollen assemblage biozone (samples C 6, C 70 cm), so the macroscopic assemblage is likely to have been formed in a maritime situation.

(vii) *Cromerian IIIb*

An *Abies–Carpinus* p.a.b. is recorded in laminated tidal sediments (bed k) at SSTA 55 cm (younger than the channel filling SSTC), SSV 6, and SSR 5–150 cm. In the latter section the *Abies–Carpinus* frequencies are much lower than at the other sites, but the assemblages seem sufficiently alike the others to be included in this pollen assemblage biozone, an allocation supported by the stratigraphy. The AP frequencies are high, principally *Pinus*, *Picea*, *Abies*, *Alnus* and *Betula*, with *Carpinus* and *Quercus*. The frequencies of Ericales taxa, including *Calluna*, *Empetrum* and Ericales, are high. Pollen of *Bruckenthalia spiculifolia* is recorded in SSV 6. The NAP is mostly Gramineae and Chenopodiaceae, with *Armeria*, *Plantago maritima* and a limited variety of other NAP taxa. The assemblage is characteristic of the tidal sediments of Cr IIIb along the coast.

(viii) *Cromerian IVa*

A *Pinus–Alnus–Picea–Betula* p.a.b. is recorded in tidal sediment of bed k at SSV 8. There are low frequencies of thermophilous trees, with higher frequencies of Ericales, and pollen of *Bruckenthalia spiculifolia*. Gramineae and Chenopodiaceae form the bulk of the NAP. The assemblage shows the continuance of tidal conditions and boreal woodland into the late Cromerian at this site.

(ix) *Cromerian IVb*

A *Pinus–Betula* p.a.b. is recorded in stiff stony silty clay of bed l at SST. The AP frequency is high, and *Picea*, *Alnus* and *Salix* are also present in addition to *Pinus* and *Betula*. Ericales frequencies are much lower and Gramineae percentages are higher than in the underlying zone. Chenopodiaceae shows a continuous pollen curve. The assemblage indicates continued revertence towards the end of the Cromerian temperate stage, with *Pinus* and *Betula* becoming the main trees of the woodland. At the Overstrand section OB a similar biozone is found in freshwater sediments, but it seems likely that the stiff stony clay of bed k at Sidestrand is marine in origin, since it has relatively high frequencies of Chenopodiaceae pollen.

8: Trimingham

(a) Lithostratigraphy

The sections along the Trimingham coast have been badly exposed during the period of the study. The cliffs are high and continually slumping, so that cliff falls and talus have obscured the pre-glacial sediments. In addition there have been difficulties over access, for this part of the coast was laid with mines during the war, and these were not cleared until the 1960s. As a result no long sections were recorded, but only small exposures, and it is impossible to relate these on the basis of stratigraphy. The details of these and of three boreholes made in the course of coast protection works are shown in Fig. 29. The boreholes show shelly sands and gravels overlain by sand and clay, with an organic freshwater horizon within the sequence at borehole BHT14A. In general the cliff sections show laminated silty clays, with tidal structure, overlain by or overlying stratified sands. In the upper parts of the sections at the south-east end of the stretch involutions effect stratified sand with streaks of drift mud and seams of freshwater sediment, and an ice-wedge cast is also present (TK, 1250 m). These are probably of early Anglian age. There seems evidence for tidal conditions in the lower part of the sequence, with regression and permafrost in the upper part. An interesting section with 2 m of freshwater stony silty mud, with wood fragments and large freshwater bivalves, was seen at TB (175 m), but unfortunately the stratigraphical relations of this horizon could not be seen. It is almost certainly the same horizon as that described by Reid (1882, p. 32) as the Sidestrand *Unio*-bed.

Table 20 gives a summary of the succession.

(b) Biostratigraphy

Table 18 lists the biozones identified in the pollen diagrams from Trimingham (fig. 27). The isolation of the sections make it difficult to draw up a sequence of biozones. Some, however, may be satisfactorily correlated with the regional sequence. Table 16 gives the macroscopic plant analyses and Table 20 gives a summary of the biozone sequence and vegetational history.

Biozone a. *Pinus–Picea* p.a.b. This assemblage is similar to those from Sidestrand SSV, at levels 10 and 12. This indicates that a correlation with Pa Ib is likely. However, the assemblage has some similarity with late Pastonian assemblages at Paston–Mundesley (BHB 2), so the correlation is tentative.

Biozone b. *Pinus* p.a.b. This assemblage, similar to the previous one except for the higher *Pinus* frequencies, is also correlated with Pa Ib.

Biozone c. Gramineae–*Pinus–Quercus–Ulmus–Carpinus* p.a.b. This assemblage can be correlated with Pa II, and comes from sediments similar to those of bed g at Sidestrand.

Biozone d. *Pinus–Alnus–Picea–Betula* p.a.b. This characteristic biozone, identified at two sites, may belong to either Pa or Cr IVa.

Biozone e. Gramineae–*Pinus* p.a.b.; biozone f. Gramineae–*Pinus–Betula* p.a.b. These two high NAP biozones are generally similar. They are both probably pre-Cromerian, since at TC biozone f is below a biozone clearly Cr IIIb in age. Their age is uncertain. The sediments appear to be marine and they may be Pa IV in age.

Biozone g. *Ulmus–Quercus–Pinus* p.a.b. This biozone occurs in the silty mud of section TB, and is placed in Cr IIa.

Biozone h. Gramineae–*Quercus–Ulmus* p.a.b. This biozone shows lower frequencies of *Pinus* and higher frequencies of Gramineae and Chenopodiaceae pollen than the previous biozone, but otherwise the forest constituents are similar. The assemblage is placed in Cr IIa.

Biozone i. *Abies–Carpinus* p.a.b. This characteristic biozone, identified at two sites, is correlated with Cr IIIb.

Biozone j. Cyperaceae p.a.b. The high NAP biozone from a thin horizon of mud below the till is placed in the e An.

(c) Vegetational history

(i) *Pastonian Ib*

The *Pinus–Picea* p.a.b. from the boreholes at Trimingham BHT 10 and BHT 16 is very similar to that at the base of the SSV section at Sidestrand (SSV 10 and 12) in its content of AP and NAP. AP frequencies are very high, with *Pinus* and *Picea* dominant. Other tree taxa are sparsely represented. The Ericales taxa frequencies are low, and the Gramineae percentages very low. Other NAP taxa are scarce. Filicales, *Lycopodium*, *Osmunda* and *Sphagnum* spores are present. The *Pinus* p.a.b. of borehole BHT 10 180 cm is similar in constitution to the *Pinus–Picea* p.a.b. except for the higher frequencies of *Pinus* and lower *Picea* frequencies. Both p.a.b.s indicate the presence of coniferous woodland at a time of marine sedimentation (at −1·5 to −8 m OD) of sands and silty clays, with a fauna with *Macoma balthica* at BHT 16 750 cm.

(ii) *Pastonian II*

The Gramineae–*Pinus–Quercus–Ulmus–Carpinus* p.a.b. comes from a silt horizon on cross-bedded sands at section TA, similar to sediments of bed g at Sidestrand. It is similar to spectra from that bed at SSV, indicating the presence of temperate woodland, with a strong local NAP representation, mostly Gramineae and Chenopodiaceae. There is an aquatic element, with *Azolla*, *Nuphar*, *Nymphaea*, *Sparganium* and *Typha latifolia*, and spore taxa include Filicales, *Lycopodium* and *Osmunda*.

(iii) *Pastonian IVa/Cromerian IVa*

A *Pinus–Alnus–Picea–Betula* p.a.b. is recorded in laminated silty clays with tidal structure at TF and TM. The assemblage, with high AP, indicates regional boreal woodland with a lack of thermophilous trees. Such vegetation occurs in the late Pastonian and the late Cromerian, and in the absence of a more detailed stratigraphy it is impossible to say whether the p.a.b. is Pastonian or Cromerian at each of the two sites.

(iv) *Pastonian IV?*

Two pollen assemblages from silty clay laminae in sand, probably marine, at TC 300 (Gramineae–*Pinus–Betula* p.a.b.) and TJ (Gramineae–*Pinus* p.a.b.) show high NAP frequencies, particularly Gramineae and Cyperaceae, with *Pinus* and *Betula* the best-represented trees. The stratigraphical relationships of the sediments are not clear, but at TC the assemblage appears to be pre-Cr IIIb, and both assemblages are tentatively assigned to Pa IV, since the AP assemblage and marine environment are similar to other Pa IV assemblages on the coast. However, it is also possible they may be Beestonian, but there is no record of Beestonian marine sediments on this part of the coast, so the former correlation is preferred.

(v) *Cromerian IIa*

An *Ulmus–Quercus–Pinus* p.a.b. is recorded at section TB in two metres of silty mud, with large freshwater bivalves. The AP is high, with high frequencies of *Ulmus* and *Quercus*, and lesser frequencies of *Pinus*, *Alnus*, *Betula* and *Picea*. There is no record of Ericales taxa. The low NAP is predominantly Gramineae and Chenopodiaceae, but the following are also recorded: Cyperaceae, Caryophyllaceae, Compositae Liguliflorae and Tubuliflorae, Cruciferae, *Filipendula*, *Plantago*, *Polygonum bistorta/viviparum*, *Ranunculus*, Rubiaceae and Umbelliferae. *Azolla*, *Sparganium* and *Typha latifolia* are present, and Filicales spore frequencies are high. The assemblage indicates regional temperate woodland in the immediate neighbourhood.

A macroscopic analysis was made from TB at 50 cm depth, a horizon with large freshwater bivalves. The assemblage shows a predominance of the aquatic and fen/reedswamp helophyte groups, but with a good representation of dry-ground herbs and shrubs. Shrubs are represented by *Salix* and *Thelycrania* cf. *sanguinea*, dry-ground herbs by *Aphanes microcarpa*, *Chenopodium glaucum*, *Chenopodium polyspermum*, *Helianthemum chamaecistus*, *Marrubium vulgare*, *Plantago major*, and *Scleranthus* sp., damp-ground herbs by *Carduus* cf. *acanthoides*, and *Selaginella selaginoides*, and *Alisma* and *Rumex maritimus* are present in the muddy substrate/shallow-water group. The fen/reedswamp helophytes present are *Hypericum tetrapterum*, *Lycopus europaeus*, *Myosoton aquaticum*, *Oenanthe* cf. *aquatica*, *Ranunculus sceleratus* and *Urtica dioica*, and the aquatics *Chara*, *Hottonia palustris*, *Hydrocharis morsus-ranae*, *Najas minor*, *N. tenuissima* (*see* Wilson, 1973), *Ranunculus* subg. *Batrachium*, *Stratiotes aloides*, *Typha latifolia* and *Zannichellia palustris*.

The assemblage shows the local existence of herb communities, in particular those of dry ground, both on basic (*Helianthemum*) and more acid (*Aphanes*) soils. In the aquatic group, the presence of *Najas tenuissima*, a species of eutrophic waters, is notable.

An analysis of non-marine molluscs by B. W. Sparks from the shelly level (see fig. 29) of site TB is given in Table 19. He comments on this list: the *Pisidium henslowanum* are very trigonal and might be *P. supinum*. This seems to be the fauna of a stream with appreciable movement, shown by a *Pisidium* group typical of such conditions. It includes five taxa of land snails and was probably deposited in the warmer part of an interglacial. The presence of a land fauna corresponds with the good representation of terrestrial plants in the macro-flora.

The assemblage from the organic mud of borehole BHT 14A, forming a Gramineae–*Quercus*–*Ulmus* p.a.b., is also placed in Cr IIa. It shows higher frequencies of Gramineae and Chenopodiaceae than the assemblage from TB, but *Quercus* and *Ulmus* are present and it is probably a variant showing much increased representation of local herb vegetation, probably maritime. The level of this assemblage is at −3·5 m OD, some 7 m below the TB level. The strata along this part of the cliff section are much disturbed by glacial action, with irregularities in the level of the Chalk (Reid, 1882, Peake & Hancock, 1970). It is possible that the level in borehole BHT 14A is undisturbed, and if so, as the pollen assemblage shows a strong maritime influence, the level, −3·5 m OD, may indicate the same approximate level for sea-level during Cr IIa (see Chapter 16).

(vi) *Cromerian IIIb*

The *Abies–Carpinus* p.a.b. of Cr IIIb is recorded in tidal silty clays at TC 0–150 cm and at TH. There are relatively high Chenopodiaceae frequencies, with Gramineae, as might be expected in tidal sediments.

(vii) *Early Anglian*

A Cyperaceae p.a.b. is recorded in silty mud at TL, a metre below till of the Anglian glaciation. The AP is low, with *Pinus*, *Betula* and *Salix* showing the highest frequencies. The NAP variety is low, with Cyperaceae at very high values, accompanied by Gramineae, Compositae Liguliflorae and Tubuliflorae, Cruciferae, *Ranunculus* and *Thalictrum*. The assemblage shows the limited occurrence of tree and shrub genera, absence of thermophilous trees and abundance of local sedge vegetation.

A macroscopic analysis from this horizon shows a restricted flora with a preponderance of the aquatic group (Fig. 28). Trees, shrubs and fen/reedswamp helophytes are not represented. *Eriophorum vaginatum* sclerenchyma spindles are present. *Myosurus minimus* may represent a dry-ground or damp-ground element (*see* Wilson, 1973). The muddy substrate/shallow water group comprises only *Eleocharis palustris* subsp. *palustris* (Cyperaceae pollen ?). Aquatics include *Chara*, *Hippuris vulgaris*, *Myriophyllum verticillatum*, *Nitella/Tolypella*, *Potamogeton filiformis*, *P. gramineus*, *P. praelongus*, *Ranunculus* subg. *Batrachium* and *Zannichellia palustris*. This restricted flora, with its northern *Potamogeton* assemblage, tallies with the pollen evidence for herb vegetation of a cold stage type.

9: Paston and Mundesley

(a) Lithostratigraphy

The sections east and west of Mundesley were well exposed in the early 1960s, before the present coast defences were built. Figs. 30–34 show sections and boreholes extending 1695 m south-east of datum 1 (Fig. 2, inset 3) and 600 m north-west of datum 2. The boreholes were either made especially for the purpose of the study or in the course of site investigation for the coast defences.

A useful marker level along this part of the coast, at or just below the base of the cliff, is an iron-cemented horizon ('pan'), either gravel or sand, showing a sharp and clean contact with overlying unconsolidated sediments (Plate 22). Clay conglomerate may also be associated with this horizon. Below this level the sediments, seen in boreholes, excavations (Plate 21) and at low tides are mainly sandy with thin silty laminae, thicker beds of laminated silty clay, with or without organic debris, and clay conglomerate. The Chalk underlies these sediments in the borehole MAT (15 m south-east, Fig. 33) at a depth of −10·5 m OD.

Above the marker level there are freshwater muds, thick blue laminated silty clays with tidal structure, beach gravels and basins of freshwater silty clays, with subordinate beds of sand, all seen in the cliff section below the till and totalling a maximum thickness of *c.* 6 m.

We shall first consider the sediments at and below the marker level. These show considerable lateral variation, but can be grouped into beds as follows. Table 23 summarises this succession.

a. Above the Chalk (top of Chalk–10·5 m OD) in borehole MAT (15 m south-east, Fig. 33) there is 45 cm of grey silty clay with stones, representing the stone bed. Marine shells were found in this sediment.

b. 6·25 m of laminated grey/dark grey silty clay with seams of grey sand lie on the stone bed in borehole MAT. Marine shells were present at various levels in this sediment.

c. In borehole MAT 2·3 m of grey sand with laminae of grey silty clay containing drift mud overlies the laminated grey silty clay.

d. 60 cm of grey sandy gravel succeeds the grey sand in borehole MAT. These gravels may also be represented in borehole C3 (270 m north-west, Fig. 33), where sandy gravels 2·5 m thick are seen at depth.

e. Above the gravel and below the modern beach sediment in borehole MAT is a further 1·7 m of grey sand with laminae of grey silty clay. At this level (about OD) in excavations (Plate 21) and other boreholes occurs more massive laminated silty clay, with clay conglomerate and driftwood (as at ML, 1370 m south-east, Fig. 30). In parts of the section the surface of this horizon is ferrugineous and forms the marker level.

f. This bed (Plate 21) consists of ferrugineous sand and gravel, clay conglomerate and cross-bedded sand, up to a thickness of 1·6 m. Wood and bone occur in the sediment. This horizon may rest unconformably on the silty clays of bed e. It has not been seen in contact with the muds of bed g. This horizon is thickest at the south-east end of the section. It is clearly distinguishable at PH (1295 south-east, Fig. 31), PD

(1515 m south-east, Fig. 30) and PG (1695 m south-east, Fig. 30). At PD it is divisible into a lower clay conglomerate, a cross-bedded sand (apparent dip north-west) and an upper ferruginous gravel. At PG, the ferruginous gravel cuts across the bedding of bed e and is a metre thick.

g. At MPA and PE (662 and 1322 m south-east, Fig. 31), up to 30 cm of brown mud overlaid the grey silty clays of bed e, with a sharp contact between them.

The beds above the marker level are as follows:

h. Detritus mud occurs immediately above the marker level at MAK (1 m south-east, Fig. 33), MU (75 m south-east, Fig. 32; Plate 23) and MF, MG, MP (139–147 m south-east, Fig. 32). The maximum thickness of these freshwater sediments is 45 cm at MU, where the exposure shows that the detritus muds form a narrow column with sides and surface eroded before and during deposition of the next bed. The thin mud horizons at the other sites, MAK, MF, MG, and MP, are in shallow depressions in the underlying ferruginous sands and gravels. At MZ (171 m south-east, Fig. 32), rolled cakes of this detritus mud are found at the base of the next horizon in a depression in the underlying ferruginous gravel.

i. This horizon is a complex of sediments, associated with marine transgression and interdigitating with one another. Four facies can be distinguished:

i(b) Pebbly sandy gravels (e.g. 62–124 m south-east, Fig. 32; Plate 22), up to 1·5 m thick, with an intercalated silt up to 20 cm thick containing *Mytilus* shells (e.g. MR, 123 m south-east, Fig. 32). These gravels are interpreted as beach gravels. In the base of the gravels at MZ (172 m south-east, Fig. 32), a number of mud cakes, reworked from bed h, were seen accumulated in a hollow in the underlying ferruginous gravel.

i(s) Sand with occasional silty laminae, with erratics of clay and mud, up to 1 m thick. This is seen below the beach gravel at MAO (57 m south-east, Fig. 33), MU (75 m south-east, Fig. 32), MW (92 m south-east, Fig. 32) and MP (147 m south-east, Fig. 32).

i(lc) Laminated silty clay with pale sand laminations (Plates 24, 25, 30), with a tidal structure and up to 2·5 m thick occurs at the north-west end of the section (0–2 m south-east, 49–345 m north-west, Fig. 33), and at the south-east end of the section (1290–1692 m south-east, Figs. 30 and 31). The lamination is usually horizontal and there are no signs of bioturbation. In MC (134 m north-west, Fig. 33), the bedding of the silty clays, here containing much clay breccia, is oblique and forms a channel filling.

i(o) Organic horizons, up to 27 cm thick, composed of detritus mud and *Phragmites* peat, occur within the laminated clays at MAM (50 m north-west, Fig. 33; Plate 24) and MAR (181 m north-west, Fig. 33). They contain remains of freshwater and brackish plants and of foraminifera, implying a change in tidal regime and organic deposition in brackish or freshwater pools.

The distribution of these facies is as follows. At the north-west end of the section the laminated clays occur directly on the ferruginous marker horizon, with the intercalated organic horizon at two sites. At MAK and MAM on either side of Mundesley (Fig. 33) pebbly gravels of the beach facies are associated with the laminated clays. South-east of this the beach gravels occur, as far as MY (182 m south-east, Fig. 32), with the sands of bed i(s) below them at certain sites. The transgression sediments are then cut out by the overlying sands until PH (1295 m south-east, Fig. 31), and at that site and south-east to the end of the section the laminated clays are again seen, with thin sand of horizon i(s) below, all resting on the ferruginous marker level.

j. This horizon is of cross-bedded sand with silty laminae and streaks of drift mud (Plate 27). It fills two channels cut in the underlying transgression sediments, one at 296 m north-west (Fig. 33), where it is up to 2 m thick, the other at 300–896 m south-east (Figs. 31 and 32), up to 2·9 m thick. The upper part of the sand (bed j i) spreads as a clearly cross-bedded unit (apparent dip north-west) over the transgression sediments, up to 60 cm thick, along the section from MAK (34 m south-east, Fig. 33) to MAD (390 m south-east, Fig. 32). A drifted stump is submerged by this upper sand at MAA (166 m south-east, Fig. 32; Plate 26). Between the upper cross-bedded unit and the lower sands occur thin silt horizons (MAH 1, MAD). At MAU an organic horizon occurs in the same position, below the cross-bedded sands of j i. This is separated as bed k.

k. At MAU (at the south-east end of Mundesley; Fig. 34; Plate 25), up to 40 cm of dark blue stony organic clay with freshwater shells and seeds was observed resting unconformably on the tidal sediments of bed i(lc). This section, length 35 m, was described by Reid (1890, *p.* 167). He referred the freshwater horizon to his Upper Freshwater Bed.

l. At MS (155 m south-east, Fig. 32) and MAH (460 m south-east, Fig. 32; Plate 28), shallow channels are cut into the sand of bed j and filled with laminated silt and black stony mud respectively.

m. A horizon of horizontally laminated silty clay with plant debris in seams and up to 60 cm thick, stretches along much of the section (Plates 26, 27). It is seen at MB (296 m north-west, Fig. 33), and MAK (1 m south-east, Fig. 33) to MX (566 m south-east, Fig. 31). At ME (48 m south-east, Fig. 33; Plate 29), a shallow channel filled with silty mud and laminated silty clay is associated with this horizon.

n. A cross-bedded sand with streaks of drift mud and thin silty laminae overlies bed m, as at MB (296 m north-west, Fig. 33) and MY (182 m south-east, Fig. 32) to MAG (467 m south-east, Fig. 32).

o. This horizon is of laminated silty clays with mud seams, occurring in narrow shallow channels cut in the sand of bed n. Such channels are seen at MM and MX (491 m and 563 m south-east, Fig. 31) and PC (1675 m south-east, Fig. 30; Plate 30).

p. This is the till which seals the succession along the Paston–Mundesley coast, the Second Till of Banham (1968).

(b) Periglacial stratigraphy

Ice-wedge casts

1. 299 m north-west (Fig. 33), within bed j.
2. 33 m south-east (Fig. 33), penetrating bed f, sealed by bed i(b).
3. 93 m south-east (Fig. 32; Plate 22), penetrating bed f, sealed by bed i(s).
4. 161 m south-east (Fig. 32), penetrating bed i(b), sealed by bed ji.
5. 181 m south-east (Fig. 32), penetrating bed f, sealed by bed i(b).
6. 391 m south-east (Fig. 32), penetrating ji, sealed by bed m.
7. 391 m south-east (Fig. 32), within bed n.
8. 464 m south-east (Fig. 32; Plate 27), within bed n, penetrating beds m and j.
9. 893 m south-east (Fig. 31), within bed n.
10. 1691 m south-east (Fig. 30), penetrating bed i(lc), sealed by bed n.
11, 12. 9 m and 25 m, MAU (Fig. 34), penetrating bed ji, sealed by gravels at base of bed m.

Involutions

181 m north-west (Fig. 33), associated with bed ji.
464–7 m south-east (Fig. 32), within and at top of bed j.
485 m south-east (Fig. 31), a. within and near top of bed j.
485 m south-east (Fig. 31), b. at base of bed n.
564 m south-east (Fig. 31), within bed o.
1672 m, 1678 m (Fig. 30), at margins of channel of bed o.

(c) Sedimentary history

Marine transgression of the Chalk surface (at −10·5 m OD) produced the stone bed (bed a), containing marine molluscs, and was followed by the accumulation of laminated silty clays, probably under shallow marine conditions (bed b). Aggradation of these clays took place to −4·25 m OD. Subsequently sands with thin silty laminae and gravels were deposited, with marine conditions still obtaining (beds c, d and e). The gravels (bed d) may imply a period of regression and shallowing. The upper sand contains much laminated silty clay, and levels with clay conglomerate and drifted wood and stumps, indicating shallow marine or tidal conditions. A period of regression followed this, with the freshwater muds of bed g formed on the surface of silty clays.

Gravel, sand and clay conglomerate (bed f) were probably also deposited during this regression, with reworking of the older sediments. This horizon was either deposited locally or heavily dissected by erosion, since in many sections it is absent, the ferrugineous marker level being represented by the iron-cemented topmost sediments of bed e (e.g. MAO, 57 south-east, Fig. 33). The ice-wedge casts (2, 3 and 5 above) penetrating this marker level are truncated, indicating some erosion of an old land surface which suffered permafrost (Plate 22). This surface was flooded by rising base level and freshwater muds (bed h) accumulated. This was followed by marine transgression (sediments of bed i) which first eroded the freshwater mud (see MU, 75 m south-east, Fig. 32; Plate 23) and redeposited part as cakes in the basal transgression sediments. First sands (horizon i(s)), then beach gravels with *Mytilus* (i(b)) and tidal laminated silty clays (i(lc)) were deposited, and are seen over much of the coast section. The transgression sediments reach a height of about 3·5 m OD, with a transgression contact (MU, 74 m south-east, Fig. 32) at about 2 m OD.

Surface textures of sand grains from the gravels of bed i(b) at ME show an early episode of glacial action followed by beach action (Krinsley & Funnell, 1965, sample E 32).

Regression followed, and the transgression deposits were channelled. Aggradation of sands with silty laminae and freshwater muds occurred in the channels (beds j, k). Involutions are found in this horizon and an ice-wedge cast (1 above) in the upper part. Ice-wedge casts penetrating the transgression sediments (4 and 10 above) may also belong to this permafrost period. The aggrading sands of bed ji spread over the surface of the transgression sediments and the freshwater muds of bed k. This sheet of sand was subject to permafrost (ice-wedge casts 6, 11 and 12 above, involutions at top of the horizon), and channelling, in which sediments of bed l accumulated. Further aggradation of silty clays (bed m) then occurred along much of the section as a horizontal sheet. Further cross-bedded sands accumulated (bed n) and were subjected to permafrost (ice-wedge casts 7, 8 and 9 above). Involutions affected the base of this sand and the underlying silty clays. The sand was later channelled, with the channels filled by laminated silty clays. Finally the succession was sealed by the deposition of till.

Table 23 summarises the sequence described above.

(d) Biostratigraphy

Table 21 lists the biozones distinguished in the pollen diagrams from Paston–Mundesley (Figs. 35 to 38). Table 22 records the macroscopic analyses from this part of the coast, and Table 23 gives a summary of the biozone sequence and vegetational history.

The stretch of coast shows much variation in facies, and the sections are not continuous, so that assembling an order of biozones is not straightforward. However, two of the temperate biozones stand out as being useful marker horizons. The lower is the *Pinus–Quercus–Carpinus–Ulmus* biozone of the Pastonian (Pa II) which was found in excavations and boreholes, roughly at and a few metres below OD. The upper is the *Abies–Carpinus* biozone of the Cromerian (Cr IIIb), represented in many sections along the coast, above OD. In isolated sections and boreholes, especially those with few or single pollen analyses, it is difficult to place the biozones concerned in the stratigraphical sequence. Nevertheless, the general succession of biozones is clear. The basal one (a, *Pinus*–Gramineae–Ericales) is followed by the development of temperate forest (c, *Pinus–Quercus–Carpinus–Ulmus*), with a subsequent reversion to largely coniferous forest (e.g. i, *Pinus–Picea*). These sediments are marine, but regression occurs at the top of this part of the succession, with the formation of freshwater sediments (bed g). Above the unconformity between beds g and h the Cromerian biozones begin. There are no difficulties here in the formulation of the biozone succession.

Biozone a. *Pinus*–Gramineae–Ericales p.a.b. This is the basal biozone and is correlated with Pre-Pa a. The borehole MAT shows 6·25 m of clayey sediments (beds a and b) included in this biozone, resting on the Chalk. Other boreholes show single pollen spectra at depth of the same type.

Biozone b. *Alnus–Betula–Quercus* p.a.b. This assemblage is recorded in the mud erratic of MN 1, where the local influence of *Alnus* and *Betula* is seen. The biozone, with an assemblage with low frequencies of *Quercus* and *Ulmus*, is considered a variant from the principal p.a.b. (biozone c) of the Pa II temperate woodland assemblage.

Biozone c. *Pinus–Quercus–Carpinus–Ulmus* p.a.b. (beds c, e and f). The assemblage shows the characteristic expression of temperate woodland in the Pastonian as we have already seen it in the Beeston sections, and is placed in Pa II. One analysis from the freshwater mud erratic within marine sediments at MN1, shows a local p.a.b. with high frequencies of *Alnus* and *Betula*.

Biozone d. *Alnus–Betula–Quercus*–Ericales p.a.b. (bed e). The spectra from section PH are placed in this biozone, which differs from the previous biozone in lower frequencies of *Ulmus* and *Carpinus* and higher NAP frequencies, particularly Ericales and *Artemisia*.

Biozone e. *Pinus–Betula–Quercus* p.a.b. (bed e). This biozone is separated on the basis of a spectrum in section MP at a level higher than silty clay containing a Pa II assemblage.

Biozones d and e are intermediate in content between the Pa II temperate forest assemblages and those characteristic of the late Pastonian, Pa IVa, with boreal forest.

Quercus is the only thermophilous tree represented firmly, and the other most abundant tree genera are those of the assemblages of Pa IVa. The biozones are placed in Pa III.

Biozone f. *Pinus–Alnus–Picea–Betula* p.a.b. (bed e). This biozone is well expressed at several sections, and its stratigraphic position is clear. It is similar to biozones in the late Pastonian at Beeston, Overstrand, Sidestrand and Happisburgh and is placed in Pa IVa.

Biozone g. *Pinus–Betula–Alnus* p.a.b. (bed e). This biozone is identified in a single sample (MPA 10). It is similar to spectra from biozone f, but shows lower frequencies of *Picea*. It is placed in Pa IVa.

Biozone h. *Betula–Alnus* p.a.b. (bed g). This biozone is recorded in freshwater sediments directly overlying marine sediments (bed e) at section MPA. It is placed in late Pa IVa, as a local p.a.b.

Biozone i. *Pinus–Picea* p.a.b. (bed e). This biozone is identified in two boreholes, BH B1 and BH B2. In the former borehole it overlies a Pa IVa biozone; in the latter it overlies a Pa II biozone. The biozone is similar to a biozone at Overstrand later than a *Pinus–Alnus–Picea–Betula* biozone, and is placed in Pa IVb.

Biozone j. *Pinus–Betula* p.a.b. (bed e). This biozone is only identified from one sample (MAP). It is placed in Pa IVc, though its stratigraphic position is uncertain.

Biozone k. *Myrica–Pinus–Betula* p.a.b. (bed g). This biozone, recorded in section PE in freshwater mud, is placed in Pa IVc.

Biozone l. *Quercus–Ulmus–Pinus* p.a.b. (bed h). This biozone is identified at three sections with freshwater sediments. It is correlated with Cr IIa.

Biozone m. Gramineae–*Betula–Pinus* l.p.a.b. (bed h). This biozone follows biozone l and is identified in three sections. It is placed in Cr II, since, as will be discussed later, it is interpreted as a local p.a.b.

Biozone n. *Pinus–Betula*–Gramineae l.p.a.b. (bed h). This biozone is identified in freshwater sediments from a single site (MAK). It is placed in Cr II and interpreted as a local p.a.b.

Biozone p. *Quercus–Ulmus–Tilia* p.a.b. (bed i(b)). A single spectrum (ME 1) shows an assemblage which is correlated with Cr IIb.

Biozone q. *Carpinus* p.a.b. (beds i(b/s), i(lc)). This biozone is identified from single samples at two sections. It is placed in Cr IIIa.

Biozone r. *Abies–Carpinus* p.a.b. (beds i(lc), i(b)). This biozone is identified in many sections along the Paston–Mundesley coast, principally in the laminated clays of bed i, but also in the beach gravels of bed i. It is placed in Cr IIIb.

Biozone s. *Pinus–Alnus–Picea–Betula* p.a.b. (beds i(lc), i(o), j). Again this biozone is identified from many sections, all in the laminated clays of bed i, and the freshwater silt of bed j, except for a local p.a.b. (Gramineae–*Pinus–Betula* l. p.a.b.) from mud (bed i(o)) identified at one site (MAM) at the lower boundary of the biozone. It will be noted that this biozone is similar to biozone f (Pa IVa), and that their stratigraphical separation is certain. It is placed in Cr IVa.

Biozone t. *Pinus–Alnus–Betula* p.a.b. (bed j). This biozone is only identified from freshwater sediments at one site (MA), as are the two subsequent biozones. It is placed in Cr IVb.

Biozone u. *Betula* p.a.b. (bed k). This biozone, recorded in freshwater mud, is placed in Cr IVc, since it resembles the Cr IVc *Betula* p.a.b. at West Runton.

Biozone v. Gramineae–Cyperaceae–*Betula–Pinus* p.a.b. (bed l). This biozone, with high NAP, is placed in an early Anglian substage, as an early part of the cold period subsequent to the Cromerian temperate stage.

Biozone w. Gramineae–Cyperaceae p.a.b. (bed m). This final biozone, with high NAP, is younger than biozone v and is placed in a later part of the early Anglian.

(e) Vegetational history: Pastonian

(i) *Pre-Pastonian a*

The *Pinus*–Gramineae–Ericales p.a.b. recorded in the boreholes BHC 7, BHC 3, MAT 680–1340 cm and BHA 1 640 cm has already been discussed in Chapter 2.

(ii) *Pastonian II*

The *Pinus–Quercus–Carpinus–Ulmus* p.a.b. of this substage is recorded in many sections near and below OD along the Paston–Mundesley coast in marine sediments of beds c, e and f. The sediments are often laminated clays with tidal structure. The pollen diagram from section PG shows the characteristic assemblage of this substage. The AP frequency is about 70%. *Quercus*, *Ulmus* and *Carpinus* are well represented, with *Pinus*,

Picea, *Alnus* and *Betula*. Ericales taxa reach about 10%, with Gramineae and Chenopodiaceae the most frequent of the other NAP taxa. Pollen of aquatics is scarce, but spores of Filicales, *Osmunda* (*regalis* and *claytoniana*-type) and *Sphagnum* are frequent. Regional temperate woodland is indicated, with a substantial local NAP element, including Gramineae, Chenopodiaceae and *Osmunda* spores, all probably derived from coastal communities.

A variation of assemblages within the substage is seen in the *Alnus–Betula–Quercus* p.a.b. of the mud erratic (MN 1) within marine sediments at MN. Higher frequencies of *Alnus* and *Betula* will be derived from local fen carr communities, and the lower frequencies of *Carpinus*, Ericales and Chenopodiaceae are a result of freshwater rather than tidal deposition.

A macroscopic analysis from the tidal silty clays at PG shows the mixed assemblage usual in marine sediments, with the maritime group preponderant (Fig. 28). Tree remains comprise fruits of *Alnus*, shrubs *Betula* cf. *nana* and *Salix*, and in the maritime group *Salicornia* seeds are abundant. Dry-ground herbs include *Chenopodium rubrum/botryodes* and *Urtica urens*, damp-ground herbs *Hypericum* cf. *hirsutum* and *Ranunculus* cf. *acris*, and taxa of muddy substrates/shallow water *Alisma* and *Eleocharis*. The fen/reedswamp helophytes include *Filipendula ulmaria*, *Osmunda* and *Ranunculus sceleratus* and aquatics *Callitriche intermedia*, *Chara*, *Elatine hydropiper*, *Hippuris vulgaris*, *Najas tenuissima*, *Potamogeton friesii*, *P.* cf. *praelongus*, *Ranunculus* subg. *Batrachium*, *Stratiotes aloides*, *Typha latifolia* and *Azolla filiculoides*. As usual with tidal sediments *Dryopteris* type sporangia are abundant. The record of *Betula* cf. *nana* is unexpected, but its origin may lie in local ombrogenous mires.

(iii) *Pastonian III*

Two pollen assemblages intermediate between those of Pa II and Pa IV are included in this substage. They are recorded in tidal silty clays at PH (*Alnus–Betula–Quercus*–Ericales p.a.b.) and MP (*Pinus–Betula–Quercus* p.a.b.). Both show rather low AP frequencies, with *Quercus* the only well-represented thermophilous tree, accompanied by *Pinus*, *Picea*, *Alnus* and *Betula*, the trees characteristic of the younger substage Pa IVa. At PH, the higher NAP frequencies include higher Ericales and *Artemisia* frequencies, but these are not shown at the section MP. The assemblages indicate a restriction of thermophilous trees in the forest and the increased representation of heath and herb communities.

A macroscopic analysis was made from marine silt with a content of drifted organic detritus of bed e at section PH. The remains were exceptional in being very poorly preserved. The analysis (Fig. 28) shows a mixed assemblage. Tree remains include *Alnus* and *Betula* fruits; and shrubs include *Salix* and *Viburnum opulus*. The maritime group comprises *Glaux maritima* and cf. *Salicornia*, and the heath group includes only *Juncus squarrosus*. There is a substantial dry-ground group with *Aphanes arvensis*, *Arenaria*, *Draba* cf. *muralis*, *Polygonum aviculare*, *Silene alba* and cf. *Urtica urens*. The damp-ground herb group includes only *Selaginella selaginoides*, the muddy substrate/shallow-water group only *Alisma*, and the fen/reedswamp helophyte group only *Osmunda*. The aquatic group is also restricted, with *Callitriche*, *Najas*, *Potamogeton pectinatus*, *P.* cf. *praelongus*, cf. *Sagittaria*, and abundant *Azolla filiculoides*. *Potentilla* achenes were abundant.

The assemblage bears out the pollen evidence for local herb communities. The low fen/reedswamp and aquatic frequencies are exceptional. Most of the macroscopic plant remains seem to have been washed in from terrestrial situations, perhaps a sign of approaching regression.

(iv) *Pastonian IVa*

The *Pinus–Alnus–Picea–Betula* p.a.b. of this substage is recorded in shallow marine or tidal sediments of bed e at several sites. The AP frequency is high, with only traces of pollen of thermophilous trees. The NAP is predominantly of Ericales taxa, including *Calluna*, *Empetrum* and Ericales, Gramineae, Cyperaceae and Chenopodiaceae. Otherwise the variety of NAP is low. Filicales and *Sphagnum* show low constant curves. The assemblage indicates boreal forest, with heath and herb communities.

A macroscopic analysis was made from marine sand with drifted organic fragments (bed e) at MAQ. The mixed nature of the assemblage is shown in Fig. 28. Tree remains include only fruits of *Betula pendula/pubescens* (but a *Pinus* log was found near a stump (not *in situ*) in the tidal silty clays at ML). The maritime group at MAQ includes abundant seeds of *Salicornia*, and the heath group spindles of *Eriophorum vaginatum*. The dry-ground herb group includes only cf. Chenopodiaceae, and the muddy substrate/shallow water group *Alisma plantago-aquatica* and *Isoetes hystrix*. *Urtica diocia* is the only fen/reedswamp helophyte, and the aquatics are *Chara*, *Stratiotes aloides*, *Typha angustifolia* and *T. latifolia*. Amongst the unclassified group there are *Carex*,

Compositae, abundant *Dryopteris*-type sporangia, Gramineae, *Rumex* and *Solanum*. The assemblage is a restricted one and shows the mixed nature of assemblages of shallow marine sediments.

A second group of p.a.b.s included in Pa IVa comprise the *Pinus–Betula–Alnus* p.a.b. and the *Betula–Alnus* p.a.b. of the section MPA. The former p.a.b. is recorded in laminated silty clay (bed e) and the latter in the coarse detritus mud with *Pinus* and *Picea* cones (bed g) overlying the silty clay. The *Pinus–Betula–Alnus* p.a.b. only differs from the p.a.b. described above by its low frequencies of *Picea*. The *Betula–Alnus* p.a.b. shows the local presence of these taxa, and Cyperaceae, in quantity, indicating regression and the local growth of woodland.

More information on the vegetation of these two p.a.b.s is given by macroscopic analyses from the sediments of each of them. Taking first the analysis from the older of the two p.a.b.s, the *Pinus–Betula–Alnus* p.a.b., trees, shrubs and aquatics form the major groups represented (Fig. 28). Trees include *Alnus glutinosa*, *Betula pendula/pubescens* and Coniferae (leaves), and the shrubs *Rubus idaeus* and *Salix*. The heath group includes only *Eriophorum vaginatum*, and there is a montane group with *Oxyria digyna* and *Thalictrum alpinum*. The dry-ground herb group includes only *Diplotaxis muralis*, and the muddy substrate/shallow-water group cf. *Alisma* and *Eleocharis*. The fen/reedswamp helophytes recorded are *Carex* cf. *rostrata*, *Filipendula ulmaria*, *Ranunculus lingua*, *Swertia* cf. *perennis* (see Wilson, 1973), and *Urtica dioica*. The aquatic group comprises *Ceratophyllum submersum*, *Chara*, *Hippuris vulgaris*, *Menyanthes trifoliata*, *Najas marina*, *N. tenuissima*, *Nitella/Tolypella*, *Potamogeton compressus*, *Ranunculus* subg. *Batrachium*, *Sparganium angustifolium*, *Stratiotes aloides*, *Typha latifolia* and *Zannichellia palustris*.

This assemblage indicates that herb-rich communities of boreal type (*Filipendula*, *Oxyria*, *Swertia*, *Thalictrum*) were present in the neighbourhood, though tree pollen is dominant. There is substantial representation of fen/reedswamp and aquatic plants, and it is likely that *Alnus* and *Betula* were present in local fen carr.

The analysis from the *Betula–Alnus* p.a.b. shows a richer representation of trees, fen/reedswamp helophytes and aquatics. The trees include *Alnus glutinosa*, *Betula pendula*, *B. pendula/pubescens* and *Picea abies* (leaves). The only shrub recorded is *Rubus idaeus*, and *Eleocharis* is the only taxon in the muddy substrates/shallow water group. The fen and reedswamp helophytes include *Bidens tripartita*, *Carex aquatilis/bigelowii*, *C.* cf. *pseudocyperus*, *Filipendula ulmaria*, *Oenanthe aquatica*, *Potentilla palustris*, *Ranunculus lingua* and *Urtica dioica*. The aquatics comprise *Hippuris vulgaris*, *Menyanthes trifoliata*, *Nymphaea alba*, *Sparganium* and abundant leaf spines of *Stratiotes aloides*. In the unclassified group there are abundant *Carex* remains.

The assemblage shows the local abundance of *Betula* (*pendula*, *pendula/pubescens*), *Alnus glutinosa* and *Picea abies*, though the last is sparsely represented by pollen. The fen/reedswamp helophytes and aquatics are well represented, though restricted in variety. Presumably the high Cyperaceae pollen frequencies correspond to the presence of sedge communities with *Carex aquatilis/bigelowii*.

(v) *Pastonian IVb*

A *Pinus–Picea* p.a.b. is recorded in sediments of bed e in boreholes BHB 1 230–450 cm and BHB 2 240 cm. The AP frequencies are high and the assemblage differs from the *Pinus–Alnus–Picea–Betula* p.a.b. of Pa IVa by the higher percentages of *Picea* and lower of *Alnus* and *Betula*. Whether this difference results from a variation in the vegetation or the taphonomy or from any post-depositional processes affecting the two types of assemblage it is not possible to say.

(vi) *Pastonian IVc*

The *Myrica–Pinus–Betula* p.a.b. of the compressed mud of bed g at PE is placed in this subdivision of Pa IV. *Pinus* and *Betula* are the only trees well represented. There is little *Picea* and no *Alnus*. The assemblage is dominated by *Myrica*. Ericales taxa are sparse, and the other NAP is mostly Gramineae, with few other NAP taxa. The macroscopic analysis from this horizon gave a very poor flora, with only remains of *Carex*, Cyperaceae and indeterminate wood fragments. The pollen and macro floras indicate that *Myrica*-rich communities occupied the surface of the laminated silty clays of the PE section, after regression, at a time when *Pinus* and *Betula* were important in the regional woodland.

The *Pinus–Betula* p.a.b. at MAP may be also related to Pa IVc, although it is found in a different stratigraphical context, in sands allocated to bed e. *Pinus*, *Betula* and Gramineae are the important taxa in this assemblage, with low frequencies of *Picea* and *Alnus*.

(f) Vegetational history: Cromerian

(i) *Cromerian IIa*

The *Quercus–Ulmus–Pinus* p.a.b. of this subdivision is recorded in the detritus mud of bed h at MU,

MF and MG. The AP frequency is high. There is a rich representation of *Quercus* and *Ulmus*. *Pinus* and *Alnus* pollen is less frequent, and there are very low frequencies of *Betula*, *Corylus*, *Taxus*, *Fraxinus*, *Salix* and *Rhamnus cathartica* pollen. The NAP is predominantly Gramineae, with low frequencies of pollen of Cyperaceae, *Armeria*, Chenopodiaceae, *Plantago maritima*, Compositae Liguliflorae and Tubuliflorae, *Artemisia*, Cruciferae, *Filipendula*, Labiatae, Ranunculaceae, Rubiaceae, *Rumex*, *Thalictrum*, Umbelliferae and *Valeriana officinalis*. In the aquatic group *Sparganium* and *Typha latifolia* pollen is frequent, with *Myriophyllum verticillatum*, *Nuphar*, *Azolla* and *Salvinia*. Filicales spores have relatively low frequencies. The assemblage indicates *Quercus–Ulmus* forest in the near neighbourhood, with the presence of a variety of other trees also indicated. The NAP indicates local fen communities with reedswamp and also the presence of maritime communities not too distant.

Macroscopic analyses were made from detritus mud at MU 25–45 cm and MG 0–10 cm. Fen/reedswamp helophytes form the largest group represented (Fig. 28), together with aquatics and water ferns. There is an insubstantial representation of other groups. Taking both analyses together, trees recorded are *Alnus* and *Fraxinus*, and shrubs only include *Salix*, even though the AP pollen frequency is so high. A large, quiet water body is indicated. The small dry-ground herb group, with only *Polygonum lapathifolium* and *Sonchus oleraceus*, indicates a similar conclusion. The fen/reedswamp helophyte group is rather restricted and only includes cf. *Cicuta virosa*, *Mentha* cf. *aquatica*, *Oenanthe aquatica*, *Ranunculus sceleratus* and *Urtica dioica*. The aquatics recorded are *Ceratophyllum demersum*, *C. submersum*, *Hydrocharis morsus-ranae*, *Lemna minor*, *Menyanthes trifoliata*, *Scirpus lacustris*, *Sparganium angustifolium*, *Typha latifolia* and *Salvinia natans*. This group thus includes free-floating, submerged rooted and emergent rooted aquatics, and eutrophic quiet water of shallow depth is indicated.

(ii) *Cromerian II*

Two l.p.a.b.s, a Gramineae–*Betula–Pinus* l.p.a.b. and a *Pinus–Betula*–Gramineae l.p.a.b., are recorded in detritus muds at four sites. At two of them, MU and MF, the muds overlie those of the Cr IIa *Quercus–Ulmus–Pinus* p.a.b. At MP and MAK, they overlie the ferruginous sands and gravels of the marker level.

a. *Gramineae*–Betula–Pinus *l.p.a.b.*

The NAP, *Betula* and Gramineae frequencies rise very suddenly at the base of this biozone at MU and MF, and at MP 0 cm. *Betula* replaces all other trees, except *Pinus*, with *Alnus*, *Ulmus* and *Quercus* falling to very low frequencies. Low frequencies of *Carpinus* and Ericales pollen begin to appear, and *Salix* is also more frequent. The NAP element, apart from the abundant Gramineae, is better represented than in the previous zone, with increases in Caryophyllaceae, Compositae Liguliflorae and Tubuliflorae, Cruciferae, *Filipendula*, and Umbelliferae. The aquatic pollen frequency also rises strongly, especially *Sparganium* and *Typha latifolia*. Filicales spore frequencies also rise sharply.

Macroscopic analyses in this biozone were made at MU 0–15 cm, MU 20–30 cm (some overlap with the Cr IIa biozone below) and MF 0–20 cm. Taking the MU analyses first, these show declining values of fen/reedswamp helophytes and a corresponding increase in aquatics. In the MU 20–30 cm analysis, trees, shrubs, the heath and dry- and damp-ground herb groups are absent. Fen/reedswamp helophytes are represented by *Carex rostrata*, *Cicuta virosa*, *Ranunculus sceleratus* and *Urtica dioica*. Aquatics recorded are *Ceratophyllum demersum*, *Hydrocharis morsus-ranae*, *Myriophyllum spicatum*, *Ranunculus* subg. *Batrachium*, *Scirpus lacustris*, *Typha latifolia* and *Salvinia natans*. In the MU 0–15 cm sample, the fen/reedswamp helophyte group is reduced, and the tree and aquatic groups are increased. The trees recorded are *Alnus* (catkins), *Quercus* (wood), *Q.* cf. *robur* (acorn) and *Prunus avium*. Fen and reedswamp helophytes only include *Ranunculus sceleratus*, and the aquatics recorded are *Hydrocharis morsus-ranae*, *Nuphar lutea*, *Typha latifolia* and *Salvinia natans*.

The MF 0–20 cm analysis gave a much more substantial list, with aquatics best represented, but many other groups present. Trees include *Betula* and *Prunus*, dry-ground herbs *Chenopodium rubrum/botryodes*, and damp-ground herbs *Carex pendula*. The muddy substrates/shallow-water group records *Polygonum minus*, *Rumex hydrolapathum* and *R. maritimus*, with the fen/reedswamp helophytes *Cicuta virosa*, *Ranunculus sceleratus* and *Urtica dioica*. The aquatic list is more substantial, with *Ceratophyllum demersum*, *C. submersum*, *Hydrocharis morsus-ranae*, *Lemna minor*, *Potamogeton natans*, *P. trichoides*, *Ranunculus* subg. *Batrachium*, *Sagittaria sagittifolia*, *Scirpus lacustris*, *Sparganium angustifolium*, *Stratiotes aloides*, *Typha* cf. *latifolia* and *Salvinia natans*.

b. Pinus–Betula–*Gramineae l.p.a.b.*
This assemblage from MAK (10 cm) again shows high *Betula* and Gramineae frequencies, with reduced frequencies of *Ulmus* and *Quercus*, but *Pinus* frequencies are higher, and the aquatic pollen and Filicales spore frequencies do not rise as in the previous biozone. The macroscopic analysis (6–11 cm) shows a restricted list of taxa. *Betula pendula/pubescens* is recorded in the tree group, and *Rubus* cf. *idaeus* in the shrub group. In the dry-ground herb group cf. *Euphorbia cyparissias* is the only taxon present, with *Alisma plantago-aquatica*, the most abundant species present in the list, is the only member of the muddy substrates/shallow water group. *Urtica dioica* is the only member of the fen/reedswamp helophyte group and the aquatics recorded are *Chara*, *Myriophyllum spicatum*, *M. spicatum* v. *pectinatum*, *Najas minor*, *Ranunculus* subg. *Batrachium*, *Typha latifolia* and *Salvinia natans*.

The change from the *Quercus–Ulmus–Pinus* p.a.b. to the two l.p.a.b.s characterised by *Betula* and Gramineae is abrupt, occurring in the course of detritus-mud sedimentation. At first sight the change might be considered to have a climatic cause. But the thermophilous aquatics (*Hydrocharis*, *Lemna*, *Salvinia*) persist through both biozones and there is evidence, both from pollen and macroscopic remains, of a temperate climate in the later biozones. The change is probably related to a rise in water level, leading to the local development of fen carr of a eutrophic type, with ensuing dominance of *Betula* (*pubescens*). The increased *Salix* frequencies will also result from this development. Evidence supporting this explanation is found in the enrichment of the aquatic group of macroscopic plant remains in the upper part of MU, and the sharp increases in aquatic pollen and Filicales spores (*Thelypteris*?). The presence of *Alnus* and *Quercus* remains in this upper part shows the continued presence of elements not well represented in the pollen rain, indicating the local nature of the pollen assemblage. The MAK macroscopic assemblage indicates shallower and drier conditions (*Alisma plantago-aquatica*), and the different pollen assemblage, with decreased aquatic pollen and Filicales spores, may then represent pollen rain not affected by overmuch local representation.

The rising water level eventually resulted in marine transgression, for at the top of the section MU (sample T) laminated tidal silty clay occurs, with a pollen assemblage indicating a Cr III age.

(iii) *Cromerian IIb*

A *Quercus–Ulmus–Tilia* p.a.b. similar to that at West Runton is recorded in section ME in a thin horizon of silty clay near the base of the beach gravel of bed i(b). This is the only record of the *Quercus–Ulmus–Tilia* p.a.b. of Cr IIb, which at West Runton is the major biozone of the West Runton Freshwater Bed. The pollen spectrum includes 5% Chenopodiaceae pollen, suggesting maritime conditions of deposition. If this is correct, even if the grey silty clay is partly reworked during the formation of the beach gravel, marine transgression in this area would appear to have started in Cr IIb times.

(iv) *Cromerian IIIa*

A p.a.b. with *Carpinus* is recorded at MU (top sample, T) in laminated silty clay of bed i(lc) and in a silty clay horizon with *Mytilus* at the junction of beds i(b) and i(s) at MW. In the assemblage, *Quercus* and *Ulmus* and a very low frequency of *Tilia* are present. The pollen spectra resemble those of Cr IIIa at Corton, and show the addition of *Carpinus* to the temperate woodland at about the time of the Cromerian marine transgression. High frequency of Chenopodiaceae pollen tallies with the tidal nature of the sediment.

(v) *Cromerian IIIb*

An *Abies–Carpinus* p.a.b. is recorded at many sites along the Paston–Mundesley coast in the tidal laminated silty clays of bed i(lc) and in clayey horizons within the beach gravel of bed i(b). The assemblages show a high AP frequency, with *Pinus* having the highest frequency, and lower percentages of *Picea*, *Abies*, *Alnus*, *Betula*, *Ulmus*, *Quercus* and *Carpinus*, and traces of *Corylus*. The tree assemblage shows the addition of *Abies*, the reduction of *Ulmus*, *Quercus* and *Tilia*, and the expansion of *Picea*. There may be over-representation of conifers in these tidal sediments. There are low frequencies of *Calluna*, *Empetrum* and Ericales. Gramineae and Chenopodiaceae are the best represented of the NAP taxa, which are, as with the aquatic pollen, of limited variety and sparse occurrence. Filicales and *Sphagnum* are regularly present in low frequencies.

Two macroscopic analyses from tidal sediments of Cr IIIb were made, at MAR 180–190 cm and MAK +9–15 cm. They show a poor and mixed macroscopic plant remains content with a restricted range of species and a strong maritime representation (Fig. 28). The only tree represented is *Betula pendula/pubescens*. The maritime group includes *Salicornia* and *Triglochin maritima*, and *Alisma plantago-aquatica* occurs in the muddy substrate/shallow-water group. The fen/reedswamp helophyte group includes *Epilobium palustre* and *Urtica dioica*, and the aquatic group *Callitriche hermaphroditica*, *Myriophyllum*, *Ranunculus* subg. *Batrachium*, *Typha latifolia* and *Azolla filiculoides*.

(vi) *Cromerian IVa*

A *Pinus–Alnus–Picea–Betula* p.a.b. is recorded in many sections of the tidal laminated silty clays of bed i(lc), superimposed on the sediments containing the previous biozone (e.g. MA, MAM, MAR), and in silty streaks of the bedded sands of bed j (MAD, MAH 1). The tree pollen frequencies are rather lower than those of the previous biozone. *Abies* pollen largely disappears, and there is a very reduced representation of pollen of the thermophilous trees. The Ericales taxa show higher frequencies, with Gramineae and Chenopodiaceae pollen remaining well represented.

A number of macroscopic analyses were made from sediments of Cr IVa (MAR 100–110 cm, 130–145 cm; MAM 2 cm; PA 30 cm, 200 cm). The sample from MAR 130–145 cm was from a horizon within the tidal sediments of stony mud with wood; the other samples are from tidal sediments. The macroscopic floras are again restricted and mixed (Fig. 28). Trees recorded include *Alnus glutinosa*, *Betula pendula/pubescens* and Coniferae (leaves). *Rubus idaeus* is the only shrub represented, and the maritime group includes *Juncus* cf. *gerardii* and *Salicornia*. The heath group has only *Eriophorum vaginatum*, and the dry-ground herb group *Plantago major*. The muddy substrate/shallow-water group contains *Alisma plantago-aquatica* and *Eleocharis mamillata*, and the fen/reedswamp helophytes *Carex aquatilis/bigelowii*, *Filipendula ulmaria*, *Lycopus europaeus* and *Urtica dioica*. The aquatic group is best represented in the detritus mud of MAR 130–145 cm, and includes *Elatine hydropiper*, *Lemna* cf. *trisulca*, *Menyanthes trifoliata*, *Ranunculus* subg. *Batrachium*, *Scirpus lacustris*, *Stratiotes aloides* and *Typha latifolia*. In the unclassified group occur *Carex* and *Dryopteris*-type (sporangia) and Gramineae. Like other assemblages in tidal sediments, this is clearly derived from a variety of plant communities.

In the section MAM, a Gramineae–*Pinus–Betula* l. p.a.b. is recorded in humified mud with *Phragmites* at 25–52 cm. This sediment must have been deposited in a brackish lagoon in an area of tidal flats. The assemblage shows an increase in *Betula*, Gramineae, Cyperaceae and aquatic pollen, with decreases in *Picea* and *Alnus*, changes indicating plants locally present (increases) and locally absent (decreases). Macroscopic analyses from this organic sediment (MAM 30–40 cm, 40–50 cm) show a restricted list of taxa. Taxa recorded are *Betula* cf. *pendula/pubescens*, Coniferae (leaves), *Carex aquatilis/bigelowii*, *Phragmites communis*, *Menyanthes trifoliata*, *Scirpus lacustris*, *Sparganium* cf. *angustifolium*, *Carex* and *Dryopteris*-type (sporangia).

The pollen assemblage has similarities with the local Gramineae–*Betula–Pinus* l.p.a.b. of Cr II, indicating the facies nature of the assemblage.

(vii) *Cromerian IVb*

A *Pinus–Alnus–Betula* p.a.b. is recorded in a horizon of silty clay with a detritus-mud fraction in the sand of bed j at MB 0 cm. The assemblage only differs from that of the previous biozone by the lower *Picea* pollen frequency and the higher frequencies of Gramineae and Cyperaceae pollen. The difference is minor and may reflect a change to freshwater conditions (high Cyperaceae frequency, *Chara* oospores present) on the regression after the deposition of the tidal sediments of Cr IVa. The regional vegetation at this time must have been boreal woodland wtih local grass communities.

The regression is confirmed by an analysis of non-marine molluscs by B. W. Sparks from the level marked s of site MB, Fig. 33, immediately above the laminated clays of Cr IIIb, given in Table 19. He comments that little can be deduced from this restricted and badly preserved freshwater fauna. However, it is significant in demonstrating that a regression took place in post-Cr IIIb and pre-Cr IVb times.

(viii) *Cromerian IVc*

A *Betula* p.a.b. is recorded in freshwater mud of bed k at section MAU (Fig. 34, Plate 25), lying unconformably on the tidal sediments of bed i(lc). *Pinus*, *Picea*, *Juniperus* and *Salix* pollen is present in low frequency. The NAP frequency is around 40%, and is predominantly Gramineae, with Cyperaceae, Chenopodiaceae, *Artemisia*, *Ranunculus* and Umbelliferae showing low continuous curves. Aquatic pollen includes *Myriophyllum spicatum*, *Nuphar*, *Sparganium* and *Typha latifolia*, and spores of Filicales, *Equisetum*, *Lycopodium annotinum* and *Sphagnum* are present. The assemblage indicates regional birch woodland, with *Juniperus* and *Salix* present, and herb communities with Gramineae and *Artemisia*.

A macroscopic analysis from this horizon shows a predominant aquatic element (Fig. 28). *Betula pendula/pubescens* fruits are recorded, together with *Chenopodium* cf. *album*, *Atriplex patula*, *Thalictrum minus* and *Valerianella* in the dry-ground group, *Ranunculus repens* in the damp-ground group, and *Alisma plantago-aquatica*, *Eleocharis palustris* and *Rumex maritimus* in the muddy subsstrate/shallow water group. The fen/reedswamp helophyte group is larger, with *Cicuta virosa*, *Filipendula ulmaria*, *Lycopus europaeus*, *Oenanthe aquatica*, *Ranunculus lingua*, cf. *Selinum carvifolia*, *Thalictrum flavum* and *Urtica dioica*. The aquatics comprise an assemblage with varying trophic requirements at the present day: cf. *Caltha*, *Ceratophyllum demersum*, *Chara*, *Hippuris vulgaris*, *Isoetes lacustris*, *Myriophyllum* cf. *alterniflorum*, *M. spicatum*, *Nuphar lutea*, *Potamogeton* cf. *alpinus*, *P. filiformis*, *P. friesii*, *P. gramineus*, *P. pectinatus*, *P. praelongus*, *P. trichoides*, *Ranunculus* subg. *Batrachium*, *Scirpus lacustris*, *Sparganium* cf. *angustifolium*, *Stratiotes aloides*, *Typha latifolia* and *Zannichellia palustris*.

The assemblage is a rich one derived mainly from local fen, reedswamp and aquatic communities.

An analysis of non-marine molluscs by B. W. Sparks from the freshwater mud of MAU is given in Table 19. He comments on the list: the molluscs are in a poor state, indicating water with movement, but otherwise rather uninformative both ecologically and climatically; although not positively confirming particular ecological conditions, the fauna is a pretty tolerant one and does not contain the species typical of the middle of an interglacial.

(g) Vegetational history: early Anglian

Two p.a.b.s with high frequencies of NAP are recorded in the uppermost part of the pre-glacial succession of the Paston–Mundesley coast. At MAH a channel filled with stony mud records a Gramineae–Cyperaceae–*Betula–Pinus* p.a.b. *Betula*, *Pinus* and *Salix* are the predominant tree/shrub pollen types. Low frequencies of *Picea*, *Alnus* and *Juniperus* also occur. The NAP is predominantly Gramineae and Cyperaceae, but the following are also recorded: Chenopodiaceae, *Plantago maritima*, Caryophyllaceae, Compositae Liguliflorae and Tubuliflorae, *Artemisia*, Cruciferae, *Filipendula*, *Plantago major/media*, *Ranunculus*, *Potentilla*, Rubiaceae, *Thalictrum* and Umbelliferae. *Myriophyllum verticillatum* and *Sparganium* pollen is recorded, and Filicales and *Sphagnum* spores. The assemblage indicates regional grassland with a variety of open ground herbs and scattered woodland.

A macroscopic analysis (MAH 10–30 cm) shows a fairly rich flora, mainly of fen/reedswamp helophytes and aquatics. The trees include *Betula pendula/pubescens*, cf. *Betula* tree sp. × *B. nana*, Coniferae (leaf), and the shrubs *B. nana* and *Salix*. *Luzula* cf. *spicata* is placed in the montane group, and the muddy substrate/shallow water group includes *Eleocharis palustris* subsp. *palustris* and *E. uniglumis*. The fen/reedswamp helphytes are *Carex aquatilis/bigelowii* and *Potentilla palustris* and the aquatics *Callitriche platycarpa*, *Chara*, *Hippuris vulgaris*, *Myriophyllum spicatum*, *M. verticillatum/spicatum*, *Potamogeton* cf. *friesii*, *P. vaginatus*, *Ranunculus* subg. *Batrachium* and *Typha latifolia*. The assemblage indicates a substantial aquatic flora, with local shrub or woodland and fringing fen/reedswamp.

The second p.a.b. ascribed to the e An substage is at ME, in sediment of bed m in a channel filled with laminated silty clay with organic laminations. The assemblage, a Gramineae–Cyperaceae p.a.b., shows about 80% NAP. *Pinus*, *Betula* and *Alnus* are present in the AP. Gramineae and Cyperaceae reach high frequencies, with few other NAP taxa, which include *Plantago major/media* and Umbelliferae. The assemblage indicates the regional prevalence of grassland communities, with scarce woodland.

Three macroscopic analyses were made in the ME channel (Plate 29). ME 5 is in the marginal part of the channel, ME 2 is from the centre in the most organic part of the filling, at the level of the pollen analysis, and ME 4 is from a less organic horizon in the centre. The group analysis of these three samples is shown in Fig. 28. It will be seen that all groups except heath and water ferns are represented, but there is considerable variation of content from one sample to another, e.g. in the frequency of the montane group. We may consider the three lists together. cf. *Betula* is recorded in the marginal sample. Shrubs are best represented in the centre organic sample (ME 2), with *Betula nana*, *Salix* cf. *herbacea*, *S.* cf. *polaris*, *S.* cf. *viminalis/phylicifolia*. *Salix* buds or budscales are recorded in all samples, though no *Salix* pollen is recorded. In the maritime group *Armeria maritima* is recorded in ME 2. In the two centre samples, ME 2 and 4, *Oxyria digyna* and *Saxifraga oppositifolia* are recorded in the montane group. In the dry-ground herb group, present in all samples, there are present cf. *Atriplex*, *Cerastium arvense*, *C. holosteoides*, *Linum anglicum*, *Minuartia* sp., and *Polygonum aviculare*. The damp-ground herb group contains only *Selaginella selaginoides*, and the muddy substrate/shallow-water group *Alisma plantago-aquatica* and *Rorippa microphylla*. The fen/reedswamp helophyte group comprises *Carex* cf. *aquatilis*, cf. *Glyceria* sp., *Iris pseudacorus*, *Parnassia palustris*, *Ranunculus scleratus* and *Urtica dioica*. The aquatics are *Callitriche hermaphroditica*, *Chara*, *Isoetes lacustris*, *Myriophyllum*, *Nitella/Tolypella*, *Potamogeton alpinum*, *P. filiformis*, *P. gramineus*, *P. natans*, *P. pectinatus*, *P. vaginatus*, *Ranunculus* subg. *Batrachium*, *Sagittaria sagittifolia*, *Stratiotes aloides*, *Typha* cf. *latifolia* and *Zannichellia palustris*. The unclassified group includes *Carex*, cf. *Alyssum*, Cruciferae (seeds and capsule valves), Gramineae, *Potentilla*, *Viola* and numerous derived megaspores.

The assemblage bears out the evidence for regional herb vegetation indicated by the pollen content, but adds much detail about species concerned. The flora contains many taxa characteristic of the 'full-glacial' floras of the Devensian cold stage (Godwin, 1975). Low shrub communities of *Salix* spp. are clearly indicated. Rich herb communities are also indicated (*Cerastium*, *Linum*, *Oxyria*), with a rich fen and aquatic flora, with a particular variety of *Potamogeton* species, all of which range north of the Arctic Circle today in Scandinavia, except for *P. vaginatus* which has a limit just south of the Circle.

Three other samples have been analysed for macroscopic plant remains, one in the laminated silty clay of bed m at MAJ, one in the basin of laminated silty clay with organic seams of bed o at PC, and one in stony clay in bed m at PH. The MAJ analysis gave a short list. It contains cf. *Betula*, and the muddy substrate/shallow-water group *Alisma* and *Eleocharis*. The fen/reedswamp helophyte group includes *Hydrocotyle vulgaris* and *Urtica dioica*, and the aquatics *Callitriche*, *Littorella uniflora*, *Nitella/Tolypella*, *Potamogeton berchtoldii*, *Ranunculus* subg. *Batrachium*, cf. *Sparganium*. The unclassified group includes *Carex*, cf. *Rumex*, *Viola* and many derived megaspores. The assemblage indicates the presence of shallow water and muddy substrates with a restricted flora, with deposition of laminated silty clays under freshwater conditions. Judging by the extensive outcrop of the sediment along the coast, the pool must have been large. Unfortunately no identifiable pollen was recovered from the sediment.

The second analysis, from stony clay in bed m at PH, gave an even shorter list, with *Betula nana*, *B. pendula/pubescens*, and *Salix* amongst the trees or shrubs, *Salicornia* in the maritime group and in the aquatic group *Myriophyllum spicatum*, *Scirpus lacustris* and *Typha latifolia*. The flora indicates predominantly freshwater conditions, but the sediment is stony and elements of it may be derived from the underlying tidal silty clays of bed i(lc).

The third analysis is from sediment at the shallow channel at PC. Cf. *Alnus* and *Salix* are recorded in the tree and shrub groups, *Eriophorum vaginatum* in the heath group, Chenopodiaceae in the dry-ground herb group, *Selaginella selaginoides* and *Senecio aquaticus* in the damp-ground group, *Callitriche platycarpa*, *Chara*, *Potamogeton filiformis*, *Ranunculus* subg. *Batrachium* and *Zannichellia palustris* in the aquatic group, and *Carex*, *Dryopteris*-type (sporangia), Gramineae, and cf. *Rumex* in the unclassified group. The assemblage is a restricted one, showing the presence of shallow-water, muddy and damp-ground communities in the neighbourhood.

A small moss flora was obtained from this sample. It included *Calliergon giganteum*, *C.* cf. *cordifolium*, *Campylium* sp., *Drepanocladus aduncus*, *Eurynchium praelongum*, *Pohlia/Bryum*, and *Scorpidium* cf. *scorpioides*. The *Calliergon* spp., *Drepanocladus* and *Scorpidium* are plants of rich fens, shallow eutrophic waters or wet woodland (carr) and have been recorded in cold stage floras of the Wolstonian or Devensian (Dickson, 1973). *Eurynchium* is rather out of place in this list, since it has been recorded fossil from interglacial or Flandrian woodland environments.

At MAA a large drifted stump of *Taxus baccata* was recorded in the early Anglian sediments (Fig. 32, Plate 26), probably reworked from the earlier Cromerian temperate stage.

10: Bacton

(a) Lithostratigraphy

Sections at Bacton, including two recorded from excavations, are shown in the bottom half of Fig. 39. The positions of the datum at BE and section BC are shown in Fig. 2. The sections here are separated from those at Paston by a length of coast where till cuts down below the level of the base of the cliff. But in general the stratigraphy is similar to that seen at Paston.

The sediments can be grouped into the following beds. Table 26 summarises this succession.

a. Grey silty sand with silt laminae and laminated silty clay is seen at the base of the excavations BI and BH (358 m and 494 m, Fig. 39), up to 70 cm thick.

b. Ferrugineous sandy gravel occurs above the grey silty sand in BI and BH, and at the base of the section as a prominent ledge in BJ (67 m).

c. Red sand, 75 cm thick, with apparent dip north-west, occurs above the gravel in BH, and is capped by a ferrugineous stony layer, of the same type that forms the marker level at Paston and Mundesley.

d. Pale and reddish coarse sand, with silty laminations at BI and BH, is seen at these sites and at the base of BJ, resting on a ferrugineous horizon. This horizon is up to 1 m thick.

e. This bed is the most prominent sediment seen along the Bacton coastal section. It is up to 2·5 m thick and consists of sediments showing a tidal structure – a pale sand, rich either in wavy laminations of grey and brown silt (BC, BE, BJ, BI, BH, BA), or in thin seams of drift mud (BB, BD). At BJ, BI, BH and BA a more clayey sediment is found near the base of the horizon (e (bc)), and, in BI and BH, at the top (e (tc)).

f. In BI and BH a series of laminated clays and sands, with a thin gravel at BI, are up to 1·9 m thick. They rest on the sediments of bed e, and underlie the till. Freshwater horizons occur near the top of this horizon in both sections. At BI silty mud occurs 20 cm from the top of the section; at BH a stiff grey clay with freshwater shells occurs 70 cm below the till. The topmost sands at BC, BB and BD may also belong to this horizon.

g. The till sealing the succession is the Second Till of Banham (1968).

(b) Periglacial stratigraphy

Ice-wedge casts

BC (Fig. 39), penetrating horizon e, sealed by bed f (see West & Donner, 1958).

Involutions

BC, BD, in the lower part of bed f (see West & Donner, 1958).

(c) Sedimentary history

The grey silty sands (bed a) at the base are marine, equivalent to bed e at Paston and Mundesley. The overlying ferrugineous gravels (bed b) and sands (bed c) may indicate regression with subsequent weathering. Beds d and e indicate that a transgression followed, with the deposition of tidal sediments, first much more sandy (bed d) and then with silt or drift-mud laminations (bed e). The transgression sediments reach a thickness of up to 2·5 m. The gravel, clay and sand of bed f contains in its upper part freshwater sediments, indicating regression with continued deposition. The ice-wedge cast at BC indicates the same, with permafrost penetrating the tidal sediments. Sand was deposited at a later date and involutions then affected this sand and the uppermost tidal sediments (BC, BE).

Table 26 gives a summary of this sequence.

(d) Biostratigraphy

Table 24 lists the biozones identified in pollen diagrams from Bacton (Fig. 40). Table 25 gives the macroscopic analyses, and Table 26 gives a summary of the biozone sequence and vegetational history.

Biozone a. *Pinus–Betula*–Gramineae p.a.b. This basal biozone, in bed a, is represented by a single sample. It is placed in Pa I.

Biozone b. *Pinus–Quercus–Carpinus–Ulmus* p.a.b. This biozone includes analyses from silty laminae in the sand of bed a in the excavations at BH and BI. The spectra are grouped together and placed in Pa II. But the assemblages were poorly preserved and variable in content and may have been affected by weathering.

Biozone c. *Pinus–Alnus–Picea–Betula* p.a.b. This biozone is found in the tidal sediments of bed e and the clay of bed f along the cliff base at Bacton. It is placed within Cr IVa.

Biozone d. Gramineae–*Betula–Pinus–Alnus* p.a.b. This high NAP biozone is from freshwater mud near the top of the excavation BI and is placed in the early Anglian substage.

(e) Vegetational history

(i) *Pastonian I*

A *Pinus–Betula*–Gramineae p.a.b. is recorded in the basal analysis of section BH. The NAP frequency is considerable, and there are only low frequencies of pollen of thermophilous trees. The assemblage is therefore placed in Pa I. Gramineae and Cyperaceae comprise most of the NAP, with low frequencies of spores of Filicales, *Osmunda* and *Sphagnum*.

(ii) *Pastonian II*

A *Pinus–Quercus–Carpinus–Ulmus* p.a.b. is recorded in sediments of bed a in section BH above the Pa I biozone, and at the base of section BI. The AP frequency is high, and together with *Pinus*, *Picea*, *Alnus* and *Betula*, *Ulmus*, *Quercus* and *Carpinus* are present. The NAP is mostly Gramineae, Cyperaceae and Chenopodiaceae, with higher frequencies of Filicales and *Osmunda* (*regalis*) than in the previous biozone.

A macroscopic plant analysis of the marine sediments of bed a at BH 545–575 cm gave a small poorly preserved flora with the maritime and aquatic groups best represented (Fig. 41). The only tree remain found was a cf. conifer leaf, and the only shrub buds of *Salix*. The maritime group includes *Salicornia* and the heath group cf. Ericaceae seeds. The dry-ground herb group includes cf. *Urtica urens* and the damp-ground group cf. *Selaginella selaginoides*. The fen/reedswamp group includes a sporangium of *Osmunda*, and the aquatics *Hippuris vulgaris*, cf. *Najas*, *Ranunculus* subg. *Batrachium*, *Typha latifolia* and *Azolla filiculoides*. The flora is the usual mixed type found in marine sediments.

(iii) *Cromerian IVa*

A *Pinus–Alnus–Picea–Betula* p.a.b. is recorded in the tidal sediments of bed e and in the more clayey sediments of bed f, at sections BB, BI, BE and BH. The AP frequency is high, and the genera named above form the bulk of the AP, with only low frequencies of pollen of thermophilous trees. The Ericales taxa form low pollen curves with the percentage highest in the most organic of the tidal sediments at section BB. Gramineae, Cyperaceae and Chenopodiaceae comprise most of the NAP, with Filicales, *Osmunda* (*regalis*) and *Sphagnum* spores.

(iv) *Early Anglian*

A Gramineae–*Betula–Pinus* p.a.b. is recorded in the freshwater mud at the top of section BI. The NAP frequency is relatively high, mostly Gramineae, with substantial frequencies of *Betula*, *Pinus* and *Alnus*. The allocation of this biozone to a substage is problematical. The AP composition, apart from the *Alnus* frequencies, resembles other biozones ascribed to Cr IVb (e.g. at Overstrand, Sidestrand, Paston–Mundesley and Ostend), but the Gramineae percentages are considerably higher. The biozone has therefore been tentatively placed in the e An substage, the assemblage indicating grassland and scattered woodland.

A restricted macroscopic flora is recorded from the freshwater mud (BH 70–100 cm), and shows a predominance of aquatics (Fig. 41). There are no members of the tree, shrub, maritime, heath or damp-ground herb groups. The dry-ground herb group contains *Potentilla crantzii/tabernaemontani*, the muddy substrate/shallow-water group *Eleocharis palustris* subsp. *palustris*, and the fen/reedswamp helophyte group *Carex bigelowii/ aquatilis*. The aquatics comprise *Chara*, *Nitella/Tolypella*, *Potamogeton* cf. *lucens*, *P.* cf. *perfoliatus* and *Ranunculus* subg. *Batrachium*. The limited assemblage supports a correlation with the e An substage and indicates a shallow pool, surrounded by *Eleocharis* and *Carex* communities.

11: Ostend

(a) Lithostratigraphy

Only a single section, 80 m long and position shown in Fig. 2, was observed at Ostend. This is drawn in Fig. 39 and showed a channel, filled with clay-mud and detritus mud, cut into a massive grey silty clay with rootlets penetrating from the surface, and sealed by till. A boring with a 1-inch single spiral auger to 3·5 m depth was made at the south-east end of the section.

The following succession of sediments was observed. Table 26 summarises this succession.

a. 1·5 m of grey sand, with a 20 cm thick horizon of grey clay 80 cm down, was found in the bore-hole OD.

b, c. A massive grey clay, non-calcareous and 2 m thick (bed b), was seen in section in its upper part (Plate 31), and in the auger hole to its full depth. Wood was present in the top 20 cm of this sediment (bed c), both as rootlets penetrating vertically and as horizontal branches (Plate 32). Race was present at 60–90 cm depth. This rootlet bed was described by Reid (1882, p. 42), who noted a section nearly a quarter of a mile south-east from Ostend Gap as follows:

	Feet
First Till (*base below high-water*)	
? (hidden)	8
Carbonaceous soil with a little lignite	½
Greenish unstratified loam, with small roots, bedding obscure passing into	1
Laminated carbonaceous silt and green clay, with a little lignite and calcareous concretions	2
Green clay, drift wood and fir cones to low water	1

This rootlet bed is lithologically similar to the rootlet bed, bed e, at Corton, where it is considered to be an alluvial deposit.

The sediment may therefore be of alluvial origin in part, or as considered by Reid (*loc. cit.*), may be a weathered surface of the 'estuarine beds', the laminated clays (of Pastonian age), recorded by Reid below the rootlet bed. If alluvial, it is likely to be of Cromerian age. No such sediments are recorded in the Pastonian elsewhere on the coast.

d. This horizon is the filling of the channel cut into bed b. The stratigraphy of the channel filling at OA was:

Till

0–52 cm Grey clay-mud, with freshwater shells at 12–16 cm.

52–88 cm Brown detritus mud, with a sharp boundary below.

88–104 cm Blue-grey clay, rootlets present (bed b).

At OC 25 cm of compressed brown detritus mud with wood lay between the till and bed b.

e. The channel filling was sealed by the First Till (Banham, 1968), which south-eastward rested on the woody surface of bed b.

(b) Sedimentary history

Clay (bed b), possibly of alluvial origin, was laid down on the sand of bed a. This clay, with its upper part containing wood (bed c), must be the horizon described by Reid (*see above*) as unstratified loam with small roots, and, according to his sections, it also overlay laminated carbonaceous silt, probably equivalent to bed j at Happisburgh. Weathering of this sediment produced calcareous concretions and surface tree growth resulted in rootlet penetration at the surface. The surface was later channelled and in the channel freshwater detritus mud and clay-mud (bed d) with wood in the basal part of the sediments, was deposited.

Table 26 gives a summary of this sequence.

(c) Biostratigraphy

Table 27 lists the biozones distinguished in the pollen diagrams from Ostend (Fig. 40). Table 25 gives the macroscopic plant analyses, and Table 26 summarises the sequence of biozones and the vegetational history.

Biozone a. Gramineae–*Betula–Pinus* p.a.b. This is identified from a single sample in the borehole OD, from the basal sediment, grey sand with grey clay intercalation (bed a). The pollen spectrum is similar to that from Happisburgh C 457 cm (biozone f). It is placed in the Pre-Pastonian, and a correlation with Pre-Pa d is suggested.

Biozone b. *Alnus–Pinus–Picea–Quercus–Ulmus* p.a.b. This assemblage occurs in the upper part of the rootlet bed (bed c). There was no well preserved pollen in the lower part of the bed b, where the sediment was decalcified and contained race, indicating a period of weathering before the deposition of this biozone. The correlation of this biozone is discussed in the next section, as it appears related stratigraphically to biozone c.

Biozone c. *Pinus–Alnus–Picea–Betula* p.a.b. This biozone is seen in the freshwater mud of bed c, lying in a channel cut in bed b. The importance of *Pinus*, *Alnus*, *Picea* and *Betula* in the spectra suggest the biozone should be correlated with Pa IVa or Cr IVa. The heavy weathering of bed b, which itself lies on laminated beds of the type seen at Happisburgh, suggests a significant discontinuity of deposition and the biozone is correlated with Cr IVa rather than with Pa IVa. Freshwater sediments of Cr IVa were also observed at Paston–Mundesley (MB 60, MAD, MAH 1) at about the same level. The pollen (biozone b) in the top of bed c would then be associated with reworking of sediments at the surface of bed b, with tree growth taking place before the freshwater episode bed d, biozone c, times. It is possible that biozone b, with low frequencies of *Quercus* and *Ulmus* pollen, may be a local assemblage relating to phase in Cr IVa times when freshwater deposition was starting subsequent to the Cromerian regression. Thus the assemblages of biozones b and c may be a marginal quietwater equivalent of assemblages from the sands of bed j at MAH and MAD in the Paston–Mundesley area.

Biozone d. *Betula–Pinus–Alnus–Picea* p.a.b. This biozone is identified from an erratic of coarse mud containing *Picea* cones and bored by *Pholas*, found on the beach at Ostend. It might be correlated with Pa IV or Cr IV. At Mundesley, *Picea* cones were found in Pa IV (MPA 0) in a freshwater sediment, and Reid also noted fir cones at Happisburgh in sediments now dated as Pastonian. Boring by *Pholas* indicates that the *in situ* position of the mud was low in the intertidal area or below LWM. A correlation with Pa IV therefore seems more probable.

(d) Vegetational history

(i) *Pre-Pastonian*

A Gramineae–*Betula–Pinus* p.a.b. is recorded in a clay horizon within the sand of bed a at the borehole OD. The AP frequency is low, with *Betula* and *Pinus* the best represented trees, accompanied by lower frequencies of *Picea*, *Alnus*, *Salix*, *Ulmus* and *Quercus*. The NAP is predominantly Gramineae, with very low frequencies of Cyperaceae, Ranunculaceae and Umbelliferae pollen. The assemblage indicates regional grassland with sparse woodland and, since it has similarity in content and level with Pre-Pastonian, high Gramineae assemblages at Happisburgh, it is placed in the Pre-Pa substage, either Pre-Pa b or d of the Beeston succession. In the absence of *Artemisia*, perhaps the latter seems more probable.

(ii) *Cromerian IVa*

Two assemblages ascribed to this zone (biozones b and c) both show high AP frequencies with *Alnus*, *Pinus* and *Picea* the most frequent tree pollen types. In the older of the two biozones, associated with the rootlet bed (bed c), there are rather higher frequencies of *Ulmus*, *Quercus* pollen and Filicales spores and lower frequencies of *Betula* pollen (*Alnus–Pinus–Picea–Quercus–Ulmus* p.a.b.). The changes associated with the start of freshwater mud deposition, rises in *Betula*, *Pinus*, *Picea* and *Carpinus*, falls in *Alnus*, *Ulmus*, *Quercus*, *Filipendula* and Filicales, are probably caused by change from local pollen deposition to a more regional origin of the pollen rain. The assemblages both indicate the character of the regional woodland, the older p.a.b. giving insight into the woodland growth associated with the surface sediment of the rootlet bed, and the younger p.a.b. indicating regional boreal woodland.

Two macroscopic analyses were made from the freshwater sediment of biozone c (*Pinus–Alnus–Picea–Betula* p.a.b.) at section OA 70–80 cm in the middle of the channel and section OC 20–25 cm at the edge of the channel. Both show (Fig. 41) a preponderance of the tree group, with lesser representation of the fen/reedswamp helophyte and aquatic groups (both represented least in the marginal OC sample). Taking both analyses together, the tree group includes *Acer campestre*, *Alnus glutinosa* (very abundant), *A.* cf. *incana* (see Wilson, 1973), *Betula pendula/pubescens*, *B.* cf. *pubescens*, and in the marginal mud of OC rich in wood, Coniferae (leaf), cf. *Corylus avellana*, *Picea abies* and *Taxus baccata*. The shrubs *Thelycrania sanguinea* and *Viburnum opulus* are only recorded in the marginal OC sample.

The dry-ground herb group, as might be expected, is best recorded in the central sample (OA), with *Ranunculus* cf. *bulbosus*, *Rumex acetosa* and *R. acetosella*, with only *Rumex* cf. *acetosa* in the marginal sample (OC). The damp-ground herb group includes only *Carex* cf. *remota*, *Solanum dulcamara* and *Ranunculus repens*, and the muddy substrate/shallow-water group *Alisma* and *Eleocharis*. The fen/reedswamp helophyte taxa are more abundant, with *Bidens*, *Cicuta virosa*, *Filipendula ulmaria*, *Lycopus europaeus*, *Lysimachia vulgaris*, *Lythrum salicaria*, *Oenanthe aquatica*, *Solanum dulcamara*, *Urtica dioica* and *Valeriana dioica*. The aquatic group is also substantial, with *Callitriche*, *Chara*, *Elatine hydropiper*, *Najas flexilis*, *N. minor*, *Nuphar*, *Nymphaea alba*, *Ranunculus* subg. *Batrachium*, *Sparganium*, *S. erectum*, *Trapa natans*, *Zannichellia palustris* and *Azolla filiculoides*. The assemblage indicates the presence of alder fen carr with fen bordering a slowly flowing water body bringing in terrestrial elements of the flora. The woodland (fen or terrestrial) taxa are more varied than in most other assemblages analysed and show a range of taxa associated with fen carr to terrestrial communities. Though the pollen assemblage indicates a boreal type of woodland, a number of relatively thermophilous taxa are present including *Acer campestre*, *Thelycrania sanguinea*, *Trapa natans* and *Valeriana dioica*, indicating summer warmth was still considerable at this time.

An analysis of non-marine molluscs by B. W. Sparks from the shelly level marked s in the freshwater mud of site OA is given in Table 19. He comments on the list: only opercula of *Bithynia* are present, and these all fall within the range of *inflata*, though it is not possible to be certain that there are no *tentaculata*; this entirely freshwater fauna represents good stream or large pond conditions, and indicates the continuing presence of species more typical of the middle of an interglacial, with typical Cromerian species.

(iii) *Pastonian IVa?*

The *Betula–Pinus–Alnus–Picea* p.a.b. recorded is from an erratic of coarse mud and shows a very high AP frequency, with very low frequencies of other AP taxa, and only low frequencies of *Calluna* and Gramineae in the land pollen. The macroscopic assemblage is largely of trees and shrubs (Fig. 41). The trees are *Alnus glutinosa*, *Betula pendula/pubescens*, *B.* cf. *nana* × *B.* tree species and *Picea abies*, and the shrubs only *Betula* cf. *nana*. A cypsela of cf. *Aster trifolium* indicates nearly maritime conditions. *Eleocharis palustris* ssp. *palustris* is the only member of the muddy substrate/shallow-water group and *Menyanthes trifoliata* the only aquatic. *Dryopteris*-type sporangia were also recorded. This assemblage must have originated in a shallow pool within woodland.

(iv) *Pollen analyses of samples from the Green Collection 1843*

Stuart & West (1976) described a *Pinus–Picea–Betula* p.a.b. and a *Pinus–Betula*–Gramineae p.a.b. from clay-muds from Ostend associated with rodent remains in the Green Collection. These are placed in Cr IVa and IVb respectively, and appear to be part of the post-regression vegetational sequence, deposited in a freshwater environment.

12: Happisburgh

(a) Lithostratigraphy

At Happisburgh sections on the foreshore and in the cliff base were recorded, and a borehole (HC, Plate 33) was carried down to the Chalk. The stratigraphy at these sites is shown in Fig. 39, with the location of the datum (HC) in Fig. 2.

The sediments above the Chalk (surface at −27·66 m OD) are grouped into the following beds (beds a to h only in borehole HC). Table 29 summarises this succession.

a. 6·9 m of grey sand with occasional laminae of grey silty clay lies on the Chalk in borehole HC. The sand contains fragments of *Mytilus* shell at −23·8 m OD and −25·6 m OD.

b. 80 cm of sandy gravel.

c. 2·0 m grey and pale silty clay, laminated in places and containing small flint pebbles, with a more sandy middle horizon with pebbles 60 cm thick, and with *Mytilus* shell fragments in the lower part.

d. 40 cm gravelly sand.

e. 11·2 m of grey sand with occasional laminae of grey silty clay.

f. 30 cm of grey and pale silty clay, laminated in places, with thin seams of grey sand.

g. 1·1 m of grey sand with angular fragments of cemented silty clay and occasional laminae of grey silty clay. The siltstone fragments resemble the underlying silty clays in colour and content.

h. 0·7 m of grey sand, with erratics of detritus mud, and grey silty clay seams.

i. 15 cm of gravel, with grey silt fragments. A gravel also occurs at a similar level at HD and HG (326 m and 356 m, Fig. 39). In these sections, seen at low tide, there was ferrugineous gravel with some clay-breccia.

j. Above the gravel in the borehole HC, in the low tide sections HD, HE, and HF and in the basal part of the cliff section HC, HH, and HA to the north-west of the borehole, occur a series of laminated grey silty clays with tidal structure, with seams of grey sand, and near the base, thin seams of detritus mud. The maximum thickness of this horizon is 4 m, in HC. Muddy seams in the laminated silty clays were seen in HA, 2 m beneath the base of the till, and at HC at −1·2 m OD: at HF 8 cm of detritus mud occurred between grey silty clays, and at HE 20 cm of compressed detritus mud with much wood occurred above a grey silty clay (Plate 34). These seams of mud are separated as bed j(o). Around the sections HE and HD were seen several logs, up to 2 m long at least and 30 cm in width. Reid (1882, p. 42) observed logs in the same area at a similar level. Conifer wood also came up with samples at +50 cm to −50 cm OD in borehole HC.

k. The First Till (Banham, 1968) overlies the laminated grey silty clays along this part of the coast. This clay is contorted near the junction with the till at HA and HC. Sand and freshwater sediments do not intervene between till and laminated clays, as they do further north-west at Bacton, Paston and Mundesley.

(b) Sedimentary history

Deposition of marine sands (bed a) from −27·7 m OD to 19·7 m OD took place on the Chalk surface, followed by a series with gravels (bed b), laminated clay and sand (bed c) and gravelly sand (bed d) to a height of −17·3 m OD. This series, containing fragmentary marine shells, a high NAP assemblage and macroscopic plant remains of maritime, freshwater and terrestrial plants, indicates a regression with shallow water conditions (*see* Chapter 2, *p.* 12). But deeper conditions again obtained with the deposition of the grey sand of bed e, accumulated to −6·33 m OD. This was followed by the deposition of a thin grey silty clay (bed f) and grey sand (beds g and h), with hard cemented silty clay fragments, possibly derived by reworking from the underlying silty clay in bed g, after a change in water level. These sands, with thin laminae of grey silty clay accumulated to a level of −1·67 m OD. The succeeding thin gravel (bed i) may indicate a regression. Certainly the muds and the laminated silty clays of bed j indicate a coastal environment with tidal conditions and deposition of laminated sediments and intercalated highly organic muds with much wood. These conditions continued to the junction with the till, at 2·5 m OD.

Table 29 gives a summary of this sequence.

(c) Biostratigraphy

The sequence of biozones at Happisburgh (Table 28) is clear. The evidence comes from the deep borehole (HC) to the Chalk, supplemented by exposures on the foreshore and at the cliff base. The pollen diagrams are given in Figs. 40 and 42, and the macroscopic analyses in Table 25. Table 29 summarises the biozone sequence and vegetational history.

Biozone a. *Pinus–Picea* p.a.b. This basal biozone is represented by the lowermost spectrum in the borehole taken from bed a.

Biozone b. *Pinus*–Gramineae–Ericales p.a.b. This is a substantial biozone, covering a considerable depth (4·3 m) of the sand of bed a. The sample at 2651 cm contains 8% *Pterocarya* pollen and this and the succeeding level (2561 cm) show higher percentages of *Alnus* (7–8%) than higher in the biozone. As the frequencies of these fall, so the NAP frequency rises during the biozone.

Biozone c. Gramineae–Ericales p.a.b. This biozone is found in the clays of bed e.

Biozone d. *Pinus*–Gramineae–Ericales p.a.b. This biozone covers 11·5 m of beds e and f. It is similar to the upper part of biozone b.

The correlation of biozones a to d has already been discussed in Chapter 2(a). Biozone d is placed in Pre-Pa a, but the correlation of the earlier biozones is problematical.

Biozone e. *Pinus*–Gramineae p.a.b. This biozone is identified from a single sample at 601 cm depth (bed g) in the borehole. Ericales frequencies have greatly decreased, and Gramineae frequencies increased, compared with the previous biozone.

Biozone f. Gramineae–*Betula–Pinus* p.a.b. This biozone, identified from a single sample at 457 cm in bed h, shows an increase in Cyperaceae and *Artemisia* pollen and a decrease in *Pinus* frequencies. It is possibly correlated with the Pre-Pastonian b biozone (Gramineae–*Artemisia* p.a.b.) or Pre-Pa d at Beeston.

Biozone g. *Betula*–Gramineae p.a.b. This biozone is again represented by a single sample at 427 cm, taken from a ball of mud near the top of the sand of bed h. This biozone shows high NAP frequencies and is placed in Pre-Pa d.

Biozone h. *Betula–Pinus*–Gramineae p.a.b. This biozone is again represented by a single sample at 381 cm, in the laminated clay of bed j. It is placed in Pa Ia.

Biozone i. *Alnus–Quercus–Carpinus–Ulmus* p.a.b. This biozone occurs in the lower part of the laminated clays of bed j. It is placed in Pa II.

Biozones j, k. In two foreshore exposures (HE 20, HF 10) horizons of mud show spectra with high frequencies of *Alnus*. These are separated as local p.a.b.s, with an *Alnus–Quercus–Pinus–Carpinus–Osmunda* p.a.b. (biozone j) in section HE, and an *Alnus–Quercus–Osmunda* p.a.b. (biozone k) in section HF.

Biozone l. *Pinus–Carpinus–Quercus–Ulmus* p.a.b. This biozone, also in bed j, is separated from the previous one by a rise in *Carpinus* and *Pinus* frequencies, with very low frequencies of *Tsuga*. The separation between these two biozones may seem marginal, but as the composition of this biozone is intermediate between the Pa II and Pa IVa biozones, it is separated and placed in Pa III.

Biozone m. *Pinus–Alnus–Picea–Betula* p.a.b. This biozone is again in the laminated clays of bed j. It is placed in Pa IVa.

(d) Vegetational history

(i) *Pre-Pastonian*

Biozones a to d at Happisburgh have already been described and discussed in Chapter 2. Biozone d is a *Pinus*–Gramineae–Ericales p.a.b. which covers 11·5 m of sand with laminated silty clay in the borehole HC. This is correlated with substage Pre-Pa a, as already described. Three further p.a.b.s are distinguished in borehole HC before the higher AP assemblages of the Pastonian commence. Both are only recorded in single samples. The first is a *Pinus*–Gramineae p.a.b. at 601 cm. Compared with the previous biozone, the Ericales pollen frequency is greatly reduced, and frequencies of *Betula* and Gramineae rise. There are more substantial traces of *Ulmus*, *Quercus*, *Carpinus* and *Corylus* pollen and of the aquatics *Sparganium* and *Typha latifolia*. The change in pollen assemblage may indicate, with the Ericales reduction, a change in the taphonomy of the assemblage, perhaps with the beginning of an element of freshwater deposition. The presence of angular fragments of cemented silty clay and erratics of detritus mud in beds g and h indicate reworking, perhaps under shallower water conditions. The assemblage indicates grassland with *Betula* and *Pinus*, and while it is to be placed in the Pre-Pa substage by reason of the high NAP frequencies, its exact correlation to the Pre-Pa substages at Beeston is unclear. It is possibly Pre-Pa c (Gramineae–*Pinus*–*Betula*–*Alnus* p.a.b. at Beeston), but it may also be a variant of the Pre-Pa a assemblage with loss of Ericales because of change in taphonomy. If it is the latter, the reworking may be related to the regression noted at Beeston between Pre-Pa a and Pre-Pa b.

The second Pre-Pa assemblage is a Gramineae–*Betula*–*Pinus* p.a.b. at 457 cm. *Pinus* pollen frequencies fall and Ericales pollen is absent. There is an increase in NAP notably Cyperaceae and *Artemisia* pollen, and of *Salix*. There is also a record of *Ephedra* pollen. Otherwise the assemblage is similar to the previous one. There is an increased representation of herbaceous communities and again the assemblage is to be correlated with a Pre-Pa substage, probably either Pre-Pa b or d. The *Artemisia* frequency is similar to that found in Pre-Pa b at Beeston.

A third assemblage, a *Betula*–Gramineae p.a.b., is recorded from an erratic of detritus mud at 427 cm. This assemblage must be taphonomically distinct from the previous assemblages, recorded in silty clay laminae. *Pinus* falls to very low values and *Betula*, Gramineae and Compositae Liguliflorae frequencies increase. The assemblage resembles that from Pa Ia (e.g. site BTA), but the Gramineae frequencies are twice as high. The assemblage, indicating the presence of grassland and *Betula* woodland, is placed in Pre-Pa d.

(ii) *Pastonian Ia*

A *Betula*–*Pinus*–Gramineae p.a.b. is recorded at 381 cm in the laminated clay and sand of bed j. AP frequencies have risen, marking the beginning of the Pastonian stage. *Betula* frequencies rise and Gramineae frequencies fall, indicating the spread of *Betula* woodland. The only macroscopic remains recorded from this sample were megaspores of *Azolla filiculoides* with massulae.

(iii) *Pastonian II*

A distinct pollen assemblage, with *Quercus*, *Carpinus* and low frequencies of *Ulmus*, is recorded in the laminated clays of bed j in the borehole HC and the sections HD, HE and HF. This is the *Alnus*–*Quercus*–*Carpinus*–*Ulmus* p.a.b. of substage Pa II. *Alnus* frequencies are high, associated with local fen woods, as indicated by the pollen spectra from mud horizons within the laminated clays. *Betula* and *Pinus* frequencies are low, and there are records for *Salix*, *Corylus*, *Taxus*, *Acer* and *Hedera*. Pollen of Ericales taxa is scarce. The predominant NAP is that of Gramineae and Cyperaceae, but there are low frequencies of pollen of Chenopodiaceae, Caryophyllaceae, Compositae Liguliflorae and Tubuliflorae, *Artemisia*, *Plantago media/major*, *Polygonum persicaria*-type, *Ranunculus*, Rosaceae, *Potentilla*, *Thalictrum*, Umbelliferae, *Urtica* and the Monocolpate grain X (*see* Chapter 18, *p.* 112). In the aquatic group, *Azolla*, *Nuphar*, *Potamogeton*, *Sparganium* and *Typha latifolia* are present, with spores of Filicales, *Equisetum*, *Lycopodium*, *Osmunda* (*claytoniana*-type and *regalis*) and *Sphagnum*.

The assemblage indicates regional temperate woodland, with local *Alnus* fen carr, in an area of tidal sedimentation.

A number of macroscopic plant remains were recovered from the laminated clays containing the *Pinus*–*Quercus*–*Carpinus*–*Ulmus* p.a.b. At HE 20 cm abundant megaspores and massulae of *Azolla filiculoides* were recorded. At HF 15 cm, in a stony silty clay with wood fragments, a more substantial list was obtained, mainly aquatics and fen/reedswamp helophytes (Fig. 41). The only tree recorded is *Betula pendula*. *Salicornia* is present in the maritime group, and *Alisma* cf. *plantago-aquatica* in the muddy substrate/shallow-water group. The fen/reedswamp helophyte group includes *Hydrocotyle vulgaris*, *Osmunda* and *Urtica dioica*, and the aquatic group *Callitriche stagnalis*, *Elatine hydropiper*, *Najas flexilis*, *Najas minor*, *Potamogeton* and *Typha latifolia*. The unclassified group contains cf. *Cyperus*, abundant *Dryopteris*-type (sporangia) and cf. Gramineae. The assemblage indicates a near-tidal situation, with a substantial contribution from nearby brackish, freshwater or terrestrial communities, both carr and herb communities.

Three macroscopic analyses from the laminated clays of bed j in borehole HC (HC 317–321 cm; 306–309 cm; 265–268 cm) show a variable representation of the different groups, with aquatics the best represented (Fig. 41). Taking the three analyses together, the only tree recorded is *Alnus* cf. *glutinosa*, *Salicornia* is present in the maritime group and *Eriophorum vaginatum* in the heath group. *Thalictrum minus* is recorded in the dry-ground herb group, and cf. *Selaginella selaginoides* in the damp-ground herb group. The muddy substrate/shallow-water group includes *Alisma plantago-aquatica*, *Eleocharis* cf. *acicularis*, *E. palustris* subsp. *palustris*, *Polygonum hydropiper* and *Rumex palustris*. The fen/reedswamp helophyte group includes *Bidens* cf. *tripartita*, *Cyperus longus*, *Lycopus europaeus*, *Oenanthe aquatica*, *Osmunda*, *Ranunculus sceleratus* and *Urtica dioica*. The aquatic group comprises *Callitriche platycarpa*, *Ceratophyllum*, *Hippuris vulgaris*, *Hydrocharis morsus-ranae*, *Lemna* cf. *trisulca*, *Nitella/Tolypella*, *Potamogeton trichoides*, *Sagittaria sagittifolia*, *Stratiotes aloides*, *Typha latifolia*, *T.* cf. *angustifolia* and abundant *Azolla filiculoides*. The unclassified group includes *Carex*, Gramineae (abundant), *Mentha arvensis/aquatica*, *Rumex* and Umbelliferae. These assemblages are chiefly derived from local brackish or freshwater and fen communities, with little evidence of terrestrial communities.

Horizons of detritus mud with wood in the sections HE and HF show local p.a.b.s. At HE (0–10 cm; *Alnus–Quercus-Carpinus–Ulmus–Osmunda* p.a.b.), the *Alnus* frequencies are high, the frequency of *Osmunda claytoniana* type spores rises very sharply and there is a fall in Gramineae and Chenopodiaceae pollen. The local presence of fen carr with *Osmunda* is clearly indicated. At HF (7–12 cm; *Alnus–Quercus–Osmunda* p.a.b.) the same trends are shown, but with higher frequencies of *Alnus* and *Quercus* and lower frequency of *Pinus*. Both p.a.b.s show the presence of the Monocolpate grain X, whose parent plant must be associated with the *Alnus–Osmunda* community. The identity of this grain is discussed in Chapter 18, *p.* 112. It is possibly that of a species of *Sabal*.

A macroscopic analysis from coarse detritus mud with wood at HE 0–10 cm gave a list with preponderant fen and reedswamp helophytes (Fig. 41). The only tree recorded is *Alnus* cf. *glutinosa*. The occurrence of the dry-ground herb *Trinia glauca* is discussed by Wilson (1973). *Eleocharis palustris* subsp. *palustris* is the only member of the muddy substrate/shallow-water group. The fen/reedswamp helophyte group includes *Carex aquatilis/bigelowii*, *C. laevigata*, *C. riparia*, *Cicuta virosa*, *Cyperus longus*, and very abundant *Osmunda* (sporangia and annuli). The aquatics comprise *Potamogeton perfoliatus*, *Typha latifolia* and *Azolla filiculoides*.

(iv) *Pastonian III*

A biozone, *Pinus–Carpinus–Quercus–Ulmus* p.a.b., intermediate in composition between the Pa II and Pa IVa assemblages, is recorded in the laminated clays of bed j in the borehole HC. It is tentatively separated from these other biozones and ascribed to substage Pa III. The AP frequency is high. *Carpinus* and *Pinus* frequencies are higher and *Alnus* frequencies are lower than the previous biozone. *Tsuga* (*canadensis*-type) occurs very sparsely. Pollen of the Ericales taxa begin to occur consistently. The character of the NAP content remains as in the previous biozone, except that *Sphagnum* spores increase in frequency. The assemblage shows a change in woodland content and spread of heath.

(v) *Pastonian IVa*

A *Pinus–Alnus–Picea–Betula* p.a.b. is recorded in the upper part of the laminated clays of bed j in the borehole HC and in section HA. The AP is high, and in addition to the named genera, there are low frequencies of *Salix*, *Ulmus*, *Quercus*, *Carpinus*, *Corylus*, *Acer* and *Hedera*. The Ericales taxa frequencies are higher than in the previous biozone, and the Gramineae frequencies lower. The other NAP taxa are few, with *Armeria*, Chenopodiaceae, Compositae Liguliflorae and Tubuliflorae, *Artemisia*, Cruciferae, *Ranunculus* and *Thalictrum*. The Filicales frequencies are low, and *Osmunda* is absent. The assemblage indicates a retreat of the thermophilous elements of the woodland, and the spread of boreal woodland and heath.

A macroscopic analysis from HC 223–233 cm shows a poor flora, with significant maritime, heath and aquatic groups present (Fig. 41). *Salicornia* is present in the maritime group, *Erica* and *Eriophorum vaginatum* in the heath group, *Campanula glomerata* in the dry-ground herb group, *Elatine hexandra* and *Eleocharis* in the muddy substrate/shallow-water group, and *Urtica dioica* in the fen/reedswamp helophyte group. The aquatic group includes *Nitella/Tolypella*, *Scirpus tabernaemontani*, *Stratiotes aloides* and *Typha latifolia*. There are abundant *Dryopteris*-type sporangia in the unclassified group. The assemblage is a mixed one of the usual type found in tidal sediments.

13: Corton

(a) **Lithostratigraphy**

The cliff section at Corton was studied by J. H. Blake in the course of the geological survey of the area in the 1880s and a detailed horizontal section was published by the Geological Survey (Blake, 1884). Recently a number of sections have been observed south of Corton (Corton Woods) and north of Corton towards Hopton (Corton Cliffs) (Plates 35, 36). These are shown in Fig. 43, with distances based on datum points north and south of Corton (Fig. 2). The OD level shown at 178 m south of the northern datum was calculated by determination of high tide on the day the section was seen and related to HWM (tidal range (springs) is 1·95 m). Fig. 43 also shows data from levelled boreholes made in the course of site investigation for coast defences.

The sediments can be grouped into the following beds. Table 32 summarises this succession.

a. Brown and yellow sand with occasional grey silty clay laminae and occasional fine gravel occurs at the base of boreholes 3/11, 3/10, 3/6, 3/5, 3/4, 3/3, 3/1 and 2/6. The greatest thickness seen, 5·1 m, was in borehole 3/3. In borehole 3/11 this sand reaches to −8·0 m OD.

b. Brown and grey laminated silty clay occurs above bed a in boreholes 3/10 (where there are fine silty laminae in the clay), 3/6, and 3/5, and, dark grey and more massive, in borehole 2/6. This last borehole shows the maximum thickness of this horizon, 1·7 m, with pollen evidence for marine deposition.

c. Red-brown and yellow sand, usually with small gravel, occurs below the Rootlet Bed (bed e) in all the boreholes. Its thickness is very variable, from 1·5 m in borehole 3/3 to 7·8 m in borehole 2/4. Gravels were also found below the Rootlet Bed in three boreholes made by the Geological Survey at the time of the survey (Blake, 1890, *pp.* 85, 86). Sand with ferrugineous gravel was also seen in section CE beneath the Rootlet Bed south of Corton and at sections CD and CF. This sand and gravel is included in this horizon. At sections CF and CG there were involutions and thin silty laminae associated with the sand.

d. In sections CD and CF, 130 m and 180 m south of Corton Woods datum, a brown sandy coarse detritus mud, up to 20 cm thick, was seen resting on ferrugineous gravel of bed c and covered by sand. In section CD the mud was in discrete lumps contorted with purple hard sand containing wood fragments and resting on the ferrugineous sand and gravel. A very similar purple hard stony sand resting on ferrugineous sand was seen at the base of section CE (100 m to the north) beneath the Rootlet Bed. This mud horizon is therefore placed below the Rootlet Bed, though it was not seen in section beneath it, and it may then be a lateral facies of part of bed b or c.

e. This bed is the Rootlet Bed described and drawn in detail by Blake (1884, 1890) (Plates 35, 36). It was seen both north and south of Corton. Section CKG, 70 m north of the Corton Cliffs datum, showed the Rootlet Bed in most detail, 2·3 m thick. The upper part was a massive blue-grey clay, the middle part a mottled brown and pale clay with race, and the lowest 0·6 m a laminated pale and brown clay with much race.

This pale base of the Rootlet Bed was noted by Reid in a borehole and was also seen in section CC south of Corton. After a scour in January 1971 rootlets were clearly seen penetrating the surface of the Rootlet Bed along much of the section then exposed (CK to CJ, 80–253 m south of the Corton Cliffs datum). Blake reported that the Rootlet Bed contained many bones, and freshwater molluscs (see analysis, Table 19, discussed later, *p.* 84) were seen in it about 100 m to the north of section CA at section CAB, at a place near where Blake (1884) noted their presence. The Rootlet Bed thus appears to be a sheet of alluvium.

f. Organic sediments rest on the Rootlet Bed in the region 70 m north to 253 m south of the Corton Cliffs datum. They consist of either a coarse detritus mud with much wood, as at CA, 16 cm thick with a sharp contact with the Rootlet Bed, or a finer mud, as at CK, 15 cm thick with a transition through to the Rootlet Bed. Rootlets were observed at CJ passing from a coarse mud with much wood down into the Rootlet Bed through a sharp contact.

g. Ten cm of ferruginous gravel and 30 cm of pale sand were seen to overlie the organic sediment of bed f at CJ, and 30 cm of grey silty sand were seen above the Rootlet Bed at CC and CE, 0 m and 30 m south of the Corton Woods datum.

h. Above the sand of bed g occur grey laminated silty clays and silts with tidal structure. They were seen thus in sections CE and CC south of Corton, and CJ north of Corton. The same sediment unit occurs above the organic sediments of bed f at sections CB, CI, CH and CK. At CI and CK there is a transition to the more organic sediments below. At section CC, which showed the maximum observed thickness of this horizon, 1·4 m, vertical sand-filled burrows were present in the sediment. Blake (1884) shows up to 3 m of this horizon in his section.

i. In section CC 0·5 m of contorted pale sand with streaks of drift mud occurred above the tidal sediments.

j. This bed is the brown till which seals the preglacial sediments along this part of the coast, described variously as the North Sea Drift, Norwich Brickearth and Cromer Till.

(b) Periglacial stratigraphy

A striking series of ice-wedge casts seen in section and plan view were seen north of Corton in January 1971 (Plate 36). They have been described by Gardner & West (1975). They were also seen in the same place by Blake (1884) and drawn in his horizontal section. According to this section they penetrate beds h and g (Blake's Pebbly Series), the mud of bed f and the Rootlet Bed (e.g. section CH) and they are sealed by the North Sea Drift and, where this is missing, by the Corton Beds.

(c) Sedimentary history

The sands with thin silt laminae of bed a form the basal unit along the Corton coast. The environment of deposition appears to have been marine, as indicated by the pollen assemblage from borehole 3/11. They were strongly channelled before the deposition of the laminated silty clays of bed b. In borehole 3/6 they contain pollen of maritime plants suggesting a near-marine or marine environment. These silty clays are of irregular occurrence, and appear to have been subject to erosion before the sand with gravel of bed c was deposited. At the north end of the section this sand and gravel is found at greater depth, with beds a and b missing. It is not known whether the sand and gravel is fluviatile or marine. It is possibly a marine aggradation following tidal silty clays of bed b below but preceding the alluviation of bed e.

Towards the end of the period of sand and gravel deposition thin freshwater sandy muds (bed d) were formed at the southern end of the section, resting on a gravel. The overlying sands may be part of bed c, and if so may suggest a fluviatile origin for this horizon. The alluvium of the Rootlet Bed was then widely deposited over the sand and gravel of bed c at about OD. The disposition of this alluvial sheet is well seen in Blake's horizontal section. The alluvium passes up into the organic muds of bed f, indicating a flooding of the alluvium surface with a rise in water level. The rootlets in the Rootlet Bed originate from vegetation associated with this rise. Before much mud could accumulate, however, marine transgression occurred and beach gravel and sand (bed g) and tidal silts and silty clays were deposited. The transgression contact is at 0·5 m OD in section CI. There followed the deposition of the sands of bed i, with streaks of drift mud. These sediments are probably fluviatile and may follow regression. At some time later than the regression an extensive ice-wedge system developed as a result of permafrost. The ice advance which deposited the North Sea Drift followed.

Table 32 gives a summary of this sequence.

(d) Biostratigraphy

Table 30 lists the biozones distinguished in the pollen diagrams from Corton (Figs. 44 and 45). Table 31 gives the macroscopic analyses, and Table 32 summarises the biozone sequence and vegetational history.

Biozone a. *Pinus–Quercus–Picea–Ulmus–Carpinus* p.a.b. This is identified from a single sample in the basal bed (a) of borehole 3/11. This temperate assemblage is tentatively correlated with Pa II.

Biozone b. *Betula–Alnus* p.a.b. This single pollen spectrum, below biozone c in section CG, was very poor in pollen. Its age is uncertain.

Biozone c. *Alnus–Quercus–Ulmus–Pinus* p.a.b. This biozone occurs in bed b in boreholes 3/10 and 3/5, and in the muds of bed d at CD and CF. It is correlated with Cr IIa.

Biozone d. *Pinus–Alnus–Quercus–Ulmus* p.a.b. This biozone, identified in a single sample from bed b in borehole 3/6, shows a composition intermediate between biozones c and e.

Biozone e. *Quercus–Ulmus–Tilia* p.a.b. This biozone is seen in the silty clays of beds a and b in borehole 2/6. It is correlated with Cr IIb.

Biozone f. *Pinus–Picea* p.a.b.; Biozone g. *Alnus–Picea–Tilia* p.a.b.; Biozone h. *Pinus–Alnus–Picea–Tilia* p.a.b. These three biozones are found in the Rootlet Bed (bed e) at sites borehole 2/6, CC and CK respectively. Pollen is not well preserved in the silty clays of bed e, and weathering is likely to have played some part in determining the constitution of these assemblages. Because the first of these assemblages is sandwiched between horizons assigned to Cr IIb and the other two contain *Tilia*, all are tentatively placed within the period covered by Cr IIb.

Biozone i. *Quercus–Ulmus–Tilia* p.a.b. This biozone is seen in the mud with wood fragments (bed f) resting on the Rootlet Bed. It is correlated with Cr IIb.

Biozone j. *Carpinus* p.a.b. This biozone is seen in the upper part of the mud (bed f) at section CA and CK, and in the lower part of bed h, the tidal laminated clay, at CK. It is correlated with Cr IIIa.

Biozone k. *Abies–Carpinus* p.a.b. This biozone occurs in the clays of beds g and h, and is correlated with Cr IIIb.

(e) Vegetational history

(i) *Pastonian II*

The *Pinus–Quercus–Picea–Ulmus–Carpinus* p.a.b. recorded in a silty clay lamina in the sand of bed a at borehole 3/11 shows high AP frequency. The thermophilous trees *Ulmus*, *Quercus*, *Carpinus* and *Corylus* are well represented. The NAP is principally of Gramineae, *Armeria*, Chenopodiaceae and Compositae Liguliflorae. The assemblage indicates the presence of temperate woodland, deposited in a silty clay of marine origin.

(ii) *Cromerian IIa*

An *Alnus–Quercus–Ulmus–Pinus* p.a.b. is recorded in coarse detritus mud at sections CD and CF and, at lower levels to the north, in laminated silty clay in boreholes 3/10 (381 cm) and 3/5 (762 cm). The AP frequency is high, with *Alnus* showing higher frequencies in the silty clay than in the detritus mud. Besides the genera named above, *Picea*, *Salix*, *Tilia* and *Corylus* occur sparsely. The NAP is mostly Gramineae, with a greater variety in the detritus mud samples, including Cyperaceae, Caryophyllaceae, Compositae Liguliflorae and Tubuliflorae, *Artemisia*, Cruciferae, *Ranunculus*, *Polygonum bistorta/viviparum*, *Rumex*, Umbelliferae and *Valeriana officinalis*. Pollen of aquatics is present in both sediment types, including *Nuphar*, *Nymphaea*, *Sparganium* and *Typha latifolia*. *Azolla filiculoides* was found in the detritus muds. The assemblage indicates temperate woodland with *Quercus* and *Ulmus*, with local *Alnus* giving a high representation in the silty clays. The assemblage is thus present in two facies, a shallow water organic facies (CD, CF), and, to the north at depth, a deeper water facies, possibly marine, in the boreholes 3/10 and 3/5.

(iii) *Cromerian IIb*

A *Quercus–Ulmus–Tilia* p.a.b. is present in the laminated clays of bed b in borehole 2/6, in the uppermost part of the Rootlet Bed (bed e) at section CA, and in the lower part of the detritus mud (bed f) overlying the Rootlet Bed at sections CA, CB and CK. AP frequencies are high, and *Quercus*, *Ulmus*, *Tilia* and *Corylus* are relatively well represented in all these pollen spectra, and the assemblage indicates regional temperate woodland with these genera. There are variations within the assemblage, related probably to facies. In borehole 2/6 the spectra from the laminated clays show higher frequencies of *Pinus*, and the NAP is characterised by higher frequencies of Chenopodiaceae pollen, with

Armeria present. All these differences are likely to be related to marine conditions of deposition. The detritus mud spectra show very high frequencies of Filicales spores, related to local fen carr.

Other high AP pollen assemblages are also placed in Cr IIb, either because they are characterised by *Quercus–Ulmus–Tilia*, or because they occur in the Rootlet Bed, which is sandwiched between sediments with the Cr IIb *Quercus–Ulmus–Tilia* biozone described above. All these occur within the Rootlet Bed and their variation from the *Quercus–Ulmus–Tilia* assemblage is probably related to peculiarities of sedimentation in the alluvial environment or to post-depositional weathering. At borehole 2/6 (45–137 cm) a *Pinus–Picea* assemblage is present in the Rootlet Bed, with the NAP showing high frequency of Compositae Liguliflorae. The pollen was badly preserved and the assemblage may have been much altered by weathering. At section CC, the Rootlet Bed showed an *Alnus–Pinus–Tilia* assemblage with much *Azolla*, and at CK a *Pinus–Alnus–Quercus–Tilia* assemblage, similar to the Cr IIb assemblage in the mud above, but with higher frequencies of *Pinus* and *Picea* and lower frequencies of *Quercus*. Two pollen diagrams from the Rootlet Bed were described by Duigan (1963), one at Corton (section FF) and one at Hopton (section EE). The Corton section is labelled SD in Fig. 43. The pollen diagrams show high *Alnus* frequencies, with low frequencies of *Pinus*, *Picea* and *Tilia*.

In the inorganic sediments of Cr IIb there are sparse occurrences of several AP taxa, including *Abies*, *Tsuga*, *Carpinus*, *Ilex*, *Fraxinus* and *Acer*. The origin of such pollen is not certain; it may be due to reworking or to derivation from vegetation upstream of the alluvial area represented by the Cr IIb sequence. The extent of the alluvium, occurring as it does south to Pakefield and Kessingland, indicates a very extensive alluvial basin, with strong possibilities for reworking and long distance river transport.

An analysis of non-marine molluscs by B. W. Sparks from the Rootlet Bed at site CAB is given in Table 19. He comments that it is a small fauna of marshy ground, but no climatic inferences can be drawn from it. Further non-marine mollusc fragments were recorded in borehole 2/6 at Corton, at the level in the Rootlet Bed marked s (Fig. 43). B. W. Sparks recorded an operculum of *Bithynia* sp. and a valve of *Pisidium* sp., together with crushed debris of freshwater shells in this sample.

A macroscopic analysis was made from the upper part of the Rootlet Bed at section CA 16–22 cm. The assemblage shows a preponderance of the fen/reedswamp helophyte group (Fig. 41). *Alnus* and cf. *Betula* are present in the tree group, *Solanum dulcamara* in the damp-ground herb group, and the fen/reedswamp helophyte group includes *Carex riparia*, *Cyperus longus*, *Lycopus europaeus*, *Ranunculus sceleratus* and *Urtica dioica*. *Stratiotes aloides* and *Typha* are present, together with abundant *Azolla filiculoides* and *Salvinia natans*. *Azolla filiculoides* was recorded in the Rootlet Bed at section CC, and with *Salvinia natans* and *Typha* at sections CI (0–10 cm) and CK (63 cm). The assemblages indicate fen and fen carr vegetation with shallow aquatic conditions towards the end of the deposition of the alluvium of the Rootlet Bed.

A macroscopic analysis was also made from the detritus mud with wood at section CA 0–13 cm. The assemblage shows strong representation of the tree and aquatic groups (Fig. 41). The tree group includes *Alnus glutinosa* (very abundant) and *Betula pendula/pubescens* and the shrubs *Frangula alnus* and *Salix*. *Solanum dulcamara* is present, and in the fen/reedswamp helophyte group there are *Bidens tripartita*, *Cicuta virosa*, *Lycopus europaeus*, *Osmunda*, cf. *Phragmites communis*, *Ranunculus lingua*, *R. sceleratus* and *Urtica dioica*. The aquatics include *Hippuris vulgaris*, *Hydrocharis morsus-ranae*, *Potamogeton*, *Stratiotes aloides*, *Typha*, *Azolla filiculoides* and *Salvinia natans*. The unclassified group includes *Carex*, *Cyperus*, *Dryopteris*-type (sporangia), Gramineae, and cf. *Ranunculus* subg. *Ranunculus*. The assemblage indicates the occupation of the immediate area by fen carr (*Alnus glutinosa*, *Betula pendula/pubescens*, *Frangula alnus*, *Dryopteris*-type) and fen with shallow water.

In the detritus mud at CB, a more restricted flora was recorded with *Alnus glutinosa*, *Osmunda*, *Najas marina*, *Stratiotes aloides*, *Typha latifolia*, *Azolla filiculoides*, *Salvinia natans* and *Dryopteris*-type (sporangia). In the detritus mud at CI (12 cm) and CK (42 cm) megaspores and massulae of *Azolla filiculoides* and *Salvinia natans* were recorded with *Typha*, and *Carex* at CK 42 cm.

(iv) *Cromerian IIIa*

A *Carpinus* p.a.b. is recorded in the upper part of the detritus mud of bed f at sections CA and CK and in the immediately overlyjng tidal silty clays of bed h at section CK. The beginning of the biozone is marked by a rise in *Carpinus* pollen frequencies and a fall in *Tilia* frequencies, seen well in the section CK pollen diagram. The AP frequencies remain high, with *Quercus*, *Ulmus* and *Corylus* still well represented. Filicales spore frequencies peak in the detritus mud at the beginning of the biozone, indicating local fen carr. At CA, the macroscopic analysis at 0–13 cm, already considered, covers this period. The change to tidal deposition is accompanied by increases in *Picea*, *Calluna*, *Empetrum*, Gramineae and Chenopodiaceae pollen, and *Sphagnum* spores. The relation between the pollen sequence and marine transgression is thus clearly seen in the diagram from CK. It took place subsequent to a change of the regional temperate forest with increased *Carpinus*.

Macroscopic analyses of the tidal sediments at section CI (13–17 cm, 17–20 cm) gave a restricted flora, with *Bidens*, *Osmunda*, *Ranunculus sceleratus*, *Najas marina*, *Stratiotes aloides*, *Typha*, *Azolla filiculoides*, *Salvinia natans*, cf. *Cyperus* and *Dryopteris*-type (sporangia). These remains evidently derive from the fen and aquatic vegetation which was swamped by the transgression.

(v) *Cromerian IIIb*

An *Abies–Carpinus* p.a.b. is recorded in silty horizons in the sands of bed g at section CC and in tidal silty clays of bed h at sections CC and CBb. The AP frequencies are lower than in earlier biozones. *Pinus*, *Picea*, *Abies* and *Betula* show increased frequencies, with *Alnus*, *Carpinus*, *Ulmus* and *Tilia* showing lower frequencies. There are increased frequencies of Ericales, Gramineae and Chenopodiaceae pollen, presumably related partly to a local change to marine deposition. The assemblage indicates a change towards a regional mixed conifer/deciduous woodland.

(vi) Betula–Alnus *p.a.b.*

This pollen spectrum from a thin silty lamina from section CG shows an assemblage not recorded elsewhere on the Corton coast. The assemblage was very poorly preserved and may have been subjected to weathering. Its age is unknown, but must be earlier than Cr IIb.

14: Pakefield and Kessingland

(a) Lithostratigraphy

The sections along this part of the coast have been obscured by talus for many years, as a result of beach aggradation south of Lowestoft. Sections showing the pre-glacial sediments were only visible at the base of the cliff at Kessingland (section K, Fig. 2, inset 6). Further north, towards Pakefield, excavations were made to expose sections, and two boreholes were also made. When J. H. Blake (1884, 1890) made the geological survey of this coast in the 1880s the position of the coastline was much as it is now and since coastal erosion was active at that time he was able to draw a complete section of this part of the coast (Blake, 1884) showing an extensive outcrop of the Rootlet Bed. The new excavations were sited on the basis of Blake's section to investigate the Rootlet Bed, the channel filled with organic deposits as well seen by Blake, and the underlying laminated sediments correlated with the Chillesford Beds by Blake (1890).

Fig. 46 shows the sections at the base of the Pakefield–Kessingland cliffs. The sediments can be grouped into beds, as follows. Table 34 summarises this succession.

a. Grey and red sandy silt with grey silty clay laminae (correlated with the Chillesford Beds by Blake) is the basal sediment of the succession along this part of the coast, with the bedding gently dipping northwards, as seen in Blake's section. It was seen in sections PA (218 m north), PB (186 m south), from 2·7 to 8·8 m in borehole PK (211 m south), and 5·7 to 8·8 m in PL (510 m south). At Kessingland K a similar sediment was seen at 0·6–3·0 m.

b. 0·7 m of grey and brown laminated silty clay with horizons of ferrugineous or grey silty sand and flattened wood occurs above bed a in the borehole PL.

c. Above bed b in borehole PK occurs 2 m of brown red sand with pale flint chips, with rounded small flints more abundant at the base. This may be the horizon of red sand with marine shells found below the Rootlet Bed in a boring by C. Reid (1890, *p.* 135), at a point 55 m north of borehole PL.

d. This bed is the Rootlet Bed of Blake and described fully in his memoir (Blake, 1890). It was seen by Blake to extend along the whole cliff section in this area, with a thickness up to 3·5 m. The upper part is normally a stiff brown or blue clay, often with rootlets and race, and the lower part a brown clayey silt with sand horizons. It was seen in sections PJ and PF (487 m and 507 m south) and at Kessingland K. At the latter section an organic horizon occurred within the clay of the Rootlet Bed.

e. This horizon is the peat with wood and seeds recorded by Blake (1884, 1890) lying on top of the Rootlet Bed. He recorded a thickness of up to 30 cm.

Blake's (1884) horizontal section shows this peat overlying the Rootlet Bed south of our excavations at Pakefield, where it underlies laminated clay and sand. Unfortunately excavations to recover these sections were not successful, but the peat and laminated clay and sand are in exactly the same stratigraphical situation as similar sediments at Corton, and they are probably of Cr II and Cr IIIb age respectively. The peat is included in the subsequent list of beds, and the laminated clay and sand can be considered as part of the bed h below.

f. This and the next two horizons form parts of a filling of a channel cut through the Rootlet Bed to the south and into bed a at the north. Bed f is a red ferruginous sand and gravel, up to 25 cm proved, at the base of the channel in sections PC and PH (208 m and 465 m south). This sand and gravel is shown at the base of the channel in Blake's section. Blake mentions that at the southern margin of the channel the gravel was rich in freshwater shells, especially *Unio* (the *Unio*-bed). Presumably this horizon is later than the ferruginous sand and gravel containing bones which Blake described from immediately below the Rootlet Bed. This latter horizon may be part of bed c above.

g. This horizon is of brown silty mud with laminae richer in silt, and forms the chief part of the channel filling. The upper part is more silty, with clearer laminations, and the basal part is richer in sand and organic debris, including wood, twigs and leaves. Blake observed a flattened trunk of oak 22 feet in length at the southern margin of the channel at the base of this bed. The silty mud is 1·8 m thick in section PE (435 m south) and 1·6 m thick in section C (208 m south). In section PJ (487 m south) it is seen near its southern margin, 0·8 m thick and resting on the Rootlet Bed.

h. This horizon is of laminated brown and red silty clays, and shows a transitional contact with the underlying more organic sediments. It is thickest (0·8 m) in section C (208 m south) and thins towards the south, so that at PJ (487 m south) it is 30 cm thick. In section PC it is more massive and not laminated in the topmost 20 cm.

i. This horizon of buff cross-bedded sands overlies the Rootlet Bed and channel filling along the length of the coast in this area. These sands are the Corton Beds of Baden-Powell (1950). The distribution and thickness of this horizon and the other related glacial deposits (North Sea Drift, Lowestoft Till) are seen well in Blake's section.

(b) Sedimentary history

The basal bed a is part of a sediment unit known to occur from Pakefield southwards to Kessingland (*see* Chapter 15). The laminated sandy and clayey silt appears to have been deposited under marine conditions, as indicated by its microfossil content. This sediment must have been eroded and channelled south of section PA (218 m north), possibly in the period of regression between Pre-Pa a and Pa noted at Beeston and elsewhere. In the depression so formed first the laminated sediments of bed b were deposited (probably under shallow marine conditions, indicated by contained fossils), and then the 2 m of red sands (bed c). Subsequently the Rootlet Bed (bed d), which must be alluvial in origin, with its contained organic and more silty laminated horizons, was deposited over the length of the coastal section south of Crazy Mary's Hole, as shown in Blake's section. Plant growth on the surface of the alluvium is shown by rootlets and by the thin (30 cm) wood peat (bed e) seen by Blake over the Rootlet Bed. The Rootlet Bed must have then been channelled at its northern margin and a sequence of freshwater sand and gravel, laminated silty mud and silts filled the channel (beds f, g, h). The organic sediments contain a flora with maritime elements, suggesting a tidal origin for the laminations. Tidal laminated clays and sands sealed the freshwater peat further south. Subsequently sediments were deposited over a level surface of the underlying pre-glacial sediments.

Table 34 gives a summary of this sequence.

(c) Biostratigraphy

Table 33 lists characteristics of the pollen assemblage biozones distinguished in the pollen diagrams from Pakefield and Kessingland (Fig. 47). Table 31 gives the macroscopic analyses, and Table 34 summarises the biozone sequence and vegetational history.

Biozone a. *Pinus*–Ericales p.a.b. This is recorded in silty clay laminae of bed a. It is correlated with Pre-Pa a.

Biozone b. *Pinus–Quercus–Ulmus–Carpinus* p.a.b. This assemblage is found in a single sample from the laminated silty clay of bed b. Since it underlies the Rootlet Bed (bed d) and 'peat' (bed e) of Cr II, it is taken to be pre-Cromerian and is correlated with Pa II.

Biozone c. *Pinus*–Gramineae p.a.b. This assemblage is recorded in the Rootlet Bed (bed d). The pollen is not well preserved in the weathered clay of this horizon. The assemblage bears comparison with some assemblages from the Rootlet Bed at Corton, and is therefore placed in Cr II.

Biozone d. *Carpinus* p.a.b. This is recorded in the silty mud of bed g from the channel filling at Pakefield. It is correlated with Cr IIIa.

(d) Vegetational history

(i) *Pre-Pastonian a*

A *Pinus*–Ericales p.a.b. is recorded in the silty clay laminae of bed a in the boreholes PK and PL. The AP frequency is around 60% with *Pinus* predominant and with much lower frequencies of *Picea*, *Betula* and *Alnus*. The pollen record of thermophilous trees is very sparse. The NAP is largely composed of Ericales, with lesser frequencies of Gramineae. Hystrichospheres, *Operculodinium* and *Tectatodinium* are relatively abundant through the sampled bed a. As discussed in Chapter 2, this assemblage is correlated with Pre-Pa a and so with substage Lu 4c in the Ludham borehole.

(ii) *Pastonian II*

A *Pinus–Quercus–Ulmus–Carpinus* p.a.b. is recorded in the laminated silty clay with wood of bed b in the borehole PL. The AP frequency is high, with *Pinus* and *Alnus* showing the highest frequencies, but with *Quercus*, *Ulmus* and *Carpinus* well represented. There are low frequencies of Ericales, *Empetrum* and Gramineae pollen. Other NAP taxa, with *Azolla*, *Sparganium* and *Typha latifolia* are sparsely represented. There are substantial frequencies of Filicales spores, with *Osmunda* (*claytoniana*-type and *regalis*), and low frequencies of Hystrichospheres, *Operculodinium* and *Tectatodinium*.

A macroscopic analysis from the laminated silty clay of this biozone showed a preponderance of the aquatic group (Fig. 41). Trees present are *Alnus* and *Betula pendula*. *Salicornia* is present in the maritime group, *Lythrum* cf. *borysthenicum* in the damp-ground herb group, and *Alisma* in the muddy substrate/shallow-water group. In the fen/reedswamp helophyte group occur *Cyperus longus*, *Ranunculus sceleratus* and *Urtica dioica*, and the aquatics present are *Callitriche truncata*, *Chara*, *Nymphaea*, *Ranunculus* subg. *Batrachium*, *Typha angustifolia*, *T. latifolia*, *Azolla filiculoides* and *Salvinia natans*. The unclassified group includes numerous *Dryopteris*-type (sporangia) (Filicales in the pollen diagram), Gramineae and Umbelliferae.

The assemblage is a mixed one of the type we have found characteristic of tidal sediments, and its presence indicates a sharp change from the conditions of the previous biozone. The pollen and macroscopic assemblages point to regional temperate woodland, with local fen carr, and aquatic communities in a maritime environment, with deposition in a tidal situation.

(iii) *Cromerian II*

A *Pinus*–Gramineae p.a.b. is recorded in the clay of the Rootlet Bed (bed d) at PF and in a more organic sediment within the Rootlet Bed in the K section at Kessingland. The pollen in these sediments was badly preserved and there were many corroded grains. The assemblage is likely to have undergone transformation in the weathered sediments of the alluvial Rootlet Bed, as discussed above in relation to the Corton Rootlet Bed. As the assemblages appear, they bear some resemblance to the Cr Ib (*Pinus–Ulmus* p.a.b.) assemblages at West Runton, but they also have resemblances to assemblages from the Cr IIb Rootlet Bed at borehole 2/6 at Corton, both showing high frequencies of *Pinus*, low frequencies of *Ulmus* or *Quercus*, and substantial frequencies of Compositae Liguliflorae pollen. There are differences, however, as there are differences in the assemblages in the Rootlet Bed at Corton. For example, there are higher frequencies of Cyperaceae pollen and Filicales spores at Pakefield and Kessingland than at Corton. However, it would seem most likely that the assemblage at Pakefield and Kessingland is a local variant within Cr II, with characteristics derived from its alluvial nature and from weathering.

(iv) *Cromerian IIIa*

A *Carpinus* p.a.b., similar in constitution to the *Carpinus* p.a.b. at Corton, is recorded in the silty mud of bed g in the filling of the channel cut in the Rootlet Bed at the north end of the section. Pollen was present in 160 cm of this filling, but heavily corroded and uncountable in the higher, more inorganic, part (bed h) of the filling. The same assemblage continues through the whole of the silty mud at section PC. It was also recorded at the top and base of the mud in section PE. The channel was thus substantially filled in this biozone. Throughout the AP frequency is high. Besides substantial frequencies of *Pinus* and *Alnus*, the thermophilous trees *Quercus*, *Ulmus*, *Carpinus* and *Corylus* are well represented, with lower frequencies of *Tilia*. *Picea* and *Betula* show low continuous curves, and there are few pollen grains of *Abies*, *Taxus*, *Fraxinus* and *Acer*. In the NAP, there are low frequencies of *Calluna*, *Empetrum* (including *hermaphroditum* type), Ericales, Gramineae and Chenopodiaceae pollen. The other NAP includes *Armeria*,

Plantago maritima, Caryophyllaceae, *Stellaria holostea*, Compositae Liguliflorae and Tubuliflorae, *Arctium*-type, *Artemisia*, *Filipendula*, *Peplis portula*, *Plantago media/major*, *Polygonum persicaria*-type, *Ranunculus*, *Rumex*, *Rumex acetosa*, *Succisa*, *Thalictrum*, Umbelliferae and *Urtica*. This extensive list implies the presence of maritime and of herb communities of open ground.

The aquatic pollen and spores includes *Azolla*, *Myriophyllum spicatum*, *Sparganium* and *Typha latifolia*, and the spores Filicales, *Lycopodium* (including *L. annotinum*), *Osmunda* (*claytoniana*-type and *regalis*), *Pteridium* and *Sphagnum* are present.

A number of macroscopic analyses were made from the sediments of this biozone at sections PC (80–90 cm, 210–230 cm), PD (one near the base, one (Da) picked from the section surface) and PH. They gave an extensive flora. The aquatics are the best represented group and there is a maritime group in all the samples. We may take all the analyses together in the following survey of the flora. The trees include *Acer campestre*, *Alnus glutinosa*, *A. incana* (*see* Wilson, 1973), *Betula pubescens*, *B. pendula/pubescens*, *Carpinus betulus*, cf. *Picea*, *Pinus*, *Populus* and cf. *Quercus*. The shrubs are *Rhamnus catharticus*, *Rosa*, cf. *Staphylea* (*see* Wilson, 1973) and *Viburnum opulus*. The maritime group includes the shrub *Corema album* and *Salicornia*. The heath group includes *Erica ciliaris/lusitanica* (*see* Wilson, Marchant & Oldfield, 1973) and abundant *Erica* seeds. There is a long list of dry-ground herbs, including *Aphanes microcarpa*, cf. *Atriplex*, *Calamintha*, *Chenopodium album*, *Herniaria* (*see* Wilson, 1973), *Hypericum montanum*, *Picris hieracioides*, *Plantago major*, *Polygonum aviculare*, *Prunella vulgaris*, *Ranunculus bulbosus*, *Senecio* cf. *sylvaticus*, *Thalictrum minus*, cf. *Tussilago farfara* and *Urtica urens*. The damp-ground herbs are *Juncus* cf. *inflexus*, *Solanum dulcamara* and *Ranunculus repens*, and the muddy substrate/shallow-water plants are *Alisma plantago-aquatica*, *Eleocharis* and *Rumex maritimus*. The fen/reedswamp helophytes are *Bidens tripartita*, *Cicuta virosa*, *Cyperus longus*, *Eupatorium cannabinum*, *Hydrocotyle vulgaris*, *Lycopus europaeus*, cf. *Lysimachia vulgaris*, *Lythrum salicaria*, *Oenanthe lachenalii*, *Osmunda*, *Ranunculus sceleratus*, *Thalictrum flavum* and *Urtica dioica*.

The aquatic group is extensive and includes *Callitriche platycarpa*, *Ceratophyllum demersum*, *Chara*, *Elatine hydropiper*, *Hydrocharis morsus-ranae*, *Myriophyllum spicatum*, *Najas marina*, *N. minor*, *Nuphar*, *Nymphaea alba*, *Potamogeton crispus*, *P. pectinatus*, *P.* cf. *perfoliatus*, *Ranunculus* subg. *Batrachium*, *Sagittaria*, *Scirpus lacustris*, *Sparganium* cf. *erectum* ssp. *neglectum*, *Stratiotes aloides*, *Trapa natans*, *Typha angustifolia*, *T. latifolia*, *Zannichellia palustris*, *Azolla filiculoides* and *Salvinia natans*. The unclassified group includes *Carex*, *Cirsium/Carduus*, Compositae, Cruciferae, Cruciferae tribe Lepidieae, *Cyperus*, *Dryopteris*-type (sporangia), Gramineae, *Juncus*, Labiatae, *Mentha arvensis/aquatica*, *Polygonum* sect. *Persicaria*, *Potentilla*, *Ranunculus* subg. *Ranunculus*, *Rumex*, *Urtica*, cf. *Veronica* and *Viola*.

This rich flora is one of the most interesting of the assemblages analysed, in respect of its rich tree flora, its list of dry-ground herbs and of the phytogeographical affinities of several of the taxa. The assemblage is clearly derived from a great variety of communities, terrestrial, aquatic and maritime. The regional woodland, with *Quercus*, *Ulmus*, *Carpinus*, *Tilia* and *Corylus*, must have been bordered near the stream associated with the channel by dry-ground herb communities requiring acid or calcareous soils, with the maritime shrub *Corema album*, and *Herniaria*, cf. *Tussilago farfara*, and *Picris hieracioides*. A rich fen and aquatic component is also present. A number of species in the list have a Lusitanian or oceanic west European type of distribution (*Corema album*, *Erica ciliaris/lusitanica*, *Herniaria*? (*see* Wilson, 1973)). The record of cf. *Staphylea* is the only one for the Cromer Forest-bed Formation, though Wilson (1973) has recorded *S.* cf. *pinnata* from the 'Cromer Forest Bed Series' in the collection of the British Museum (Natural History).

15: Southern and north-western limits

(a) Southern limit

At the south end of the Pakefield–Kessingland coastal section (Fig. 46), the Rootlet Bed (bed d, Cromerian) was seen to overlie in section K brown sand with grey clay seams, a sediment similar to that (bed a) seen in the basal parts of the boreholes at Pakefield PK and PL. Blake (1890) described this sediment as 'Chillesford Beds' and its disposition is well seen in his horizontal section (Blake, 1884). Blake (1890, *p.* 14) described the section at Kessingland as follows: 'At the extreme southern end of the cliff, near the road that leads to the village of Kessingland, there is less sand in the laminated clays of the Chillesford Beds, which are, moreover, contorted in places, and contain ferruginous concretions. In these and in other respects, they resemble those to be seen at the northern end of Covehithe Cliff, 2½ miles further south, where fine sections of the Chillesford Beds are to be seen, as well as in the next cliff further south, at Easton Bavent.' This correlation of Blake's was supported by Reid (1890).

The cliff section at Covehithe is sketched in Fig. 48, which also shows the positions and stratigraphy of five boreholes made for site investigation purposes at Covehithe. The general stratigraphy shows a laminated clay unit, underlain by shelly sands and gravels (crag) and overlain by sands and gravels. Gravel of Westleton Bed type directly overlies the clay unit (*see* Hey, 1967). Pollen analyses of this clay (positions of samples shown in Fig. 48, in boreholes LC3 and LC5) gave pollen spectra (Fig. 47) similar to those from the Baventian (Lu 4b) clay at the type site at Easton Bavents (Funnell & West, 1962), 3 km to the south. In fact, the laminated clays at Easton Bavents can be traced north in the cliffs and so correlated with those at Covehithe.

The p.a.b.s recorded, an Ericales–Gramineae p.a.b. in borehole LC 5, and an Ericales p.a.b. in borehole LC 3, are similar apart from the considerably higher Ericales and lower Gramineae frequencies in the latter, and are similar to the Lu 4b assemblages at Ludham, Easton Bavents and Aldeby, as discussed in Chapter 2, with low AP frequencies, composed of *Pinus*, *Picea*, *Alnus* and *Betula*.

As already mentioned, the pollen spectra from bed a at Pakefield show a *Pinus*–Ericales p.a.b. correlated with Pre-Pa a and so with the Lu 4c substage at Ludham, with higher *Pinus* frequencies than in Lu 4b.

The relation between the Covehithe borehole data and cliff section, the section at Kessingland K and the Pakefield boreholes are shown in Fig. 48. A correlation is suggested between them, based on sediment type and pollen analysis. There is, however, considerable difficulty in the correlation of the post-Baventian (Lu 4b) to pre-Cromerian sequence in the stretch of coast from Covehithe to Kessingland, as the following discussion of the sequence of events will show:

(i) Deposition of crag. Pre-Baventian crag at Easton Bavents is Antian in age (Funnell & West, 1962). This crag can be traced in section to the southern end of Covehithe cliffs and in boreholes one kilometre to the north of this, where it lies at −12 m OD below laminated clay of Lu 4b age.
(ii) Deposition of laminated clays of Baventian (Lu 4b) age. These clays dip northward, as shown in Fig. 48.
(iii) Erosion of Baventian sediments, seen in the Covehithe cliff section.
(iv) Deposition of Westleton Bed gravels and associated laminated tidal silty clays at Covehithe.

(v) At this point two possibilities occur:

(a) Erosion of the Westleton Beds and Baventian clays north of Covehithe, to form a basin in which the Pre-Pa a–Lu 4c sandy silt was deposited at Pakefield. There is no trace of Westleton Bed gravels in the Pakefield sequence, which suggests this is a reasonable alternative.

(b) The Pre-Pa a–Lu 4c sandy silt of bed a at Pakefield was deposited after the clays of Lu 4b, then subsequent erosion of them both was followed by deposition of the Westleton Beds at Covehithe and the laminated silty clay of bed b at Pakefield ascribed to the Pastonian.

The former alternative seems the more likely, taking into account the levels of the different horizons shown in Fig. 48. But further detailed investigation of the Covehithe to Kessingland area is necessary to clarify the succession of events.

(vi) Deposition of the clays (bed b) with wood and the sand of bed c in the channel cut in bed a at Pakefield (borehole PL). The clay of bed b is tentatively correlated with the Pastonian. If the sand of bed c is the sand with marine shells described by C. Reid at this locality, then this horizon may be Pastonian crag.

(vii) Alluviation at about 1–3 m OD in Cromerian times, to produce the Rootlet Bed at Pakefield–Kessingland. The southern limit of this alluviation must have been to the north of Covehithe and the south of Kessingland, as the Covehithe Cliffs (and the cliffs to the south) show no trace of an alluvial deposit, probably because pre-Cromerian deposits were higher than the alluvial level.

(b) North-western limit

We have so far considered sections west in cliffs as far as Beeston, just east of Sheringham (Figs. 16 to 18). From Sheringham west to Weybourne, a distance of 4·3 km, the cliffs show a more-or-less continuous, but degraded, series of sections of 3–5 m of pre-glacial sediments lying on the Chalk surface, which rises gently above HWM towards Weybourne (see Harvey *et al.* 1974). These sections were described by Reid (1882). Towards Weybourne he saw freshwater sediments overlying a Rootlet Bed, which rested on sands and loams, shelly in places, which he called Weybourne Crag. A stone bed occurred between this crag and the Chalk. Unfortunately it has not so far proved possible to relate this sequence satisfactorily to that seen east of Sheringham. West of the lifeboat house at Old Hithe (GR TG 1415 4347), the laminated clays which occur either near the base of the section or in basins higher in the section have been heavily weathered and the pollen content of these clays was too poor for satisfactory analysis.

However, sections under Skelding Hill, west of Sheringham, positions shown in Fig. 2, showed unweathered laminated clays under the great thickness of drift present here. The stratigraphy, biostratigraphy and vegetational history of these sections are now described.

(i) *Lithostratigraphy at Skelding Hill*

Sections at Skelding Hill are drawn in Fig. 49. The following beds are distinguished in these sections:

a. Stone bed of large flints at section WQ.

b. Pale grey silt, 10 cm thick in section WQ.

c. Stone bed of flints, with marine shells. This lies above the stone bed a at section WQ, but rests on the Chalk further west.

d. Pale sand, 50 cm thick at section WQ.

e. Laminated blue-grey silty clay, up to 1 m thick, at sections WQ and WP. The clay apparently lies in a shallow basin, but to the west, at sections WO and WN, it has been reworked to produce a clay conglomerate at these sections and at WK.

f. Pale cross-bedded sand, with streaks of drift mud, a black clay-mud lens at section WL, and clay pebbles.

g. Laminated silty clay, to 1·2 m thick, brown and weathered in the upper part and blue-grey below, lying in a shallow basin, descending to the stone bed at section WM. A 2 cm-thick lens of detritus mud occurs at the base of this horizon at section WM.

h. Red cross-bedded stony sand.

i. Contorted Drift.

(ii) *Biostratigraphy at Skelding Hill*

Two pollen assemblage biozones are recorded in this series of sediments, as listed in Table 33. The pollen diagrams are given in Fig. 19.

Biozone a. *Pinus* p.a.b. This is recorded in the three horizons of silt and silty clay in beds b, e and g.

Biozone b. Gramineae–*Pinus* p.a.b. This is recorded in a single sample from a 1 cm thick black clay-mud within the cross-bedded sand of bed f.

(iii) *Vegetational history at Skelding Hill*

The *Pinus* p.a.b. of the laminated clays of beds b, e and g occur in sediments both below and above the Gramineae–*Pinus* p.a.b. The assemblage shows high

AP frequencies, with *Pinus* frequencies reaching over 65%. *Betula*, *Alnus* and *Picea* are the only other trees represented at frequencies up to 10%. There are low frequencies of *Ulmus* and *Quercus* pollen, and in the uppermost laminated clay (WM) of *Carpinus* pollen. The NAP includes low frequencies of *Empetrum*, 15–20% Gramineae, and low frequencies of Cyperaceae, Chenopodiaceae pollen and Filicales spores. The pollen of *Sparganium* and Filicales spores rise sharply in the detritus mud at the base of WM section. Spores of *Lycopodium* (*alpinum*, *clavatum*), *Osmunda* (*claytoniana*-type) and *Sphagnum* are present, with the last much higher in the WM section.

This assemblage is present in laminated clays which are interpreted as marine and tidal since they contain consistent frequencies of Chenopodiaceae pollen and also foraminifers. Allowing for over-representation of *Pinus* in marine sediments, the assemblage indicates the presence of *Pinus* woodland, with a lesser presence of *Betula* and *Picea* and a low representation of thermophilous trees, *Ulmus*, *Quercus* and *Carpinus*. The frequencies of *Empetrum* and *Sphagnum* suggest the presence of heath. The pollen spectrum from the detritus mud at the base of WM indicates perhaps more closely local conditions of vegetation, with higher frequencies of *Betula*, *Ulmus*, *Sparganium* pollen and Filicales spores. This is evidently more of a freshwater horizon, in contrast to the tidal sediments above.

The Gramineae–*Pinus* p.a.b. may reflect again local conditions, since it shows in an exaggerated form the differences already observed between the freshwater and tidal sediments of the section WM. *Betula*, *Ulmus*, *Typha angustifolia* pollen and Filicales spore frequencies are again higher, and there is a great increase of Gramineae pollen at the expense of *Pinus* pollen.

The correlation of these assemblages with those found further east is not obvious. First, it seems that they fall into one biostratigraphical unit, with the differences exhibited by the WL Gramineae–*Pinus* assemblage being the result of local variation in taphonomy. This unit, with high *Pinus* frequencies and traces of thermophilous genera including *Carpinus*, suggests substage Pa Ib. The assemblages in fact have a similarity with the Pa Ib biozone of marine sediments of section SSV at Sidestrand. Their AP characters are also similar to the Pa Ib biozone at Beeston (BR/S), though this is formed in freshwater sediment just prior to the transgression.

It may be noted that the Skelding Hill marine silty clays and stone bed rest on the Chalk at 6·6 m OD, some 3 m above the level of the top of the Chalk at Beeston. No assemblages of Pre-Pa age have been identified at Skelding Hill, and the Pa Ib assemblage is there associated with the stone bed. At Beeston the stone bed is associated with Pre-Pa assemblages. Thus, as the Chalk surface rises between Beeston and Skelding Hill, the Pre-Pa a substage sediments disappear, so that Pa Ib sediments rest on the Chalk at Skelding Hill. The stone bed, as might be expected, is of different age at Beeston and Skelding Hill, so that the Weybourne Crag fauna with *Macoma baltica* must cover Pre-Pa a to Pa Ib times in the stone bed of this area. Probably the Weybourne Crag westward to Weybourne, resting on the rising Chalk surface, is also of Pastonian age.

16: Marine sediments and relative land/sea-level changes

(a) Notes on marine sediments

In the descriptions of the stratigraphy, several types of marine sediment have been recorded. Broadly speaking, three main types have been identified: beach gravels, tidal laminated fine sediments and cross-bedded sands. The following notes briefly discuss each of these sediments.

(i) *Beach gravels*

Pebbly gravels, often with oblique bedding, were recorded at Beeston (Pastonian), West Runton (Beestonian), Sidestrand (Cromerian) and Paston–Mundesley (Cromerian). These were either associated with tidal laminated clays and sands (as at Beeston, bed j; Sidestrand, bed k) or with marine shells (as at West Runton, pre-West Runton Freshwater Bed, west of Woman Hithe, bed h; Paston–Mundesley, bed i(b)). The laminated beds associated with the gravels contain a flora with a strong maritime element.

(ii) *Tidal laminated sediments*

Sediments showing flaser bedding, lenticular bedding and laminated (clay/sand) bedding were found in Pastonian and Cromerian biozones. These come mainly into the Type III lithofacies of Terwindt (1971), in his classification of inshore estuarine and tidal-inlet deposits, formed under low current conditions, though his Type II with cross-bedded sands, and flaser bedding, indicating higher current activity, was observed at a few sites (e.g. Bacton BB).

These Pastonian and Cromerian laminated sediments could be of subtidal or tidal flat origin, but in view of their disposition as wide level 'sheet' units over long stretches of coast, it seems probable that they are of tidal-flat origin. If so, they belong to the misch flat (Reineck, 1967). In certain of the sections (e.g. Paston PB) showing these laminated sediments, a sand (Paston–Mundesley, bed i(s)) occurs beneath which may have originated in the earlier sand-flat, part of the tidal-flat sequence. Mud-flat sediments were not observed. The paucity of bioturbation in the misch-flat sediments is noteworthy. It was only seen in laminated sediments of Sidestrand SSR (Cr IIIb) and of Corton CC (Cr IIIb). Presumably the primary sedimentary structures have been preserved because of fairly rapid accumulation, implying a strong source of clastic sediment. This in turn may indicate nearby river mouths.

It is not possible to determine whether these sediments were deposited under lagoonal or open conditions. No certain lagoonal sediments were observed, but there may have been a complicated coastline with offshore bars, where sheltered water allowed the deposition of extensive tidal flat deposits, as for example seen in the 'Watt' of the Friesian and NW German coast.

Laminated Type III sediments were also seen in the Pre-Pastonian in the Mundesley borehole MAT. This facies resembles that seen in the Baventian (Lu 4b) at Easton Bavents (Funnell & West, 1962) and in the Ludham borehole (West, 1961). The flora of the sediments contains a maritime element. At Ludham, the corresponding foraminifer zone (LVII of Funnell, 1970) indicates very cold conditions and accumulation in shallow marine conditions. The lateral extension of the Mundesley clays is, however, not known. It was not found in boreholes to the west (BHC3, BHC7) or in the Happisburgh borehole HC, where much more sandy sediments, with thin clay laminae, were found. It is possible that the Mundesley MAT laminated sediments are subtidal, with a local development depending on a good supply of clastic sediments.

(iii) *Cross-bedded sands*

This type of sediment was best seen in the section SSV at Sidestrand, where the Chalk had been uplifted. The sediment falls into Type I of Terwindt (1971), formed under high current action, with mega cross-bedded structures, peat lumps, detrital or drift muds, shells and clay-pebbles. The marine molluscs from this sediment at SSB, described by Norton (1967), are almost exclusively intertidal, and suggest deposition in shallow water in an extensive tidal area. The variation from laminated clays Mundesley (MAT) to cross-bedded sands (Sidestrand SSB) seen in Pre-Pa a is in accordance with the variability of lithofacies characteristic of tidal and subtidal situations, with the subtidal sediments showing different degrees of fine clastic content, depending on local sources of sediment such as river mouths.

(b) Relative land/sea-level changes

The superposition of freshwater and marine sediments in the coastal sections and boreholes has given a considerable amount of data for the study of relative land/sea-level changes. Transgression and regression contacts have been recorded and levelled at many sites, and these, through pollen analysis, can be correlated with particular biozones or with the sequence of biozones.

Evidence for marine sedimentation occurs in all the stages involved, except the Anglian. The widely distributed Pre-Pastonian a sediments are all marine, while the very restricted Pre-Pastonian b to d sediments are freshwater. The Pastonian sediments show a very substantial marine component in substages II to IVa. The Beestonian sediments are taken to include a marine horizon. The Cromerian sediments are marine in substages III to IVa, more restricted in time than in the Pastonian. These limits in the Pastonian and Cromerian apply to sections near or above present OD. At lower levels in boreholes, marine sediments occur earlier in these stages, as would be expected.

In the following discussion of relative land/sea-level changes, the compaction factor has been ignored. Since the whole area under consideration has been glaciated, compaction must have occurred, especially of biogenic deposits, but its magnitude is difficult to assess.

The stage-by-stage account which follows starts with the youngest marine sediments, those of the Cromerian, since this stage shows the most complete record of land/sea-level changes in the sequences studied. The record of the earlier stages is more fragmentary. The data significant for relative land/sea-level change are presented in Tables 35 to 41 and are summarised, stage by stage, in the coastal sections shown in Fig. 50.

(i) *Cromerian*

Transgression. The levels and ages of Cromerian transgression contacts and brackish muds are shown in Table 35. At West Runton, the contacts lie at levels between 2·0 and 5·1 m OD, and are dated to a time earlier than a part of Cr IIIa and later than a part of Cr IIb. The sharp junction between sediments at the contacts and the variable levels of the contact shows the erosive nature of the transgression in this area, which evidently was accompanied by beach gravel formation.

At Sidestrand a transgression sequence dated to Cr Ib occurs in an area subjected to glacitectonic movement, so the original levels remain unknown.

At Paston–Mundesley, a sharp transgression contact at *c.* 2 m OD was observed at MU, at the junction of Cr II and IIIa. At ME a silty clay within the lower part of the beach gravel showed a Cr IIb pollen assemblage. At Corton, the transgression contact, with a transition, is seen in Cr IIIa at 0·5 m OD, and at Pakefield also in Cr IIIa at 2·2 m OD.

The data indicate that in Cr IIIa (Corton–Pakefield) or Cr IIb (Paston–Mundesley) or by Cr IIIa (West Runton) a rising sea-level subjected to transgression large areas of freshwater sedimentation, including the alluvial sheet of the Corton–Kessingland Rootlet Bed, as well as the freshwater sediments and soil at West Runton. Levels of the transgressed freshwater sediments are given in Table 36. Earlier than this, marine Cromerian sediments are found at lower OD levels (*see* Table 36): −6 m to −4 m OD in Cr IIb at Corton; −3·5 m OD in Cr IIa at Trimingham.

Regression. The levels and age of regression contacts are shown in Table 37. The contacts are sharp in all sections observed, and they are accompanied by gaps in the biozone sequence. The data indicate regression in Cr IVa. Marine sediments of Cr IVa are present at West Runton (up to 7·5 m OD), Paston–Mundesley (up to 4·3 m OD) and Bacton (up to *c.* 4 m OD). Freshwater sediments of Cr IVa occur at Paston–Mundesley (MAD) at 3·0 m OD and at Ostend at *c.* 2 m to 3 m OD. The regression contact occurs at West Runton (Cr IVa/IVc) at 7·5 m OD, at Overstrand (Cr IIIb/IVc) at *c.* 5·3 m

OD, and at Paston–Mundesley (Cr IIIb/IVc; Cr IIIb/IVb) at 2·5 m to 3·7 m OD.

Levels of marine sediments. Table 36 shows the OD limits of Cromerian (Cr IIIa to IVa) marine sediments and the levels of associated freshwater sediments. At Beeston the marine sediments occur at *c.* 8 m OD, at West Runton at 2·0 m to 7·5 m OD, at Paston–Mundesley at 1·6 m to 4·3 m OD, at Bacton *c.* 2 m to *c.* 4 m OD, at Corton 0·5 m to *c.* 3·5 m OD, and at Pakefield 2·3 m to *c.* 4 m OD. These sediments are mostly intertidal in origin and must represent sedimentation near to mean sea-level in late Cromerian times. This is borne out by the occurrence at Paston–Mundesley of brackish muds (Table 35) at 2·8 m and 2·9 m OD within the tidal sediments at the boundary of Cr IIIb/Cr IVa.

Deformation. The late Cromerian transgression sequences recognised are all within 0·5 m to 5·1 m OD over a stretch of coast 70 km long from West Runton to Pakefield. The only sequences with transitions are at Pakefield at 2·2 m OD and Corton at 0·5 m OD, both in Cr IIIa. At Paston–Mundesley, the transgression took place, probably within a few metres above OD, by part of Cr IIb, and at West Runton by part of Cr IIIa. The levels of the freshwater and marine sediments concerned are given in Table 36, and sketched in Fig. 50. The data indicate no great net between-site deformation of the area since Cromerian times. The pattern of different levels within the limits may, however, suggest a small (not more than 2–3 m) relative deformation with the higher limits of marine sediments at the north-west end (West Runton) and south end (Pakefield) than in the Paston–Mundesley area. More data are required to confirm this indication.

(ii) *Beestonian*

Marine sediments of this age, including beach gravels, shelly sands and tidal laminated sands and clays, have been identified only at Beeston and West Runton (Table 38, Fig. 50). They occur at heights between 3·0 m and 7·8 m OD, and overlie freshwater deposits with a pollen and plant macroscopic remains content taken to characterise the stage.

The transgressions associated with the Pastonian and Cromerian temperate stages parallel those of later, interglacial, temperate stages. The Beestonian transgression, however, appears to be associated with sediments showing evidence of cold climate, though, as discussed earlier, no palaeobotanical data were obtained from the laminated clays of the marine transgression. The highest sea-level indicated by the sediments must be near 7 m OD. Such a level for cold-stage sea-level would seem unusual in the context of glacio-eustatic lowering of sea-level. Perhaps there was an analogous situation to that which obtained in the southern North Sea when the Corton Beds (glacio-marine in part?) of the Anglian stage were deposited to a height of at least 10 m OD in the Lowestoft area, though the palaeogeographical basis for this has not yet been determined. Alternatively the Beestonian marine sediment may represent a eustatic level of a cold stage. If so, is it possible that the glacial erratics found on shore platforms near OD around the southern coast of England were emplaced at this time?

(iii) *Pastonian*

Transgression and regression. A Pastonian transgression is clearly indicated in the Beeston sections (Table 39). The transgression contact shows a transition near the boundary of Pa Ib and Pa II, at heights near 5 m OD. The regression following this transgression is recorded at Beeston and Paston–Mundesley (Table 39). Pa IVa tidal sediments are present in both these areas and so are freshwater sediments of the same substage. The regression evidently took place in Pa IVa. The sharp regression contact at Beeston is at 5·1 m OD and is at the Pa III/IVa boundary. At Paston–Mundesley, a sharp contact was seen at a Pa IVa/IVc boundary at PE, and at MPA the regression had occurred by a part of Pa IVa times at −0·8 m OD.

The data indicate a rising sea-level in Pa Ib times with Pa II, III and IVa marine sediments occurring up to the heights shown in Table 40. As in the Cromerian, it is thought that the laminated sediments of this age are inter-tidal in origin and their levels indicate the approximate contemporary sea-level. Brackish muds within the tidal sequences occur at West Runton and Happisburgh. They are of Pa II age and occur at *c.* 1·5 m OD and −0·5 m OD respectively, showing that in these areas contemporary sea-level was near OD during a part of Pa II times.

Levels of marine sediments. While transgression can be demonstrated in the Beeston area, contemporary sea-level of Pa II in the West Runton and Happisburgh areas, and regression in the Beeston and Paston–Mundesley areas, there is a substantial body of Pastonian marine sediment near and below OD in the areas from West Runton south-east to Happisburgh and in the

Corton to Pakefield areas. The levels of these sediments are shown in Table 40. Thus at Trimingham Pa Ib sediments occur at −9·0 m to −1·5 m OD, at Paston–Mundesley Pa II/III sediments occur at −3·2 m to 2·4 m OD, and at Happisburgh Pa I to IV from −1·5 m to 2·4 m OD. Thus a basin of Pastonian marine sediments appears to occur between Happisburgh and West Runton, the sediments of which can be clearly seen on the uplifted Chalk in the Sidestrand area (SSV).

In the Corton to Pakefield areas Pa II sediments are recorded at −6·7 m and −5·4 m OD respectively. A further basin of thick Pastonian sedimentation may exist here, limited to the south by the laminated clays of the Baventian, rising above OD at Covehithe (Fig. 48). This distribution of basins is seen in Fig. 50, showing the OD heights of Pastonian marine sediments.

Deformation. The main body of Pastonian sediment is marine. Evidence of contemporary sea-level is only afforded by transgression contacts at Beeston, and brackish muds at West Runton and Happisburgh. The problem is whether these data are sufficient to demonstrate any great net between-site deformation.

The data from the Happisburgh to Skelding Hill areas may indicate a degree of deformation. Pastonian II sea-level was near 0·5 m OD at Happisburgh and −1·5 m OD at West Runton. Subsequently at West Runton a thickness of marine clay conglomerate was deposited to 5·0 m OD. At Beeston, the marine transgression at the beginning of Pa II is at 5·6 m OD. This suggests that the Beeston area has undergone uplift relative to West Runton since Pa II times by perhaps as much as 5 m. A further indication of the same trend is shown by marine Pa Ib sediments at Skelding Hill, with their base on the stone bed at 6·5 m OD. The Pastonian section of Fig. 50 shows the height and substage relationships of sections in these areas.

The upper limit of Pastonian laminated sediments at Paston–Mundesley and Happisburgh is 2·4 m OD and regression contacts here are not more than 2 m OD, compared with the heights reached by Pastonian sediments at Skelding Hill (9·5 m OD). This may indicate relative subsidence in the centre of the study area, compared with the north-western area.

The age of the deformation between West Runton and Beeston–Skelding Hill, if rightly inferred, remains to be considered. The height of sediments assigned to Cr IVa, marine at West Runton and probably marine at Beeston, is similar (at *c.* 7 to 8 m OD; see Table 36), as is the height of Beestonian marine sediments (Table 38). This may indicate that deformation took place in pre-Beestonian times.

(iv) *Pre-Pastonian*

The sediments of Pre-Pa a are considered entirely marine. Their disposition is given in Table 41 and in Fig. 50. At the north-west end of the coast section, they shallow as the Chalk surface rises, until they are represented by the stone bed only at Beeston. At the south of the coast section, at Pakefield, the sediments remain thick but rise above OD (3 m OD at Pakefield). Pre-Pa a sediments at Happisburgh and Mundesley have the greatest OD depth, to −16·5 m at Happisburgh. Table 41 also shows the height of Baventian (Lu 4b) sediments at Covehithe and Easton Bavents.

The freshwater sediments of Pre-Pa b–d (Table 41) show that regression occurred after the deposition of Pre-Pa a sediments. This is recorded only at Beeston (Pre-Pa b–d) and East Runton (Pre-Pa d), at 3·8 m to 5·1 m OD and −2·2 m to −1·0 m respectively. Such a difference in height of freshwater Pre-Pa d sediments between East Runton and Beeston would result from the deformation discussed under the Pastonian heading above.

(v) *General*

The transgression sediments of the late Cromerian and Pastonian record rises of sea-level in these temperate stages. It seems unlikely that sea-levels in these stages rise much above the levels indicated by the tidal sediments. The vegetational histories form entities and no deep-water marine sediments have been recorded younger than the tidal laminated sediments. If we take the highest limit of laminated sediments as giving maximum mean sea-level of each transgression,* and if we assume no deformation, the Cromerian limit is 7·5 m OD and the Pastonian limit 2·4 m OD in the Paston–Mundesley and Happisburgh areas, and 9·5 m OD in the Skelding Hill area. As already discussed there may be an element of deformation involved in the relative heights of the Pastonian sediments.

The data indicate the presence in the study area of a tract of land sensitive to sea-level change on the margin of the area of subsidence of the southern North Sea. The Crag basin (Fig. 52) is filled with Lower Pleistocene marine sediments, and this fill evidently was completed by the end of the Pastonian. Deformation may have followed and then, during the Cromerian, a large alluvial plain must have developed in the Corton–Kessingland area, towards the southern margin of the study area, with a transgression in the late Cromerian, recorded to a height of 7·5 m OD.

* The contemporary tidal range is unknown. At present it is low (4·5 m (springs) at Cromer, and 1·9 m (springs) at Lowestoft).

This low value of the Cromerian limit and the closeness of transgression contact levels over 70 km of coast are of particular interest. It is not known how much deformation by upwarping or downwarping has occurred since the Cromerian in the area. But the low variation in levels established on the coast suggests either little net movement may have taken place or that isobases of warping parallel the coast. A second point is that Hoxnian sea-level in west Norfolk reached in late Hoxnian times (Ho III) at least 23 m OD (West, 1972), considerably higher than the Cromerian transgression sediments in north-east Norfolk. It is impossible to say how much this difference is accounted for by differences of eustatic sea-levels in these temperate stages and how much by deformation. A third point is that the Pastonian and Cromerian limits approximate to the level of the wave-cut shore platform of Ireland, Wales and south-west England. This suggests the possibility that the platform may be of considerable age and have developed in pre-Hoxnian temperate stages.

17: Environmental history of the stages

In the preceding chapters we have described the stratigraphy, biostratigraphy, and the sedimentary and vegetational history of each stretch of coast. In this chapter we take a summary and overall view of each stage along the total length of the coast sections. Table 42 shows the principal pollen assemblage biozones recorded in each of the several stretches of coast and their correlation with substages and stages. Fig. 51 gives a diagrammatic survey of the marine and freshwater sediments of each stage and provides a basis for the following discussion. Fig. 52 summarises in a simplified manner the areal distribution of certain stratigraphical units and facies around the coast.

(a) Pre-Pastonian

Sediments pre-dating the Pastonian have been recorded at all the areas studied on the Norfolk and Suffolk coasts except at Corton, and at Skelding Hill, west of Sheringham, where Pastonian marine sediments rest on the Chalk. The distribution of Pre-Pa sediments is clearly seen in the diagrammatic section, Fig. 51. The basin of marine deposition deepens south-east from Beeston to Happisburgh, with sediments rising above OD in the Pakefield area of Suffolk.

(i) *Pre-Pastonian a*

The characteristics of this substage and the distribution of its sediments have already been described in Chapter 2. The distribution (*see* Table 42) indicates the presence of a marine embayment between Beeston and Covehithe, with a stone bed forming on the Chalk at the north-western margin, in the Mundesley to Beeston area. The presence of shallow-water sediments below Pre-Pa a sediments at Happisburgh indicate a major marine transgression near the beginning of Pre-Pa a times. The area of deposition is shown in Fig. 52.

The vegetation of Pre-Pa a appears to have been of a park–tundra type, with a severe climate prevailing. The marine fauna gives similar indications about climate. Probably the brecciation of the Chalk beneath the stone bed in the marginal areas of the basin is associated with the same cold climate. Evidence for a period of glaciation before deposition of Pre-Pa a sediments is found in the surface textures of the sand grains (Krinsley & Funnell, 1965).

(ii) *Pre-Pastonian b, c and d*

Sediments of these substages have been recorded at Beeston, East Runton, Ostend and Happisburgh. The most complete sequence is at Beeston, where all three substages are recorded. At Ostend and Happisburgh the evidence is more fragmentary and only few pollen analyses show high NAP frequencies in sediments between Pre-Pa a and Pa sediments. It is not clear whether these analyses are from freshwater or marine sediments, though one of the Happisburgh analyses (biozone g) comes from a freshwater mud ball.

At Beeston, however, freshwater sediments of fluviatile and pool environments show biozones attributed to Pre-Pa b, c and d and at East Runton there is a substantial bed of freshwater muddy silt correlated with Pre-Pa d.

Pre-Pa b. The fluviatile and pool sediments of this substage contain a herb-pollen assemblage biozone derived from herbaceous communities adjacent to a river, with scattered but restricted woodland of trees of boreal type. The climate is evidently severe enough to restrict the growth of more temperate trees.

Pre-Pa c. This substage is only recorded in fluviatile sediments at Beeston. The slightly higher AP frequencies in the biozone of this substage include *Ulmus* consistently, indicating an amelioration of climate, but not to a degree characteristic of temperate (interglacial) stages. Herb-pollen frequencies remain high, indicating the continued presence of a park–tundra type of vegetation.

Pre-Pa d. This substage is recorded in mud and muddy silt at Beeston and East Runton. AP frequencies are lower than in the previous substage and vegetational conditions generally resemble those of substage Pre-Pa b, with severe climatic conditions.

(iii) *General*

The sequence described above indicates transgression at the beginning of Pre-Pa a and regression at the end of Pre-Pa a. The record of Pre-Pa b to d is fragmentary, but at Beeston fluviatile and pool sedimentation leads directly into the Pastonian. A gap may then be present between the marine shelly sediments and stone bed of Pre-Pa a and the later freshwater sediments of Pre-Pa b to d, but the successive units can only be grouped into one (cold) stage in the absence of evidence of the length and climatic conditions of the period involved in the probable non-sequence.

(b) **Pastonian**

The woodland biozones following the herb biozones of the Pre-Pastonian, grouped into the Pastonian stage, are recorded around the coast from Skelding Hill to Pakefield, mostly in marine sediments (Fig. 51). The thickest sequence (*c.* 10 m) is seen in the Paston to Trimingham area. The Pastonian sediments thin to the north-west as the Chalk rises above OD, and to the south, as the Pre-Pastonian rises above OD.

Four substages of the Pastonian have been distinguished. The most widespread Pastonian sediments are those of Pa II (see Table 42) (*Pinus–Quercus–Carpinus–Ulmus* p.a.b.), a period including the transgression sediments at Beeston, and sediments such as the clay conglomerates of the East Runton to West Runton area. Sediments of Pa I are less well-recorded, and those of Pa III are fragmentary and difficult to group; those of Pa IV are better represented and have been found on many sections of the coast.

The basal sediments of the Pastonian may be Pa I or II in age, indicating unconformity between Pre-Pa and Pa. The probability of a non-sequence between Pre-Pa a and b has already been discussed, and this may have contributed to the non-sequence between Pre-Pa and Pa, but it is also probably related to erosion accompanying marine transgression in the early Pastonian.

(i) *Pastonian I*

Both marine and freshwater sediments of this substage have been recorded. Freshwater sediments occur only at Beeston. The marine sediments include tidal laminated silty clays at Skelding Hill, deeper water sands with clay laminae at Sidestrand and Trimingham (to −9 m OD) and shallower water clays and sand at Bacton and Happisburgh. The base of the sediments at Trimingham is well below the level of the Pre-Pa a laminated clay at Mundesley (MAT), indicating an unconformity and period of erosion between Pre-Pa and Pa. An unconformity between Pre-Pa a and Pa I or Pa II is also shown by the sections on uplifted Chalk at Sidestrand (Fig. 24).

The pollen assemblages of Pa I show the development of first *Betula–Pinus* woodland (Pa Ia) then coniferous forest with *Pinus* and *Picea* (Pa Ib). The earlier biozone is recorded at Beeston and Happisburgh and the latter at Skelding Hill (marine), Beeston (freshwater), Sidestrand (marine) and Trimingham (marine). The expansion of woodland and the presence of thermophilous plants (*Cladium mariscus*), by analogy with Flandrian vegetational history, indicate a temperate climate was established by this time.

(ii) *Pastonian II*

Sediments of this substage are widespread around the coast from Beeston to Pakefield. Various facies are represented: tidal laminated clays (Beeston, West Runton, Happisburgh, Pakefield), brackish or freshwater intercalations in tidal sediments (West Runton, Happisburgh), clay conglomerates with gravel (West Runton, Paston–Mundesley), and cross-bedded sand with clay laminae (Sidestrand). The facies are those to be expected in a large tidal estuary. The transgression which produced the tidal sediments at Beeston deposited laminated clays over mud with an apparently conformable transition, indicating a relative rise in sea-level which flooded coastal pools already present.

The pollen assemblages of Pa II show the presence of mixed deciduous woodland (*Pinus–Quercus–Carpinus–Ulmus* p.a.b.) with evidence for coastal swamp communities with *Alnus* and *Osmunda*. The assemblage appears to present a climax type of vegetation formed under temperate climatic conditions after the initial *Betula* and coniferous assemblages of Pa I.

(iii) *Pastonian III*

The separation of biozones grouped to this substage is, as already discussed, tentative. The sediments concerned are mainly tidal laminated sediments of very limited thickness, occurring at Beeston, West Runton, Paston–Mundesley and Happisburgh. At WRCNA, the biozone overlies the marine cross-bedded sand of bed f, indicating continued marine conditions in this substage. The pollen evidence is insufficient to characterise the substage clearly, and the basis of the segregation from Pa II and Pa IV assemblages is that the Pa III assemblages show a decreased representation of some thermophilous tree genera and some increase of Ericales pollen (Beeston, West Runton, Paston–Mundesley), indicating a change in the woodland characteristic of Pa II, or the presence of *Tsuga* pollen (West Runton, WR AA, Happisburgh HC). At West Runton assemblages of the former type are younger than the assemblage with *Tsuga*.

These changes in pollen content may indicate a change in the stable Pa II woodland, with the development of heath and restriction of thermophilous tree genera, but possibly changes in taphonomy of the assemblages may also have been significant in causing them.

Thus the occurrence of gravels in the shallow marine and tidal sequences of Sidestrand at SST and SSU between sediments of Pa II and Pa IV age (Plate 20) may indicate a regression or physiographical change in this interval and this may account for the rarity and variability of the pollen assemblages between Pa II and Pa IV.

(iv) *Pastonian IV*

The Pastonian regression occurs in this substage. Marine sediments of Pa IVa and b age occur at Beeston, Cromer, Overstrand, Sidestrand, Paston–Mundesley and Happisburgh, and freshwater sediments of Pa IVc age at Beeston and Paston–Mundesley. At Beeston the final freshwater unit is fluviatile, with sands with silt laminae, but at Paston–Mundesley it is predominantly limnic with a strong mud component.

The woodland vegetation is at first mainly coniferous, with *Pinus*, *Picea*, *Alnus* and *Betula*. The pollen assemblage is comparable to those presently obtained from the mixed coniferous woodland of the Middle Boreal region of Finland (Hicks, 1977). In certain of the sequences (Overstrand, Paston–Mundesley) an increase of *Picea* representation subsequently occurs and the assemblages showing this change are segregated as Pa IVb (*Pinus–Picea* p.a.b.), though whether the differences are caused by changes in vegetation or taphonomy is uncertain. The later assemblages of Pa IVc show an increase in the importance of *Betula* in the woodland, these assemblages marking the final woodland assemblage of the Pastonian, and indicating climatic deterioration at the end of the temperate stage.

(c) **Beestonian**

The Beestonian cold stage includes a complex of lithostratigraphical units and periglacial phenomena which are intercalated between sediments of Pastonian and Cromerian age. The sequence may be far more complex than is now recognised but, in the absence of any means of subdivision based on climatic change, we are unable to make meaningful subdivisions within it, apart from distinguishing a Late Beestonian unit.

The sediments included within the stage can be briefly listed as follows:

4. Freshwater marl (West Runton; Late Beestonian)
3. Fluviatile sand, silt or clay and mud (Beeston)
2. Marine sand and gravel with shells (West Runton, Beeston)
1. Fluviatile sand and silt or clay (West Runton, Beeston).

In addition there are ferrugineous sands and gravels at Overstrand, Paston–Mundesley and Bacton between the Pastonian and Cromerian sediments, which may be Beestonian, but whose age is impossible to determine in the absence of plant remains.

Ice-wedge casts are associated with sedimentary units 1 and 3, and possibly 2, above, and they also pierce the ferrugineous sands and gravels of Paston–Mundesley, so that in that area they provide evidence of cold conditions in the interval between the Pastonian and Cromerian temperate stages. Thus, in summary, evidence for the Beestonian is derived from sediments in the Beeston and West Runton areas and from ice-wedge casts in the Paston–Mundesley area. Further south, from Ostend to Kessingland, there is no evidence for sediments of Beestonian age.

We may now consider the Beestonian sediments and periglacial phenomena of the Beeston to West Runton areas and of the Overstrand to Bacton areas in more detail, also paying attention to evidence for weathering and down-cutting within the sequences.

(i) *Beeston to West Runton*

Incision of the Pastonian sediments at Beeston was followed by fluviatile filling of a channel under cold conditions, indicated by periglacial phenomena and by flora. Freshwater silts, sealing a post-Pastonian ice-wedge cast, occur at West Runton above the Pastonian marine silty clays. Marine gravel and sand overlie these freshwater beds at West Runton, and gravels of similar constitution overlie the freshwater beds at Beeston. A marine incursion therefore follows a period with permafrost, fluviatile deposition and a high NAP pollen flora indicating herb vegetation. At Beeston a further period with fluviatile and pool deposition, times of permafrost and a mainly herb-pollen flora followed the deposition of the gravels.

At West Runton a major period of incision occurred later than the formation of the marine gravel and sand, and in the channel so formed marl was deposited. This is the Late Beestonian sediment of immediately pre-Cromerian age, with two pollen assemblage biozones, a Gramineae–Cyperaceae–*Artemisia* p.a.b. (l Be a) and a Gramineae–Cyperaceae–*Betula* (l Be b).

This whole Beestonian sequence thus shows a marine incursion with down-cutting and fluviatile deposition occurring before it and fluviatile deposition and strong down-cutting after it. Weathering with the absence of pollen content was noted at two horizons within the sequence. The sediments were silty clays where pollen would normally be expected. These horizons were the upper part of the massive tidal Pastonian sediments at Beeston and the silty clay laminae of the Beestonian marine sands at West Runton. In the former example, the weathering, which resulted in a massive non-calcareous grey-blue silty clay like the 'Rootlet Bed' at Ostend, probably occurred in late Pastonian or early Beestonian times, while the weathering at West Runton is probably related to the period of down-cutting which followed the deposition of the marine sediments. The presence of ice-wedge casts at several points in the Beestonian succession (*see* Table 11) points also to a succession of land surfaces in the sequence. We may envisage a river plain with intermittent sedimentation, channel formation and ice-wedge polygons.

(ii) *Overstrand to Bacton*

South-east of West Runton, no sediments can be referred to the Beestonian by virtue of their contained fossils. At Overstrand, Paston–Mundesley and Bacton ferruginous sands and gravels are found between Pastonian and Cromerian sediments. There occurs above the Pastonian at Overstrand, Paston–Mundesley and Bacton clay conglomerate with gravel, then a cross-bedded sand with apparent dip west or north-west and an upper ferruginous gravel. This group of sediments may belong to the Pastonian regression. The cross-bedded sand is similar to that seen at West Runton in the late part of the Pastonian. At Sidestrand (SSV) the Cromerian is separated from the Pastonian by a thin ferruginous gravel. There is no record of fluviatile sediments of the kind seen in the Beestonian of West Runton and Beeston.

However, truncated ice-wedge casts occur at Paston–Mundesley piercing the ferruginous sands and gravels (the marker level), indicating erosion of an old land surface which suffered permafrost. This permafrost level, being sealed by Cromerian sediments, is likely to be of Beestonian age. Erosion at this level is also indicated by the absence of the ferruginous sands and gravels at Paston–Mundesley in several sections, with the junction between the Pastonian and Cromerian sediments being marked by the iron-cemented top of sands of Pastonian age.

South of Bacton no sediments have been found which might be of Beestonian age. The Cromerian at Corton lies on Pastonian sediments and at Pakefield there are sands between the Pastonian and the Cromerian 'Rootlet Bed', but their age is unknown.

(d) **Cromerian**

Cromerian sediments are widely distributed around the coast and have been recorded in all the coastal stretches studied except at Skelding Hill, East Runton, Cromer and Happisburgh. The north-western limit of preserved deposits is at Beeston and the southern limit at Kessingland (*see* Chapter 15). To the north-west of Happisburgh, the marine sediments of Cr III and IV are present over long stretches of section, with the freshwater sediments of Cr I and II being more locally distributed. At the south end of the coast section, at Corton and Pakefield, both freshwater sediments of Cr II and marine sediments of Cr III occur over long stretches of the coast. There is thus abundant evidence for the reconstruction of the landscape of the Cromerian in our area.

(i) *Cromerian I*

Sediments of this substage are recorded as the earliest Cromerian sediments of the central channel filling and in one of the swale fillings at West Runton and also in the glacitectonically disturbed sections at Sidestrand. At West Runton the sediments are freshwater, with two pollen assemblages (allocated to Cr Ia and Ib) showing the development of woodland from *Betula–Pinus* to *Pinus–Ulmus*. At Sidestrand (SST) the freshwater or brackish and tidal sediments fill a narrow channel cut in Pastonian sediments, and contain a *Pinus–Ulmus–Betula* assemblage allocated to Cr Ib. Their original height is unknown, as they must have been moved by the glacitectonic disturbances seen in the Sidestrand section. Evidence for marine transgression in the course of Cr Ib is seen in the SST section, so it is probable that the original height was much below OD.

(ii) *Cromerian II*

Sediments of this substage are recorded at West Runton, Trimingham, Paston–Mundesley and Corton. At West Runton two p.a.b.s are recorded, *Pinus–Quercus–Ulmus* (Cr IIa) and *Quercus–Ulmus–Tilia* (Cr IIb) in freshwater sediments in the upper part of the central channel filling and in one of the swale fillings. The pollen assemblages show the full development of mixed oak woodland in the Cromerian temperate stage. At Trimingham, Cr IIa is recorded in a section (TB) and in a borehole (BHT 14A). *Quercus* and *Ulmus* are important in the woodland, and brackish conditions are indicated by the borehole flora at $-3{\cdot}5$ m OD.

At Paston–Mundesley freshwater sediments of Cr IIa with *Quercus* and *Ulmus* are present in isolated sections, which indicates that the sediments were largely removed by erosion. A single pollen spectrum with a *Quercus–Ulmus–Tilia* assemblage (ME 1, Cr IIb) is associated with beach gravels, which occur widely along the coast in this area. Marine transgression evidently truncated and eroded freshwater sediments, probably in later Cr IIb times.

At Corton, Cr IIa sediments are recorded in freshwater facies near OD (CD, CF) and, possibly marine, at depth (BH 3/5, 3/10). Marine sediments of Cr IIb age are recorded at -4 to -6 m OD, filling what must be a wide channel in underlying sediments (Pre-Pastonian near OD at Pakefield). The filling of this channel ended with the deposition of an alluvial sheet (Rootlet Bed), placed in Cr II, seen along the coastal sections of Corton, Pakefield and Kessingland. This must mark the alluvial plain of a large river occupying the channel already referred to, and which may possibly be the proto-Thames. Hey (1976) suggests that the quartz and quartzite pebbles of the beach gravels associated with the transgression in Cr III at Corton are likely to have been supplied by the Thames. At Corton, a rise in water level leads to the formation of wood peat with an associated *Quercus–Ulmus–Tilia* p.a.b. over the Rootlet Bed in Cr IIb times, a rise followed by the marine transgression of Cr III.

(iii) *Cromerian III*

Sediments of this substage contain pollen assemblages which show the development of woodland first with *Carpinus* (Cr IIIa) and then with *Abies* (Cr IIIb). The Cr III sediments are brackish or marine except partly at Corton, where marine transgression is recorded in Cr IIIa.

The Cr IIIa tidal sediments are recorded very sparsely at West Runton (WR BCa) and at Paston–Mundesley (MU T; MW 1). At Corton and Pakefield brackish/tidal sediments are recorded much more substantially, giving much detail of the flora associated with the *Carpinus* woodland. In this area marine transgression occurs within Cr IIIa, as described in Chapter 16.

The sediments of Cr IIIb are marine, predominantly tidal. They occur widely along the coast at West Runton, Overstrand, Sidestrand, Trimingham, Paston–Mundesley, Corton and probably Pakefield, overlapping the freshwater sediments of earlier Cromerian times, and showing the strength of the transgression over wide tracts of low-lying land.

(iv) *Cromerian IV*

Sediments of this substage are widely distributed along the coast from Beeston to Ostend. They have not been found in the Corton to Kessingland area. Three subdivisions of this substage related to pollen assemblage biozones have been distinguished. They reflect the changing woodland composition associated with climatic deterioration towards the end of the Cromerian temperate stage. Cr IVa shows predominantly tidal sediments with a *Pinus–Alnus–Picea–Betula* assemblage, Cr IVb contains freshwater sediments with pollen assemblages including *Pinus* and *Betula* as major taxa, and Cr IVc contains freshwater sediments with *Betula*-dominated pollen assemblages. The *Pinus–Alnus–Picea–Betula* assemblage, as with that of Pa IVa, bears comparison with pollen assemblages derived from the mixed coniferous woodland of the Middle Boreal region of Finland (Hicks, 1977).

The Cromerian regression occurs in Cr IVa. The tidal sediments of the early part of Cr IVa generally overlie those of Cr IIIb and are recorded at Beeston, West Runton, Overstrand, Paston–Mundesley and Bacton. Freshwater sediments of Cr IVa are seen at Paston–Mundesley and Ostend. Regression in Cr IVa was accompanied by down-cutting with fluviatile (Paston–Mundesley) or pond (Ostend) sediments being then deposited.

Cr IVb sediments are recorded more sparsely at

Overstrand (OB), Sidestrand (SST) and Paston–Mundesley (MB). The sediments are certainly freshwater at Overstrand and Paston–Mundesley, and probably so at Sidestrand. Freshwater sediments of Cr IVa and b contain a number of thermophilous taxa (*Thelycrania sanguinea*, *Trapa natans*, *Salvinia natans*) indicating considerable summer warmth until late in the temperate stage. The change to mixed coniferous woodland in Cr IVa may then be in part associated with progressive mor development, as described for the Eemian temperate stage in Denmark by Andersen (1974), as well as with climatic change.

Cr IVc freshwater sediments are recorded at West Runton (WRAYa, WRAY) and Paston–Mundesley (MAU) where they unconformably overlie Cr IVa and Cr IIIb tidal sediments respectively.

The localised occurrence of Cr IVb and c sediments is thus associated with irregular fluviatile and pond deposition following the marine regression in Cr IVa times.

(v) *General*

The Cromerian sequence thus shows the results of marine transgression over a very extensive low-lying tract of land, in which an extensive sheet of alluvium had already been deposited in the Corton to Kessingland area, with fluviatile deposition in a channel further north-west at West Runton. The transgression, with its beach and tidal facies, best exhibited at Paston–Mundesley, was followed by regression, with the resumption of fluviatile conditions in the Paston–Mundesley area. As will be described in the next section, these conditions continued into the early Anglian in this area.

(e) Anglian

Cold conditions of the Anglian stage are recorded by herb p.a.b.s and periglacial phenomena associated with freshwater sediments lying between Cromerian sediments and Anglian tills. Freshwater sediments with pollen are recorded at Trimingham, Paston–Mundesley and Bacton. The pollen assemblages are of two types, those with very high NAP frequencies and those with high NAP frequencies, but also with *Betula* and *Pinus*. At Paston–Mundesley (MAH) the latter type occurs below the former type, the change presumably reflecting continued climatic deterioration.

Ice-wedge casts of early Anglian age and at various levels within the e An sequence have been recorded at West Runton, Paston–Mundesley and Bacton. At Corton an early Anglian ice-wedge polygon system was seen.

The sediments thus show the continuance of fluviatile and pond conditions into the e An substage, with more than one period of permafrost developing. There followed the ice advances which deposited the Anglian tills (Banham, 1975).

18: Conclusions and correlations

(a) Summary of the stratigraphy and nomenclature

The lithological variation of the pre-glacial sediments around the Norfolk and Suffolk coasts is, as we have seen, very great. However, it is possible to establish certain formal lithostratigraphical units which have a reasonably wide distribution, and a sequence of such units will serve also to summarise the stratigraphy of the areas we have studied. These lithostratigraphical units or beds can be grouped into members, which combine to form the Cromer Forest-bed Formation. The place of this formation in the East Anglian pre-glacial Pleistocene is outlined by Funnell & West (1977).

In making the classification given in Table 43, we inevitably have to omit some horizons which are scantily represented in the coastal sections, so the classification is incomplete in the sense that it does not include all sediment units described in the lithostratigraphical sections. Certain insubstantial but significant units are placed in brackets in the Table in their stratigraphical position.

Table 43 gives the proposed classification and nomenclature of beds, with type sites, and groups them into members. All members are placed in the Cromer Forest-bed Formation, except the lowest, the Sidestrand Member, which is placed in the Norwich Crag Formation of Funnell & West (1977). The members are named after stretches of coast where the component beds are well represented.

(b) Identification of breaks in the succession

The succession given above contains evidence for non-depositional unconformities at several horizons, in the form of erosional intervals and non-sequences. The non-sequences are revealed by reference to the 'standard' biozone sequence, and certain of these are indicated in the 'hiatus' column of the Tables summarising environmental history. Evidence may also stem from differential deformation of certain horizons, already discussed in Chapter 16, and from horizons with weathered sediments or with rootlets.

We will consider each of these sources of evidence in turn, first noting periods where there is, in contrast, evidence for continuous sediment or biozone sequences. Fig. 53 summarises the evidence we discuss in this chapter.

(i) *Continuous sequences*

Sequences showing continuity of sedimentation and/or of biozone succession occur in two parts of the pre-glacial sequence. They are from Pre-Pa b to Pa IV at Beeston (BR–S), and from l Be through the Cromerian to eAn. Both these periods cover 'late-glacial' time followed by the development of a temperate stage, with the addition that the stratigraphical evidence at Paston–Mundesley indicates that the Anglian stage followed closely on the Cromerian.

(ii) *Breaks in the succession*

Non-depositional unconformities or erosional intervals are present at the following points in the succession:

Prior to Pre-Pa a. The Happisburgh borehole HC shows evidence for a marine transgression at the base of Pre-Pa a, which is probably associated with the formation of the Cromer Stone Bed on the Chalk to the north-west, as discussed in Chapter 2. The presence of an old land surface on the Chalk in this area to the north-west is indicated both by the disturbance of the top of the Chalk by frost action and by the occurrence of remains of large vertebrates in the stone bed, e.g. *Archidiskodon*.

Pre-Pa a/b interval. The post-Pre-Pa a regression is followed by formation of freshwater sediments of Pre-Pa b–d. An interval of time of unknown length occurs between the Pre-Pa a and b sediments. This interval may be that which produced unconformity between Pre-Pa a and Pa along the coast at Sidestrand (Fig. 24), Paston–Mundesley (Fig. 51) and Pakefield. In the last area sediments ascribed to Pa lie in a channel in Pre-Pa a sediments.

Pa/Be interval. At Beeston, the oldest Beestonian sediments (bed n) lie in a channel cut in the Pastonian sediments and there may be a break of unknown length between them. Weathering of Pa marine sediments (bed k) along here (e.g. BAX) may indicate a considerable time interval here, if, as seems likely, the weathering is earlier than the deposition of the later Beestonian sediments.

Within Be. Periods of permafrost and aggradation occur within the Beeston Beds. Non-depositional intervals must occur related to times of permafrost interrupting deposition of the Beeston Beds and in relation to the Woman Hithe Clay, but their length is unknown. Further non-sequences may be associated with the marine gravel facies (Woman Hithe Gravel) and the freshwater sediments of the Woman Hithe Clay and Beeston Beds. Certainly an erosional interval occurs between the deposition of the Woman Hithe Gravel and the l Be calcareous sediments (bed a) of the central channel filling at West Runton. The laminated clays in sand associated with the Woman Hithe Gravel do not contain pollen and weathering of this horizon may have taken place in the erosional interval.

Pa/Cr interval. The absence of fossiliferous Beestonian sediments south-east of West Runton, and the superposition of Cr sediments on Pa sediments at Sidestrand, Paston–Mundesley and Corton at levels near and below OD speak for intervals with erosion between Pa and Cr. These intervals may be any or all of those discussed in relation to Pa/Be and within Be non-sequences.

(iii) *Deformation*

In Chapter 16 we pointed out that the height distribution of Pastonian sediments compared with those of the Cromerian indicated the possibility of a period of warping between the Pastonian and Cromerian stages, with relative uplift in the Skelding Hill and Covehithe areas, so that by Cromerian times deposition took place over a fairly level plain extending north-west from Kessingland to West Runton. If this conclusion is right, there may be a considerable interval between the Pastonian and Cromerian, perhaps associated with one or more of the intervals related to or within the Beestonian and discussed above.

(iv) *Weathered horizons, rootlet beds and tree stumps*

The Forest-bed was so named because of the abundance of tree stumps and wood in sediments exposed on the foreshore of the Norfolk coast. There was controversy over whether the stumps were found *in situ* or were drifted to their positions in the sediment. Reid (1890) concluded that no tree stump had been recorded in its rooting position and that all the ones he observed had broken roots and often were drifted into estuarine sediments. For this reason he called the horizon Forest-bed (Estuarine). My own observations confirm those of Reid. Drifted tree stumps were recorded in Pastonian sediments at ML (Fig. 30) and in early Anglian sediments (*Taxus*, MAA, Fig. 32). In the former case, the wood drifted into shallow tidal waters, and in the latter, the stump was deposited in a fluviatile environment.

Logs of wood were recorded at Beeston (cf. *Pinus*, Pre-Pa a; *Pinus*, Pa IV), West Runton (Cr), Cromer (Pa), Sidestrand (SSV, Pa), Paston–Mundesley (Pa, eAn) Happisburgh (Pa) (Plate 34). The records are in both freshwater and marine sediments.

Rootlets have been observed at many sites in the past (Reid, 1890). Their distribution gave rise to the hypothesis (Blake, 1877, 1890) that the Rootlet Bed of Corton represented a horizon which could be widely traced on the Suffolk and Norfolk coasts.

Rootlets mark the position of former land surfaces populated by plants. They may occur in inorganic sediments below biogenic sediments, as at Corton, or they may occur in inorganic sediments eroded at the surface and sealed by sands and gravels (as at Overstrand). They may be nearly contemporaneous with the sediment they are in, or may penetrate much older sediments which formed a land surface at the time. They may be found on land surfaces formed by regression or on land surfaces where plant growth is enhanced by a rise in water-level (as at Corton).

Table 44 indicates the positions of horizons with rootlets recently observed. The sediments they penetrate can usually be dated but the age of the plant growth may be more difficult to determine. The Table indicates the uncertainties. The land surfaces involved in many of the sites were certainly subjected to plant growth in the Cromerian (Plates 2 and 3), others more doubtfully in the Cromerian and one in the late Pastonian. Blake's hypothesis is therefore largely supported by my observations.

The Cromerian land-surface with rootlets may be associated with weathered horizons, indicated by decalcification, presence of calcareous concretions, and oxidation. But these horizons may occur without the presence of rootlets. Horizons showing such weathering are included in Table 44. This table also includes the Cromerian palaeosol described by Valentine & Dalrymple (1975).

(v) *Conclusions*

Breaks in the succession form the substantial evidence for non-sequences. Rootlet and weathered horizons and a palaeosol are nearly all dated to the Cromerian, and occur within known sequences, implying that they are not indicative of important breaks in the known succession.

The evidence from breaks in the succession indicates non-sequences in the positions shown in Fig. 53. The most obvious breaks appear to be between the Pre-Pa a and b and between Pa and Cr temperate stages. Evidence of considerable faunal distinction between Pa and Cr supports the existence of the second of these breaks (*see p.* 114).

(c) **Flora**

In this section we summarise and further consider the evidence concerning floras already described in the previous chapters on the successions of the coastal areas. We shall first consider the floras of the cold and temperate stages, and then go on to discuss the occurrence of particular taxa and the relationship of the floras and the vegetational history to those of the earlier parts of the Pleistocene and Neogene in Britain and northwest Europe, and to those of the later parts of the Pleistocene in Britain.

(i) *Cold stage floras*

The cold stage floras are recorded in both marine and freshwater sediments. The pollen assemblages and sparse macroscopic records in marine sediments of Pre-Pa a age and earlier have been discussed in Chapter 2. In the Happisburgh borehole HC the high NAP assemblage of biozone c, Gramineae–Ericales p.a.b., is the oldest horizon in the coastal succession showing freshwater influences and clear indication of regional grassland with a damp ground element including *Montia fontana* ssp. *fontana* and *Selaginella selaginoides*.

Within the Pre-Pa a marine sediments two erratics of mud at Sidestrand (SSF2, SSJ) show high Gramineae assemblages, with the same relatively high representation of Compositae (mainly Compositae Liguliflorae) pollen seen in the Happisburgh biozone c.

The freshwater sequence at Beeston covers the Pre-Pa b–d period, with a Gramineae–*Artemisia* p.a.b., Gramineae–*Pinus–Betula–Alnus* p.a.b. and a Gramineae–Cyperaceae–*Betula–Pinus* p.a.b. The second of these biozones, in substage Pre-Pa c, has an increased AP frequency and may indicate a slight amelioration of climate of pollen-interstadial type. Pre-Pa d is also represented at East Runton by a Gramineae–Cyperaceae p.a.b., in sediment which gave a good macroscopic flora, mainly aquatic, but also with *Ranunculus hypoboreus*.

Cold-stage floras included in the Beestonian are recorded at Beeston in the early part of the stage and at West Runton in the late part of the stage. The earlier floras show the presence of herb-rich grassland with local low-shrub communities containing *Betula* and small-leaved *Salix* spp. The herb records include *Armeria maritima*, *Cherleria sedoides*, *Saxifraga oppositifolia* and *Thalictrum alpinum*.

The Late Beestonian sediments at West Runton (WRAQ) include two p.a.b.s, a Gramineae–Cyperaceae–*Artemisia* p.a.b. and a Gramineae–Cyperaceae–*Betula* p.a.b. There is a rich macroscopic flora associated with these pollen assemblages, mostly of aquatics or plants of damp ground or shallow water. *Ranunculus hypoboreus* is again present, with *Betula nana*.

Floras of the e An substage are recorded at Mundesley–Paston and Bacton. At sites in the Mundesley–Paston area a rich macroscopic flora is present with *Betula nana*, small-leaved *Salix* spp., *Armeria maritima*, *Oxyria digyna*, *Saxifraga oppositifolia* and *Selaginella selaginoides*.

These floras from the cold stages have the same characteristics that are so well known from floras of the last, Devensian, cold stage, and which have been described in detail by Bell (1969, 1970) and Godwin (1975). They include an arctic-alpine or arctic-montane element (*Betula nana*, *Salix* spp., *Arabis alpina*, *Oxyria digyna*, *Ranunculus hypberreus*, *Saxifraga oppositifolia*, *Thalictrum alpinum*, an alpine element (*Cherleria sedoides*), a maritime element (*Armeria maritima*, *Plantago maritima*), plants of dry and damp open ground, often now weeds or ruderals (e.g. *Artemisia*, *Atriplex patula*, *Linum anglicum*, *Cerastium arvense*, *C. holosteoides*, *Plantago major/media*, *Potentilla anserina*, *Selaginella selaginoides*), and a rich aquatic flora. It is evident that the characteristic cold stage flora of the British Pleistocene is firmly recorded in pre-Pastonian times and recurred in Beestonian and early Anglian times, before the deposition of the first glacigenic sediment (Cromer Tills) in the coastal areas. In the Beestonian and early Anglian, permafrost features seem associated with these floras, and we conclude that these periods share the characteristics of the periglacial floras and frozen ground phenomena well known from the Devensian periglacial areas.

(ii) *Temperate stage floras*

The vegetational history of the Pastonian and Cromerian temperate stages has already been summarised in Chapter 17. Pollen assemblage biozones have been grouped into substages I to IV for each temperate stage. A clear division on this basis is possible in both stages, except for the Pa II and III separation, the feasibility of which has already been discussed.

Table 45 shows the biozone sequences and grouping into substages. The development of the Cromerian sequence is very different from that of the Pastonian sequence. The Cromerian shows more structure of forest development in substages Cr II and III, with a biozone with *Tilia* developing in Cr II, and *Carpinus* and *Abies* biozones in Cr III. Both *Tilia* and *Abies* pollen frequencies may be high in the Cromerian, compared with their sporadic occurrence in the Pastonian (*see pp.* 108, 111). The behaviour of *Carpinus* is also very different in the two temperate stages. The pollen of this genus first occurs in quantity in Pa II, but in the Cromerian its appearance is delayed till after the development of the mixed oak forest of Cr II. On the other hand the biozones of substage IV are similar in the two temperate stages, especially in IVa, where *Pinus–Alnus–Picea–Betula* biozones occur in both the Pastonian and Cromerian. The assemblage is similar to that found derived at the present time from the southern part of the boreal coniferous woodland in eastern Fennoscandia, and it is perhaps not surprising that towards the end of a temperate stage restriction of thermophilous tree taxa occurs with hardy tree taxa surviving to form coniferous woodland.

The extent to which taphonomy has contributed to the segregation of the biozones is not certainly known. The Pastonian sediments, apart from Pa I and late Pa IV sediments, are mainly marine and were deposited in various facies of a tidal estuary. This may contribute to the difficulties of Pa II/III zonation. The Cromerian sequence is largely freshwater in Cr I, Cr II, early Cr III and late Cr IV, while late Cr III and early Cr IV are marine. This zonation is so clear that it seems it is a true reflection of regional vegetation history regardless of taphonomy. We note, however, that in the Corton Cromerian pollen diagrams both *Abies* and *Carpinus* occur sparsely in Cr II pollen assemblage biozones. This area must have had a very large pollen catchment, larger than in Cr II times at West Runton and Paston–Mundesley, because of the alluvial situation, and it seems probable that these genera were already present at this time in the catchment, though they did not increase their frequency until Cr III times.

(iii) *Exotic taxa*

Wilson (1973) discussed the size of the exotic element in the flora of the pre-Anglian sediments of the coast and concluded that out of 281 species and 31 'extra' genera (i.e. identifications to genera not already represented in the species list), about 8% were exotic to the present British Flora. Table 46 lists exotic taxa. The distribution of certain of these taxa in the various stages is discussed in the following sections.

(iv) *Notes on taxa*

In this section we discuss identification of certain pollen taxa and the possible significance of the presence of very low frequencies of certain pollen taxa, and we compare the behaviour of some taxa in the Pastonian and Cromerian temperate stages with their behaviour in later temperate stages.

Algae

Dinophyceae. The frequency of the genera *Operculodinium*, *Tectatodinium* and *Leptodinium* is shown in the pollen diagrams. The identification and morphology of the species in these genera in the early Pleistocene marine sediments of East Anglia is described by Wall & Dale (1968). Apart from the sparse records of these genera shown in the diagrams, *Operculodinium* and *Tectatodinium* show more substantial frequencies at certain levels, as follows.

Operculodinium. This taxon in the diagrams is *O. israelianum*. Considerable frequencies occur in Pre-Pa a (Sidestrand SSF; Pakefield PK, PL), Pa II (West Runton WRM, WRAC, WRF; Pakefield PL) and Pa IVa (Sidestrand SSP). Very high frequencies were observed in crag with abundant *Macoma balthica* at West Runton (WRAS), but the age is doubtful; it may be Pa I. In the Ludham borehole *O. israelianum* is recorded throughout the sequence, but sparsely except in Lu 2 (Thurnian). Wall & Dale (1968) suggested an abundance of this taxon may indicate cool conditions. In the present study it was found, not as frequently as in Lu 2, in the cold substage Pre-Pa a and in more temperate conditions of Pa II. Its presence at these times is not, however, consistent, indicating a complex of factors determining its presence in these marine sediments.

Tectatodinium. This taxon has much the same distribution as *Operculodinium*. It occurs in Pre-Pa a (Sidestrand SSF, SSV 14; Mundesley MAT; Pakefield PK, PL), Pa II (West Runton WRF; Beeston BQ; Paston–Mundesley BHA 2; Pakefield PL) and Pa IVa (Sidestrand SSP). It also occurred abundantly in the Ludham borehole in Lu 4c (≡ Pre-Pastonian a) and Lu 5 (≡ Pastonian), as well as Lu 2.

Pteridophytes

Azolla filiculoides. The megaspores and massulae of *A. filiculoides* occur abundantly in freshwater deposits of Pastonian and Cromerian age and, less frequently, in tidal and other marine sediments of these stages. Their identification has been discussed by Wilson (1973). When they occur abundantly (as in the drift mud intercalated in tidal sediments at Sidestrand, SSTB, Pa II, and in the West Runton Freshwater Bed, Cr II), they occur as clustered masses of megaspores with massulae attached, up to 2 mm across. The massulae were counted or noted in the pollen analyses, with results shown in the pollen diagrams. The Tables show the numbers extracted during the macroscopic analyses. In some samples (e.g. Pre-Pa c, Beeston BR–S 140–155) the presence of *Azolla* megaspores with or without massulae attached occur at levels attributed to cold stages. The presence of these remains in marine sediments indicate that they can withstand considerable transport and the possibility of secondary deposition should be borne in mind. Thus Iversen (1936) found *Azolla* massulae with glochidia in the oldest Danish till, 'Moraine A' at Esjberg.

Osmunda. Annuli and spores of this genus are recorded most commonly in the temperate stages. The identification of the spores is not straightforward since the range of variation shown by reference material of the species is considerable, especially in *O. claytoniana* L. Identification has been discussed by Turner (1966, 1970) and Zagwijn (1960). In the pollen analyses, three taxa have been segregated.

Osmunda. The spores fall into classes intermediate between the next two taxa.

Osmunda regalis L. Spores thick-walled, sculpture elements fused with a coarse reticulate or rugulate pattern.

Osmunda claytoniana-type. Spores thin-walled, blunt peg-like sculptural elements, separated from one another. This taxon includes *O. claytoniana* and *O. javanica* Blume. The sculptural elements tend to differ from those of *O. cinnamomea* L., which are usually more pointed and tapering (Turner, 1966). It seems probable that this spore taxon originates from *O. claytoniana*, a species of temperate eastern North America and (a different variety) east and south central Asia, rather than *O. javanica* which has a tropical distribution in eastern Asia.

The spores of *O. claytoniana*-type are most common in Pa II, especially in sediments associated with tidal conditions (Sidestrand SSE, Happisburgh HC, HF), where they accompany lower frequencies of spores of *O. regalis*. This combination parallels the present-day occurrence of *O. regalis* and *O. cinnamomea* in coastal swamps of eastern North America (e.g. see Martin 1959). *O. claytoniana*-type spores occur less frequently in Pa I, Pa III and Pa IV, rarely in Pre-Pa a and also rarely in Cr III. *Osmunda regalis* occurs rarely or with low frequencies in Pre-Pa a, Pa I, Pa II, Pa IV, and in Cr II, III and IV. *Osmunda* spores occur throughout the succession, most frequent in the temperate stages, but also with a low representation in Pre-Pa c.

Gymnosperms

Abies. In the Pastonian pollen of *Abies* occurs sparsely, usually single grains in a particular analysis. It is consistently present in this way, however, in some of the Pastonian sections: Skelding Hill (Pa Ib, WP), Beeston (Pa II, BQ, BV; Pa III, BKA; Pa IVa, BR–S), Overstrand (Pa IVb, BHO 3B), Sidestrand (Pa IVa, SSE, SSU, SSTB) and Bacton (Pa II, BI). Again it is difficult to decide whether the tree was present in the Pastonian or whether the pollen is secondary. Pollen of *Abies* is very scarce in the temperate stages of the Ludham borehole (West, 1961), but the tree is a low pollen producer and the pollen grain is substantial and preserves well. In view of the records from Sidestrand, it seems that *Abies* may have been present in Pa IVa, but such a conclusion is very tentative.

In the Cromerian *Abies* pollen frequencies may rise above 10% in Cr IIIb, but usually the frequency is somewhat lower. There are sparse records in Cr IIIa (*Carpinus* biozone), and in the marine sediments of the Corton borehole 2/6, in Cr IIb, so it appears that the tree expanded its frequency at the beginning of Cr IIIb, being already present though rare. The pollen frequencies fall away to rare records in Cr IVa. This oligocratic behaviour of *Abies* in the Cromerian is similar to its behaviour in the Hoxnian.

Cupressaceae. Inaperturate split pollen grains without a ligule were recorded in many of the pollen analyses. Two types were recognised. A small type, corresponding to *Juniperus*-type, size *c.* 22 μm, was recognised at certain sites, including l Be at West Runton (WRAQ), Cr Ia at West Runton (WRAQ, WRCGB, WRCGC), and Cr Ib at West Runton (WRCGB).

A larger type, size 30 μm, thin-walled with scattered orbicules, was recorded in the Pastonian at Sidestrand (SSV, Pa II; SSP, Pa IVa), labelled Cupressaceae b in the pollen diagrams. In the reference collections grains of similar size and structure were found in *Chamaecyparis thyoides* (L.) B.S.P. and *Thuja orientalis* L. If the fossil is one of these two, the former seems more likely, since it is associated with lowland coastal mires in eastern North America, while the latter is a tree of some altitude in north China at present. A present-day associate of *C. thyoides*, *Osmunda claytoniana*, is recorded abundantly in the Pastonian.

Macroscopic remains of *Chamaecyparis thyoides* (L.) B.S.P. were recorded by Huckerby & Oldfield (1976) from Early Pleistocene (Bidartian) sediments of the French Pays Basque, together with Cupressaceae pollen and high *Alnus* pollen frequencies. A nearby similar palaeoecological situation to that at Sidestrand is suggested. *Chamaecyparis* macroscopic remains identified as *C. pisifera* Endl. have been recorded by Szafer (1954) in the Pliocene and Lower Pleistocene of southern Poland, and *T. orientalis* was also recorded by him in the Pliocene. A search was made in the Sidestrand sediments for Cupressaceae macroscopic remains, but none were found. In view of the association of the Cupressaceae b pollen curve with marine sediments, and curves for *Alnus* pollen and *Osmunda claytoniana* spores, it is tentatively concluded that this type of grain is probably *C. thyoides*.

In the pollen diagrams Cupressaceae pollen is also recorded at several other sites. If the records are labelled Cupressaceae, measurements were not made. There are such records for Pre-Pa a, Pa I, II, III and IV, Be, Cr I, III, IV and e An.

Ephedra. Pollen grains of *Ephedra* were recorded from three analyses, one (*E. distachya*-type) in Cr Ib at West Runton (WRCGB), two grains in a Beestonian high NAP spectrum at Beeston (BM 2) and one (*E. distachya*-type) in the Happisburgh borehole in a high NAP spectrum of Pre-Pa b/d age.

Picea. Two sizes of *Picea* pollen grains have been distinguished in Pliocene and Pleistocene pollen assemblages (Zagwijn 1960; Andersen, 1961). The smaller, with a size similar to the grain of *Pinus sylvestris*, has been referred to *P. omoricoides*, a fossil form related to *P. omorika*. The larger *Picea* type, with a length of longest axis averaging *c.* 120 μm, is referred to *P. abies*. Macroscopic remains of *P. abies* have been recorded in Pa IVa at Beeston and Mundesley and in Cr IVa at Ostend. One leaf of cf. *P. omorika* was recorded at Beeston (BR-S) in Pa II/III (Wilson, 1973).

Occasional small grains of *Picea* occurred in the pollen analyses, but the majority of grains were of *P. abies* type, and a clear separation to prove the presence of *P. omoricoides* pollen was not undertaken.

The behaviour of the *Picea* pollen curves is similar in the Pastonian and Cromerian. In the Pastonian, as seen in the Beeston pollen diagrams, substantial frequencies occur in Pa I, continuing through Pa II and III, rising in Pa IVa and b, and falling away in Pa IVc. In the Cromerian, as seen at West Runton, *Picea* pollen frequencies rise in Cr Ib, and continue strongly through Cr II and Cr III, reaching high values (*c.* 15%) in Cr IVa. The frequencies are reduced in Cr IVb and c. *Picea* (*abies*) must have been present through the temperate stages. The same seems to have occurred in the Hoxnian temperate stage in East Anglia (West, 1956; Turner, 1970), though in this stage the pollen frequencies are much lower, especially at the beginning of the stage. This behaviour contrasts with that in the Ipswichian, where the presence of *Picea* has been a subject of discussion (Phillips, 1974); no Ipswichian macroscopic remains have been found (as they have in the Hoxnian, Cromerian and Pastonian), but the low pollen frequencies recorded at some sites may indicate the presence of *Picea*. *Picea abies* is not present in Britain during the Flandrian temperate stage. These changes in the behaviour of *Picea* clearly indicate the decreasing ability of *Picea* to re-immigrate to Britain in the temperate stages in the course of the Pleistocene, possibly related to the disappearance of cold stage refugia in western Europe.

Taxodiaceae; *Sciadopitys*. A ligulate grain of *Sequoia*-type (Zagwijn, 1960) was recorded near the base of the Happisburgh borehole, site HC, in Lu 1/2? A grain of *Sciadopitys* was recorded on the near-tidal sediments, Cr IIIa, of the channel filling at Pakefield, site PC.

Taxus. See entry below under *Ilex*, *Hedera* and *Taxus*.

Tsuga. The two types of *Tsuga* pollen, *T. canadensis*-type and *T. diversifolia*-type, described by Rudolph (1935) and Stuchlik (1964), are recorded sporadically in the pollen analyses. Both types have been recorded much more abundantly in the Ludham borehole in the Ludhamian and Antian temperate stages (West, 1961), and in the Antian at Easton Bavents (Funnell & West, 1962). In both these temperate stages *T. canadensis*-type is far more abundant than *T. diversifolia*-type. In the few pollen analyses from the Coralline Crag, only sparse *T. diversifolia*-type has been recorded (Andrew & West, 1977).

In the present analyses, 23 of the samples yielded a total of 39 *Tsuga* pollen grains, 12 *Tsuga*, 17 *T. canadensis*-type and 10 *T. diversifolia*-type. All the records are from marine (or brackish) inorganic sediments where secondary derivation is likely to play a part in pollen input. If we discount as certainly secondary records from Lu 4b at Long Covert (Covehithe) and Pre-Pa (Pre-Pa a: WRQ, SSB, MAT; Pre-Pa c: BR-S), *Tsuga* pollen is also recorded in Pastonian and Cromerian sediments, as well as in Lu 1/2? in the Happisburgh HC borehole. The highest percentage, 4%, was found in marine silty clay ascribed to Pa III at West Runton, site WRAA, where six grains of *T. canadensis*-type and one of *T. diversifolia*-type were recorded. This abundance in an otherwise unexceptionable Pastonian pollen spectrum would seem to indicate the presence of *Tsuga* in the Pastonian. At Happisburgh, in the long Pastonian pollen sequence of the borehole HC, sparse *Tsuga canadensis*-type is recorded in Pa III and Pa IVa in three successive samples. At Happisburgh, site HF, two grains of *T. diversifolia*-type are recorded in Pa II, and at Sidestrand (SSTB), Beeston (BKA) and in the borehole BHB 1 of Paston–Mundesley, *Tsuga* is recorded in Pa IVa and b. A single grain of *T. canadensis*-type is recorded from Pa Ib at Sidestrand (SSV). It therefore seems likely that *Tsuga* (with taxa yielding one or the other pollen type) was present in the Pastonian, but infrequently, with *T. canadensis*-type pollen more abundant than *T. diversifolia*-type, as with the earlier Pleistocene records.

The Cromerian records are single grains from Cr II and IIIa at Mundesley (MAK and MA respectively) and seven grains (0·5%) of *T. diversifolia*-type from three levels of Cr IIb marine sediments at depth at Corton (BH 2/6). These grains are likely to be secondary, particularly in view of the good record of Cromerian vegetational history from the coastal sections.

Angiosperms

Carpinus. In the Pastonian, *Carpinus* pollen frequencies rise, with *Quercus* and *Ulmus*, near the beginning of Pa II (see Beeston pollen diagram BQ, Fig. 20) and remain at levels around 5–10% through Pa II and Pa III. The highest observed frequency was 19% at West Runton (Pa III, WRAA). In Pa IV frequencies fall to very low values. In the Cromerian, *Carpinus* frequencies do not rise until near the beginning of Cr IIIa (*Carpinus* biozone) (see Corton pollen diagram CK, Fig. 45), and values of up to 13% are reached in Cr III. Nuts of *Carpinus betulus* were recorded in Cr IIIa sediments at Pakefield, where the pollen frequency reached 12% at the most. In Cr IIIb (*Abies–Carpinus* biozone), *Carpinus* frequencies remain similar to those of Cr IIIa, but they fall to low values in Cr IVa.

Carpinus thus behaves as a mesocratic (*sensu* Iversen, 1958) in the Pastonian and as an oligocratic taxon (*sensu* Andersen, 1966) in the Cromerian, as it is in the Hoxnian, Ipswichian and Flandrian temperate stages. This change of behaviour is discussed by West (1970), and may be interpreted as a result of reduction in biotypic variation associated with climatic changes and enforced migrations or of difficulties and delays in immigration related to the position of cold-stage refuges.

Carya. A very few grains of *Carya* were recorded: one in marine sediments of Pa III at West Runton, site WRAA, accompanied by high frequencies of *Carpinus* and the highest frequency of *Tsuga* (4%) found in the analyses, and one in marine sediments of Cr IIIb at Sidestrand, site SSR. This low frequency makes it impossible to decide whether or not the presence of *Carya* pollen is secondary. That they are secondary seems likely in view of the low frequency in inorganic sediment.

Corylus. The pollen frequencies of *Corylus* remain low in the Pastonian and Cromerian temperate stage, a contrast with the much higher frequencies seen in the post-Anglian temperate stages. Frequencies rarely rise above 10% and are usually within the 2–7% range. Macroscopic remains of *C. avellana* are scarce. Reid (1890) observed that they were rare and only recorded at Ostend and Pakefield. In the present investigation nuts of cf. *C. avellana* were recorded from Ostend in Cr IVa.

The low *Corylus* frequencies become firmly established in Pa II (see Beeston pollen diagram BQ, Fig. 20) and Cr II (see West Runton pollen diagrams WRAQ, WRU, Fig. 13), accompanying the establishment of the temperate forest of Pa II and Cr II. In the Pastonian, the rise of *Corylus* at BQ appears to be contemporary with that of *Ulmus* and *Quercus*, but in the Cromerian at West Runton it is later than *Ulmus* and *Quercus*. The highest frequencies, exceptionally above 10% and never above 15%, are recorded in Pa II and Cr II, with lower frequencies continuing into Pa III/IV and Cr III/IV.

Corylus thus appears to behave as part of the temperate forest in both the Pastonian and Cromerian temperate stages, and appears to have a much less important role in the vegetation than in the later temperate stages.

Ericales. The pollen taxa recorded in this order are *Calluna*, *Empetrum*, *E. hermaphroditum*-type, *Bruckenthalia* and Ericales.

Empetrum includes all *Empetrum* tetrads including determinable crumpled tetrads. In some analyses *E. hermaphroditum*-type was separated on a basis of size, to include tetrads in the size range 30–42 μm, with *E. nigrum* including tetrads in the size range 25–29 μm. This division was based on measurements of tetrad sizes in the Cambridge reference collection, which gave averages of 27 μm for *E. nigrum* and 36 μm for *E. hermaphroditum*. *E. hermaphroditum*-type was recorded in Pa IVa wood peat at Beeston (BR–S) and, more sparsely, in Cr IIb and Cr IIIb sediments at Corton (CK) and Pakefield (PC). The highest frequency was recorded at Beeston (BR–S, 20 cm) where 39% of *Empetrum* tetrads could be referred to *E. hermaphroditum*-type.

Pollen grains referable to the ericaceous *Bruckenthalia spiculifolia* (*see* Andersen, 1973) were recorded in Pa IVa wood peat at Beeston (site BR–S, one grain at 20 cm, one at 30 cm), and in Cr IIIb and IVa in tidal sediments at Sidestrand (site SSV, four grains in sample 6, two in sample 8). The species, though poorly represented, thus appears to have an oligocratic (*sensu* Andersen, 1966) distribution in the two temperate stages, according with its present calcifuge habit in montane woods and subalpine pastures in the Balkans.

Ericales includes Ericaceae tetrads and crumpled indeterminable Ericales tetrads.

The interpretation of high frequencies of Ericales in Pre-Pa a and Ba marine sediments is discussed in Chapter 1.

Eucommia. Grains of *Eucommia* pollen were recorded at levels 3b and 13 of the Sidestrand marine sediments at SSV. The pollen grain is tricolpate, with a large polar area and boat-shaped colpi, and has been described and figured by Zagwijn (1960). The monotypic family Eucommiaceae contains the central Chinese species *Eucommia ulmoides* Oliv. Macroscopic remains of *Eucommia* occurring in the late Tertiary and Early Pleistocene of Europe have been referred to *E. europaea* Mädler, but Tralau (1963) considers the fossil to be referable to *E. ulmoides*. The pollen of *Eucommia* occurs in Europe in the Neogene and in Early Pleistocene temperate stages.

Ilex, *Hedera*, *Taxus*. Pollen of these taxa was rare. The distribution of pollen analyses with records for them is shown in Table 47. *Ilex* pollen is most frequent in Pa II and Cr II. A seed of *Ilex aquifolium* is recorded from Cr II in the West Runton Freshwater Bed. *Hedera* pollen is most frequent in Pa II and Cr II to IVa. *Taxus* pollen has a similar distribution. Both *Hedera* and *Taxus* continue well (by this standard of infrequency) into the late Cromerian, with a leaf and fruit of *Taxus baccata* at Ostend in Cr IVa. At Mundesley, a pollen record of *Taxus* occurs in the e An, together with, at MAA, a stool of *Taxus*, clearly not *in situ*.

Polemonium. A single grain of *Polemonium* was recorded in Pa IVa at Sidestrand, site SSE.

Pterocarya. 19 (8%) grains of *Pterocarya* were recorded in marine sediment ascribed to Lu 1/2? at the base of the Happisburgh borehole HC. Apart from this, single grains were recorded near the base of the Mundesley borehole MAT in marine sediment of Pre-Pa a age, in Pa IVa marine sediment of two localities at Sidestrand (SSP, SSE (two levels)) and in Pa IVb marine sediment in the Mundesley borehole BHB 1. Although few grains were found in temperate stage deposits, these are all from Pa IV, and it is possible that *Pterocarya* may have been present in the latest part of the Pastonian. We may compare this occurrence with its low frequency in the later part of the Hoxnian interglacial stage (Ho IIIb, IVa) at Marks Tey (Turner, 1970). On the continent, *Pterocarya* pollen and nuts of *P. fraxinifolia* have been recorded from Holsteinian temperate stage deposits (Turner, 1975).

Tilia. The pollen type recorded is *T. cordata*. There are infrequent and sporadic records of pollen grains, usually single grains, from the Pastonian at Beeston (Pre-Pa c,

BR-S; Pa Ib, BTB, BU; Pa II, BG, BU; Pa IVa, BR-S), Trimingham (Pa Ib, BHT 16) and Mundesley (Pa Ib, BHB 1; Pa II, MU, MW). It is difficult to determine whether *Tilia* is a rare member of the Pastonian flora or whether the pollen is of secondary origin. There was no record of *Tilia* in the temperate stages of the Ludham borehole (West, 1961), which may indicate poor possibility of reworking into the Pastonian. *Tilia* is also a low pollen producer, so it seems possible that *Tilia* may have been a member of the Pastonian flora.

The *Tilia* pollen frequencies of the Cromerian are far more substantial, reaching a maximum of around 5% in Cr IIb (see West Runton pollen diagram WRU, Fig. 13). The frequencies continue but with lower values in Cr IIIa, but fall away in Cr IIIb (*Abies–Carpinus* biozone), with only sporadic occurrences in Cr IV.

Monocolpate X. A monocolpate pollen grain, named Monocolpate X in the pollen diagrams, was recorded in frequencies up to 7% in the Pa II tidal sediments and detritus muds associated with tidal sediments at Happisburgh (HC, HE, HF). The grain is of *c.* 40 μm length (polar axis) and *c.* 25 μm width. It is tectate, with columellae clearly visible, the tectum having perforations < 1 μm in diameter. Each perforation is surrounded by a ring of columellae, so that the 'muri' are double in appearance.

The morphology of this grain could not be matched in the Cambridge pollen reference collection of taxa in the families Liliaceae, Amaryllidaceae or Iridaceae. An examination was made of pollen of various species of palm, including species of temperate regions in south-east North America. The species examined included *Chamaerops humilis* L., *C. serrulata* Michx., *Paurotis cookii* (Bartl.) Standl & Steyerm., *Roystonia regia* (L.) Cook, *Sabal adansonii* Guerns., *S. etonia* Swingle, *S. minor* (Jacq.) Persoon, *S. palmetto* (Walt.) Todd, *Serenoa repens* (Bartr.) Small and *S. arborescens* Sar. The pollen of *Sabal adansonii* and *S. etonia* showed a similar size, structure and sculpture to the fossil. These are both species of south-eastern United States. Of these two, *S. adansonii* (S. Carolina, Florida, Louisiana) is found on low ground along streams, an ecological situation possible in the sediment context, while *S. etonia* favours sandy scrubland. It is possible that the Monocolpate X grain is *S. adansonii* or a closely related taxon, associated with the *Alnus–Osmunda* communities described in Chapter 12, *p.* 80.

(v) *Phytogeographical remarks*

The bulk of the flora recorded in the pollen and macroscopic analyses is native to the present British flora. The arctic–alpine and alpine elements present in the cold stages are, as already discussed, elements in the flora which recur in the later Pleistocene cold stages. The temperate-stage floras are fundamentally floristically similar to those of later temperate stages. The Pastonian flora, however, is enriched by the presence of the East Asian or East Asian–North American taxa *Eucommia*, *Tsuga*, *Osmunda claytoniana*(-type), of the New World species *Azolla filiculoides*, of the boreal circumpolar species *Chamaedaphne calyculata* and *Swertia perennis*, of the oceanic west European species *Erica lusitanica*, *E. umbellata*, of the Balkan species *Bruckenthalia spiculifolia* and (cf.) *Picea omorika* and the European or western Asian taxa *Salvinia natans*, *Pterocarya* (*fraxinifolia*?), *Najas minor*, *N. tenuissima*, *Abies alba* and *Picea abies*.

The Cromerian shows a less rich exotic element. *Azolla filiculoides*, *Salvinia natans*, *Osmunda claytoniana* (-type), *Bruckenthalia spiculifolia*, *Najas minor*, *N. tenuissima*, *Abies alba* and *Picea abies* are present. The oceanic west European *Corema album* is also present with the boreal circumpolar *Alnus incana*, the temperate circumpolar (cf.) *Staphylea*, and the continental European species *Trapa natans*.

The presence of these taxa in the Pastonian and Cromerian floras show the former expanded ranges of many mainland European, Eurasian and North American taxa. Extinctions in Britain have thus occurred across a wide range of plants of various present-day phytogeographical distributions.

(vi) *Relation to earlier and later floras and vegetational history*

The Pastonian and Cromerian pollen assemblages differ from those of earlier Pleistocene temperate stages (Antian and Ludhamian (West, 1961)) in the scarcity of pollen of *Tsuga* and *Pterocarya*. Very low pollen frequencies of these pollen taxa occur mainly in the Pastonian, as discussed in section (iv) above. It was there concluded that, although there is a possibility that such low frequencies might result from secondary derivation in these mainly minerogenic sediments, there was a likelihood of the presence of *Tsuga* and *Pterocarya* in the Pastonian. *Eucommia* also occurs in the Pastonian, with substantial frequencies of Cupressaceae pollen. These properties of the Pastonian pollen assemblages ally them with the floras of the temperate stages of the

Early Pleistocene, even though they have a very restricted representation of the 'Tertiary-relic' taxa much more abundantly present in the Antian, Ludhamian and Coralline Crag (Pliocene) floras of East Anglia (West, 1961; Andrew & West, 1977).

In the Cromerian this component of the flora is further reduced. *Tsuga*, *Pterocarya* and *Eucommia* appear to be absent and Cupressaceae b pollen taxon has not been recorded. The *Juniperus* type of pollen is found in the preceding late-glacial, and in the early and late parts of the temperate stage, where it appears only sparsely.

The Pastonian and Cromerian share a feature not seen in the later temperate stages. This is the substantial representation of *Picea* pollen from the beginning of the temperate stages. Otherwise, the Cromerian is more closely allied in structure to the later interglacial stages, as discussed below.

These trends in the floras of the temperate stages of the Early and Middle Pleistocene are the result of restriction of range of the taxa concerned in north-west Europe, induced by climatic fluctuation and the concomitant problems of migration and of survival in refugia, with the number and extent of refugia important in determining the success of recolonisation and richness of biotype content (West, 1970).

When we come to compare the Cromerian biozone sequence with that of the post-Anglian temperate stage, the Hoxnian, we find that it has certain similarities, most noticeably in the following of the Cr II and Ho II biozones with *Quercus*, *Ulmus* and *Tilia* by Cr III and Ho III biozones with *Carpinus* and *Abies*. However, there are also considerable differences between the two. The pollen curves for *Ulmus*, *Tilia* and *Picea* are much more substantial and the curve for *Corylus* much less substantial in the Cromerian than in the Hoxnian of East Anglia. *Pterocarya* pollen, characteristic in low frequency of Ho IIIb/IVa, has not been recorded in the Cromerian; neither has the Type X tricolpate pollen grain of the Hoxnian (Turner, 1970). A *Hippophaë* biozone such as is found in the late-glacial preceding the Hoxnian is not found in the late-glacial preceding the Cromerian. Thus though at first sight the Cromerian biozone sequence has similarities with that of the Hoxnian, there is enough substantial difference between them to distinguish the two in East Anglia, providing good sequences of vegetational history are available for comparison.

(d) Fauna

(i) *Marine molluscs*

Assemblages of marine molluscs from particular horizons in the succession described here have been described by P. E. P. Norton (1967 and Appendix 2). Other assemblages have been noted by Reid (1882*a*, *b*) and Markham (1973). The following summarises the placing of particular assemblages in the succession recognised here. New analyses, with ecological comments, are listed in the Appendix 2 by P. E. P. Norton, together with further details of the provenance of all the assemblages.

Cr IIIb
Leda myalis Bed at West Runton (Reid, 1882).

Be
Bed i (shelly sand) of the pre-freshwater bed succession at West Runton, site WRCO.

Pa II
Bed b (shelly sand and drift mud), site SSE at Sidestrand (Norton, 1967).

Pa Ib
Shelly sand in Trimingham borehole BHT 16.

Pa/Pre-Pa a
Bed c (shelly sand) of pre-freshwater bed succession at West Runton, sites WRF, WRAT.
Bed b (shelly sand) in Overstrand boreholes (Markham, 1973).

Pre-Pa a
Bed b (shelly sand), Sidestrand, sites SSB, SSK (Norton, 1967).
Bed b (laminated silty clay), Mundesley borehole MAT.

Earlier than Pre-Pa a
Bed c (silty clay), biozone c, Happisburgh borehole HC.
Bed a (sand) biozone a (Lu 1/2 ?), Happisburgh borehole HC.

All these assemblages, except the scanty records from Happisburgh of pre-Pre-Pa a age, contain *Macoma balthica*. This species was considered by Reid (1890) and others to be the characteristic mollusc of the Weybourne Crag, absent from the Norwich Crag. The Weybourne Crag was therefore separated from the Norwich Crag and thought to be younger in age. The present evidence shows that assemblages with *M. balthica* are found in marine horizons from Pre-Pa a times onward to Cr IIIb. The Weybourne Crag as defined by Reid therefore has a wide time range, and it may be best to discontinue use of the term. For the Pre-Pa a sediments of crag type, the term Sidestrand Sands is suggested (Table 43).

The problem of the relative age of the Norwich Crag and *M. balthica* faunas is discussed in a later section (section (f)).

While *M. balthica* first appears in the Pre-Pa a in East Anglia, it does not occur in the Netherlands until Holsteinian (≡ Hoxnian) times (Spaink & Norton, 1967; Norton & Spaink, 1973). Since a strong regression occurs in Tiglian zone TC 4c and continental sediments are then only present until marine sediments are again found in late 'Cromerian complex' and later interglacial times (Zagwijn & Staalduinen, 1975), it is probable that Pre-Pa a is younger than TC 4c. This point will be considered further when correlations with the Netherlands are discussed in section (g) below.

(ii) *Non-marine Mollusca, by B. W. Sparks*

The non-marine Mollusca are not as illuminating as in some interglacial sections, being mostly restricted to the West Runton Frehswater Bed, where they are fully interglacial and dominantly freshwater in type. However, certain typical forms are present, *Valvata goldfussiana*, *Bithynia inflata*, *Marstoniopsis scholtzi*, *Pisidium clessini*, *Vallonia enniensis*, and *Vitrina semilimax:* these are not confined to the Cromerian but together they are indicative of this stage.

(iii) *Vertebrates*

The pre-glacial sequence is noted for the richness of its vertebrate fauna. Early descriptions of the fauna were provided by Newton (1882, 1891), but the ascription of many records to particular named horizons proved difficult because the stratigraphical sequence was complex.

In recent years, however, the fauna has been reinvestigated and re-described, with special attention paid to the stratigraphical context where this has been possible. The distribution of vertebrate taxa in time through the pre-glacial sequence and the problems of the taphonomy of the fossils have been described and discussed by Stuart (1974).

Stuart (1975) has newly described the fauna of the type Cromerian from Reid's Upper Freshwater Bed at West Runton (central channel, site AJS, Fig. 6), and related it to the vegetational history of the Cromerian. The fauna includes aquatic, amphibious and terrestrial elements, the last including woodland and grassland taxa. The assemblage appears to have accumulated largely from the remains of terrestrial vertebrates washed in as the channel eroded its banks. The fauna is distinctly different from faunas from older parts of the Cromer Forest-bed Formation in the absence of 'Villafranchian' taxa: *Mimomys pliocaenicus*, *Euctenoceros tetraceros*, *Dama nestii*, *Libralces gallicus* and 'zebrine' horse. At Ostend *Arvicola* has been recorded from sediments thought to be Cr IV in age (Stuart & West, 1976), replacing *Mimomys* (*savini*) of Cr II sediments at West Runton.

Faunas of pre-Cromerian age are much less well-known. The sediments are less well-exposed and opportunities for collecting are limited. But much of the material collected from foreshore sites and not tied to stratigraphy probably is of pre-Cromerian age. Since the Pastonian clay conglomerates and their assorted marine sediments (Forest-bed (Estuarine) of Reid) occupies much of the coastal foreshore from West Runton to Happisburgh, it is probable that much of the foreshore fauna is Pastonian, though an origin from Pre-Pastonian sediments is also possible in the Sheringham to Trimingham area. The vertebrate fauna of the Beestonian is unknown, except for a *Bison* metatarsal obtained from the gravels of bed i (pre-West Runton Freshwater Bed) 50 m west of site WRAZ (Fig. 11) at West Runton, and an elephant jaw at Beeston, site BZ, at the junction between beds k and o (Fig. 17).

Azzaroli (1953) described the deer fauna of the Weybourne Crag and Forest Bed of Norfolk and noted an abundance of species occurring at different horizons in continental Europe. He concluded that three faunas could be distinguished, and related them to the stratigraphy of Reid. The earliest was a fauna of upper Villafranchian type occurring in the Weybourne Crag, principally at East Runton; the second was a fauna of 'Mosbach–Mauer–Sussenborn' type occurring in the Forest-bed (Estuarine), and the third was a younger fauna occurring in Reid's Upper Freshwater Bed. The age of the third fauna is Cromerian and Azzaroli's evidence of this has now been enlarged by Stuart. The age of the earlier faunas is problematical. From the point of view of provenance, these pre-Cromerian faunas might be associated with the land-surface of the stone bed antedating Pre-Pa a, the Pre-Pa a crag, the land surface associated with the regression after Pre-Pa a, Pastonian sediments, a land surface related to the late Pastonian regression, or to reworked Pastonian or Beestonian remains associated with this land surface.

Much further work is required on the relation of the vertebrate finds and records to the details of pre-Cromerian stratigraphy. But it should be possible using the present stratigraphical framework to provide more accurate dating of the Villafranchian elements of the pre-glacial fauna. The East Runton area especially requires re-investigation, since the pre-Cromerian fauna is best known from this area. Present evidence suggests foreshore sediments in this area are mainly Pastonian, and it is probable that the faunal remains at East Runton are mostly of this age.

(e) Relation to the pre-glacial sequences described by previous authors

The 'Cromer Forest-bed' was studied in the last century by many workers, both amateur and professional. There was a lack of agreement on the details of the succession and its relation to the Crags. Since little original work on the stratigraphy of the pre-glacial sequence has been reported since the beginning of this century, we can refer to H. B. Woodward's (1891) summary of the state of affairs in 1890. He compares, as in Table 48, classifications by J. Gunn (1891), J. Prestwich (1890) and C. Reid (1882*a*, *b*; 1890). These classifications have a general similarity. Gunn's is perhaps the nearest to the stratigraphy now recognised. Reid's classification is the simplest and it became the accepted one, while Prestwich's is rather complex and became neglected.

The sediments which Reid included in his units are now known to have diverse ages, and it may be useful to note the allocation of sediments described by him and by Gunn to the stratigraphical sequence now known. The complexity of Prestwich's units makes a brief comment on them difficult. The equivalence of his units to those of C. Reid, as he gave them, are shown in Table 48, while the stratigraphy figured in his sections can usually be correlated easily with the sequences more recently exposed, so that a further discussion of his stratigraphy is not considered necessary.

(i) *J. Gunn*

The relation of Gunn's units (Table 48) to the sequence now known is as follows, starting at the base of the succession.

Estuarine soil: Blue clay with stools of trees and indurated ferrugineous gravel (Elephant Bed). This unit contains the upper part of the Pastonian marine sediments, with gravel and clay conglomerate, as seen in the Mundesley–Bacton area.

True Forest Bed. This included the horizon with tree trunks at Happisburgh, Pa II in age, as well as other horizons elsewhere with tree remains and of unknown age. *Reconstructed Forest Bed.* This horizon included sand and gravel with drifted tree remains, and was seen by Gunn at Paston and Happisburgh. It is equivalent to the upper parts of the sediment immediately below the marker level at Mundesley–Paston, with its sands and gravels lying below the land surface penetrated by ice-wedge casts.

Unio Bed and Rootlet Bed. These horizons, according to Gunn, included the Rootlet Bed at Corton, Pakefield and Kessingland and the 'Upper Freshwater Bed' at West Runton, sediments now included in the Cromerian freshwater suite of sediments.

Laminated Beds. These include the tidal laminated silty clays of Cr III and IV, well-developed at Mundesley and Paston and described there in their correct position by Gunn.

(ii) *C. Reid*

C. Reid in his memoir on the geology of the Cromer area (Reid, 1882*a*) described many coastal sections in considerable detail, synthesised his scheme of stratigraphy and compared it with those of other authors. He also drew the annotated coastal section (Reid, 1882*b*) and again discussed the sections and their interpretation in his Pliocene memoir (1890). The following gives the relation of certain sediments he included in his stratigraphical units to the stratigraphy now known, starting at the base of the succession. The horizons mentioned are described in either or both of Reid's two memoirs. *Weybourne Crag.* This term was applied by C. Reid (1882*a*) to shelly sand with laminated clay occurring above the Chalk and in the stone bed at Weybourne. This Weybourne Crag was characterised by the presence of *Macoma balthica* (Linné), not known from the Red and Norwich Crags. Reid (1890) later noted that the Weybourne Crag may be a littoral facies of equivalent age to the Chillesford Clay. At Weybourne, in the absence of pollen assemblages in the sediments, it has not been possible to correlate the Crag with the sequence found further east. At Skelding Hill, shelly sediments associated with the stone bed are Pa I in age. At Beeston and eastward to West Runton, shelly sediments associated with the stone bed are Pre-Pa a in age. At West Runton marine shells with *Macoma balthica* are also present in crag associated with tidal laminated clays of Pa II (*see* Fig. 8). Further south-east Reid noted typical Weybourne Crag on the Chalk at East Runton. This Crag is cut through by the mud of our section ERA and is therefore Pre-Pa a in age. The Weybourne Crag noted by

Reid at Trimingham is associated with laminated clays and may be Pa in age.

As described on *p.* 113, shelly sediments with *Macoma balthica* have been seen at Overstrand (Pre-Pa or Pa), at Sidestrand (Pre-Pa a), Trimingham (Pa I) and Mundesley (Pre-Pa a).

The Weybourne Crag, as defined by Reid, is then a shelly sandy facies of Pre-Pa a to Pa I and II age. This period covers two marine transgressions separated by a period of regression of unknown length.

Lower Freshwater Bed. This was observed in only a few places by Reid. The 'black mud' recorded by him three-quarters of a mile north-west of Cromer, cutting into the Weybourne Crag and overlain by sediments of the Forest-bed (Estuarine), is Pre-Pa d in age. The 'laminated lignite and loam' seen by him at Trimingham is probably, by analogy with the sections at Sidestrand, Pastonian in age.

'Laminated peat' recorded south-east of Mundesley by Reid (1890) is probably of early Cr age, and the cakes of peat in gravel recorded south-east of Mundesley and ascribed to the Lower Freshwater Bed in a section similar to MZ are probably Cr II in age.

The Lower Freshwater Bed thus includes the freshwater sediments formed after the post-Pre-Pa a regression, freshwater sediments associated with the rise in water-level heralding the Pastonian transgression, and probably freshwater sediments at Mundesley of early Cr age, predating the Cr marine transgression.

Forest-bed (*Estuarine*). This unit was recorded by Reid in nearly all sections around the coast. According to Reid, it is this horizon which is more especially known as the Cromer Forest-bed, and he said it consisted of alternating seams of sand, clay, and gravel, with masses of lignite. It was from this horizon that much of the forest-bed fauna originated.

At Beeston, Reid recorded the Forest-bed (Estuarine) as clays, gravel and sand from between the stone bed and 'freshwater loam' (his Upper Freshwater Bed). The age of this horizon is Pre-Pa b and c. At West Runton, the grey-blue Pa silty clays (WRAX) and the clay conglomerates of Pa age are included in the Forest-bed (Estuarine). Reid also seems to have included the shelly sand and gravels of the Beestonian seen at West Runton to the west of Woman Hithe in this unit, for the reason that they were marine sediments occurring below the Upper Freshwater Bed. The clay conglomerates seen along the coast on the foreshore to East Runton, Cromer and Overstrand, richly fossiliferous, of Pa age, were all included in the Forest-bed (Estuarine). At Mundesley and Bacton, tidal laminated clays of Cr III and IV age were ascribed to it, as were laminated clays and clay conglomerates of Pa age at Happisburgh.

Reid thus included marine sediments of Pa and Cr age in his Forest-bed (Estuarine). He found difficulty in separating sediments of this unit from those of the Weybourne Crag when the Lower Freshwater Bed was absent. This is not surprising, since the facies of both units may be similar. The same difficulty was encountered in the recent work and could only be solved by pollen analysis. Reid realised the difficulty of the sequence at Mundesley in having marine sediments above and below the Upper Freshwater Bed (Cr), and in the Cromer memoir ascribed the later of these marine sediments to the *Leda-myalis* Bed. Later, in the Pliocene memoir he changed this ascription to the Forest-bed (Estuarine).

Upper Freshwater Bed. Reid recorded this unit at many places around the coast. At Beeston 'peaty loam with opercula of *Bythinia tentaculata*' is ascribed to the Upper Freshwater Bed. The pollen content shows it is Pre-Pa d to Pa I in age. At West Runton, the Upper Freshwater Bed includes the type sediment for the Cromerian. The age of the Upper Freshwater Bed here is l Be in the lower part, and Cr I and II above this. Reid's '*Unio*-bed' at Sidestrand is probably Cromerian in age, by analogy with the Trimingham (TB) freshwater sediment. Near Mundesley, Reid described as part of the Upper Freshwater Bed a peaty clay with seeds above laminated clays very near the position of MAU, where clay-mud of Cr IVc age was found above laminated clays of Cr IIIb age. At Bacton, freshwater sediments were recorded above laminated clay and included in the Upper Freshwater Bed. The laminated clays are Cr III or IV at Bacton and the freshwater sediments may be e An by analogy with the freshwater sediments above laminated clay at section BI. The Upper Freshwater Bed at Ostend is Cr IV in age.

Most of the sediments included in Reid's Upper Freshwater Bed, including those at West Runton, are therefore Cromerian in age, either pre-dating the transgression (West Runton, Sidestrand, Mundesley) or post-dating it (Mundesley).

Leda-myalis *Bed.* This marine unit, based on the marine sands and gravel above Reid's Upper Freshwater Bed near Woman Hithe at West Runton, was recorded by him at only a few places. At Beeston, gravel and sand with marine shells of the *Leda-myalis* Bed was described by Reid above his Upper Freshwater Bed (Pre-Pa to

Pa I in age). This sediment is Pa II in age, and is related to the Pastonian marine transgression. At West Runton, the gravel with *Mya truncata* and related sands with marine shells are Cr IIIb in age. Further east, under Wood Hill, 3–5 m of sand below the till and above the Cromerian level were ascribed to the *Leda-myalis* Bed. This sand is all e An in age, post-dating ice-wedge casts. In the Cromer memoir, marine sands and gravels above Reid's Upper Freshwater Bed at Mundesley were ascribed by him to the *Leda-myalis* Bed. As mentioned above, he later changed this ascription to the Forest-bed (Estuarine) unit.

The *Leda-myalis* Bed is thus in its type area a sediment of the Cromerian transgression, Cr III and IV in age.

Arctic Freshwater Bed. This was only seen by Reid at Beeston, Mundesley and Ostend, occurring immediately below the till and characterised by a flora of northern affinity, with dwarf willows and dwarf birch. At Beeston this horizon appears to be of Be age. At Mundesley and Ostend it is e An in age.

Reid's climatic sequence. Reid could not find any evidence of any change in climate (or variation in the fauna and flora) between the periods when his Lower and Upper Freshwater Beds were laid down. He discussed at length the discrepancy between the evidence for cold conditions afforded by the Weybourne Crag molluscs in contrast to the evidence of terrestrial fauna and flora for temperate conditions. He gave a biogeographical explanation in terms of an elevation of land to the south, cutting off connection of the seas to the south, allowing only access of the northern marine fauna in a climate which permitted the existence of a temperate fauna and flora. The re-examination of the sections has, in contrast, led to the postulation of alternating cold and temperate periods.

Reid's sequence, based on several years' observations of the coastal sections, was a remarkable synthesis, considering the difficulty of correlation from section to section along the coast. The differences between Reid's sequence and the present one stem largely from the recent recognition of distinct marine transgressions, each of which may have very similar sediments (e.g. the tidal laminated clays of Pastonian and Cromerian age). The improvements now possible result largely from the application of pollen analysis to the problems of correlation, a method not available to Reid, but one he would have very much appreciated.

(f) Correlation in Britain

We have already considered in a previous section the relation of the pre-glacial floras and vegetation to those related to earlier and later stages in East Anglia. We may now consider the correlation of the coastal deposits with those elsewhere in Britain. This is a limited task, since there are few horizons or sites elsewhere where the sediments are probably of similar age. The matter of correlation may best be dealt with stage by stage. Table 49 summarises the correlations now discussed.

(i) *Pre-Pastonian*

The correlation of the Pre-Pa sediments with those elsewhere has already been considered in Chapter 2. Correlation of Pre-Pa a sediments is made with Lu 4c sediments at Ludham (West, 1961).

(ii) *Pastonian*

Sediments above the Lu 4c sediments in the Ludham borehole (West, 1961) contained a temperate pollen assemblage, placed in Lu 5, similar to the temperate pollen assemblages of the Pastonian, with *Ulmus* and *Carpinus*. The Lu 5 assemblage may then be Pastonian. A rather similar pollen assemblage, described as the Chillesford pollen assemblage, has been found in sediments of the Norwich Crag Series in south-east Suffolk (West & Norton, 1974). This assemblage was provisionally correlated with the Pastonian. The sediments were associated with the Westleton Beds (Hey, 1967), so a correlation of the Westleton Beds with the Pastonian was suggested. This correlation is highly tentative, for it is known that additional temperate stages are recorded in the Netherlands sequence (Zagwijn, 1975) and placed in the 'Cromerian complex', so that the Chillesford pollen assemblage may well be related to a stage hitherto not identified in East Anglia. In view of this doubt, and because the relation of the Westleton Beds to the Pastonian in the area north of Covehithe is problematical, it would seem best to leave this question open until further evidence is available.

If the correlation is indeed correct, then considerable height displacements seem necessary, with the relative raising of the south-east Suffolk area. It would also mean that mollusc assemblages with *Macoma balthica* (Weybourne Crag) in north-east Norfolk were the same age as those without it (Norwich Crag) in central Norfolk and Suffolk. If the correlation is incorrect, an extra temperate stage is required, older than the Pastonian and younger than the Antian, the marine sediments of which comprise Norwich Crag with the Chillesford pollen assemblage in Suffolk and Norwich Crag of the type area. The Pastonian would then be confined to east Norfolk and north Suffolk and be separated from a Norwich Crag temperate stage in a way not yet known. This latter correlation finds support in a consideration of the erratic content of the Norwich Crag Formation in the Norwich area (Funnell, 1970) and of the coastal pre-glacial sequence (Hey, 1976). In the Norwich area, an increase of the quartz-quartzite content occurs in the higher parts of the Norwich Crag Formation, with flint predominant below. Low frequencies of quartz-quartzite are also found in the Westleton Beds (Hey, 1967, 1976). On the other hand, the erratic content of Pre-Pa a gravels on the coast at Beeston and Sidestrand, described as Baventian at that time since I then considered Pre-Pa a as part of the Baventian, is considerable, as it is in the Pastonian gravels of the coast. Hey (1976) pointed out the contrasts of erratic content of 'Baventian' gravels in east Suffolk and north Norfolk, and of 'Pastonian' gravels of the Westleton Beds and north Norfolk. These difficulties might be resolved if the Westleton Beds were correlated with the Norwich Crag in the Norwich area, with the later incoming of quartz-quartzite in quantity occurring by or in Pre-Pa a times, when *Macoma balthica* is found in the marine sediments of the coast. Evidence for glacial action is found in the surface textures of sand grains in Pre-Pa a sediments at Sidestrand and in the quartz-quartzite gravels of the Norwich area. This is the earliest evidence for glaciation yet found in Norfolk.

The correlation in Table 49 shows the more likely of the two correlations discussed above.

A correlation of the Bridlington Crag with the Pastonian was made by Reid & Downie (1973) on the basis of dinoflagellate cyst assemblages. But the Crag also contains pollen of *Abies* in a tree assemblage with *Betula*, *Pinus* and *Picea*. *Abies* is absent or very scarce in the Pastonian, and a correlation with the Cromerian or Hoxnian seems more likely. The matter requires further investigation.

(iii) *Beestonian*

The complex nature of the Beestonian, with the possibility of non-sequences within and after it, and its poor fossil record, make any correlations with this stage hazardous. The Kesgrave sands and gravels of south-east Suffolk (Rose, Allen & Hey, 1975; Rose & Allen, 1977) and correlated gravels in Essex ('White Ballast'), the product of a major river draining north-east in a periglacial environment, have been placed in the Beestonian, on the basis that they belong to a pre-Cromerian and post-Westleton Beds cold stage. A further pre-Cromerian cold-stage sediment that may partly be Beestonian in age is the Plateau Drift of the Oxford area, since it appears to predate, in part at least, a fossiliferous channel filling at Sugworth, Oxfordshire, correlated with the Cromerian and associated with an old river at an elevation higher than the Hanborough Terrace but lower than areas with Plateau Drift (Briggs *et al.*, 1975; Goudie & Hart, 1975). In view of the stratigraphical problems between the Westleton Beds and the Pastonian, it may be best for the present to say that the sediments correlated with the Beestonian for the reasons described above belong to cold stages between the Chillesford pollen assemblage–Westleton Beds temperate stage and the Pastonian.

(iv) *Cromerian*

The richly fossiliferous sediments at Sugworth (see above) have been correlated with the Cromerian by reason of their flora and fauna. A further site with a Cromerian fauna is at Westbury-Sub-Mendip, Somerset, where cave sediments were exposed during quarrying of Carboniferous Limestone (Bishop, 1974, 1975). A soil (Valley Farm Soil) of suggested Cromerian age has been described in south-east Suffolk (Rose & Allen, 1977). It is described as a rubified *sol lessivé modaux*, enriched in clay and free iron, and is developed on the Kesgrave sands and gravels described above.

(v) *Anglian*

Sediments of the Cromer Forest-bed Formation on the Norfolk and Suffolk coasts clearly underlie the Middle Pleistocene Anglian tills (Turner 1973; Banham, 1975). The uppermost fossiliferous sediments of the Formation, those of the Arctic Freshwater Bed, are associated with cold floras and permafrost indicators. They are considered early Anglian in age.

(g) Correlation in the North-Sea basin and north-west Europe

Correlation with the Middle and Lower Pleistocene sequences of continental north-west Europe may be based on a variety of evidence, including vegetational, floristic and faunal history and palaeomagnetic chronology. Even so, little success can be gained in achieving reasonably firm correlations, partly because many more data are required, especially in the palaeomagnetic field, and partly because the stratigraphy is complex with numerous non-sequences, and we can only make very tentative conclusions regarding many of the stages represented. A start may be made by considering the uppermost of the correlations here attempted. There seems no doubt that the cold and glacial (Anglian) stage which follows the Cromerian temperate stage is to be correlated with the Elsterian glacial stage of the continent. In East Anglia and on the continent the vertebrate faunas of temperate stages older than these glacial stages have an archaic fauna (*see p.* 114) with some taxa not found in the later temperate stages. Moreover, the post-Anglian and post-Elsterian temperate stages, the Hoxnian and the Holsteinian temperate stages respectively, have been firmly correlated on the basis of vegetational history (Turner, 1975).

In correlation of the pre-Anglian sequence we are therefore dealing with the pre-Elsterian sequence of north-west Europe. This sequence is by no means clearly established. It is most detailed in the Netherlands (Zagwijn, Montfrans & Zandstra, 1971; Zagwijn, 1974, 1975, 1977; Zagwijn & Staalduinen, 1975) and in Schleswig-Holstein (Menke, 1975). The latter area has given long pollen diagrams from isolated sequences of sediment associated with salt structures. The diagrams are divisible into cold and warm stages and these stages have been correlated with those of the Netherlands, the correlation being chiefly based on connecting the conspicuous warm stages of Tiglian C with Menke's Ellerhoop stage.

The vegetational and floristic history of the Schleswig-Holstein area is more detailed than in the Netherlands, but there are considerable differences between them which Menke associated with the contrasting environments – the lowland alluvial environments of the Netherlands river systems and the Kaolin sands area of Schleswig-Holstein. These differences are for example seen in the highest frequencies of eutrophic and 'Tertiary-relic' taxa in the Netherlands Lower Pleistocene, and in higher frequences of oligotrophic and dystrophic taxa in Schleswig-Holstein. There also seems to be evidence in the Schleswig-Holstein sequence of stages not known from the Netherlands.

This comparison brings out the difficulties of Middle and Lower Pleistocene correlation. There have not yet been enough sites studied to assess regional variations of vegetational history and the lithostratigraphy of the areas is not well enough developed to be a realistic basis for correlation independent of biostratigraphy.

In the Netherlands the pre-Elsterian stratigraphy is reasonably well understood and the associated vegetational history may form a basis for correlations with East Anglia, across the southern part of the North-Sea basin. We will therefore restrict our consideration of correlation to this southern North-Sea area.

Fig. 54 summarises the Netherlands sequence, as it is now known. The pre-Tiglian TC 4c sequence is recorded in marine and continental facies, but TC 4c contains evidence of a strong regression and continental facies of the Rhine–Meuse–Scheldt sediments are then represented until, in the later part of the 'Cromerian complex', marine sediments are present in the northern Netherlands (Interglacial IV?) and in the western Netherlands (Interglacial III or IV?) (Zagwijn, 1977).

The division into stages in the Netherlands is based on the alternation of high NAP 'glacial' stages and 'interglacial' stages with abundant pollen of thermophilous trees. In the lower temperate stages, Tiglian and Waalian (and a recently described temperate biozone in the Menapian), the thermophilous element includes a good representation of 'Tertiary-relic' taxa, e.g. *Tsuga*, *Carya*, *Eucommia*, and *Pterocarya*. The Tiglian is characterised by the presence of *Azolla tegeliensis*, which by the Waalian is replaced by *A. filiculoides*. In the temperate stages of the 'Cromerian complex' these 'Tertiary-relic' taxa are much less well represented. Their pollen sequences are not well known, but substantial pollen diagrams have been published from a number of sites. The correlation of these diagrams in north-west Europe has been considered by Turner (1975). Interglacial I shows a mixed oak forest biozone with *Carpinus*, followed by a biozone with *Eucommia* and higher frequencies of *Carpinus*. *Abies* is sparsely present. In the Netherlands this stage is represented by a sequence at Loenermark (Polak, Maarleveld & Nota, 1962) and in Germany at the Osterholz (Grüger, 1968). Interglacial II includes a mixed oak forest biozone. *Carpinus* is only present in very low frequency in some samples and *Abies* is also sparsely present. This period is represented at Westerhoven (Zagwijn & Zonneveld, 1956), and possibly by pollen diagrams from Harreskov and Ølgod in Denmark (Andersen, 1965). Interglacial III may be represented in the Netherlands by pollen diagrams at Rosmalen (Ridder & Zagwijn, 1962) and near Eindhoven (Zagwijn *personal communication*, 1977), showing a mixed oak forest biozone with *Carpinus*, which is present from an early part of the temperate stage. *Abies* is scarce and there are very low frequencies of *Tilia*. Interglacial IV is associated with marine sediments of the Formation of Urk (Zagwijn & Staalduinen, 1975).

We may now consider this sequence in relation to that known from East Anglia. On the basis of the high frequencies of *Tsuga*, and presence of *Pterocarya* the temperate stages of the Ludhamian and Antian in the East Anglian Lower Pleistocene were correlated with the Tiglian and Waalian respectively by myself in 1961. Zagwijn (1975) preferred a correlation, shown in Fig. 54, with Tiglian TC 1–4b and TC 5–6 respectively, with the intervening Thurnian cold stage equivalent to TC 4c.

If this latter correlation is correct, then the regression of TC 4c may be marked in East Anglia by the evident non-sequence between the Thurnian and Antian of the Ludham borehole, and possibly by the regression evidence in the Happisburgh borehole HC at about the same depth OD (see Chapter 2), with the Baventian correlated with the Eburonian (Zagwijn, 1975), both the Baventian clays at Easton Bavents and the basal part of the Eburonian sediments having a normal polarity.

The transgression of the cold substage Pre-Pa a (Lu 4c), occurs later and may then be the equivalent of the Menapian or Glacial A, B or C of the 'Cromerian complex'.

To move now to the top of the pre-Anglian succession, it seems very likely that the Cromerian must be the temperate stage which immediately precedes the Anglian, and since its vegetational history, with a mixed oak forest biozone (Cr II) with substantial frequencies of *Tilia*, followed by a biozone with *Abies* and *Carpinus*, is unlike any sequence thought to be in 'Cromerian complex' Interglacials I, II and III, it is reasonable to suggest that the Cromerian may be correlated with Interglacial IV. Further afield, it also seems likely that, on the basis of the substantial development of an *Abies–Carpinus* biozone, the Cromerian is to be correlated with the Rhume Interglacial at Bilshausen near Göttingen (Lüttig, 1965; Müller, 1965), and the temperate stage in the upper part of the succession at Voigstedt, west of Leipzig (Erd, 1973).

The correlation of the Pastonian temperate stage presents more difficulty. The gap between the Cromerian and Pastonian has already been discussed. The position of possible non-sequences is shown in Fig. 53. The low frequencies of 'Tertiary-relic' taxa appear to rule out a correlation with the Waalian, which shows substantial frequencies of such taxa. A palaeomagnetic result from West Runton (section WRM), in Pa II marine sediments, showed normal magnetisation (Montfrans, 1971). In view of the palaeomagnetic results from the Netherlands sequence, this suggests that the Pastonian cannot be correlated with Interglacial I which shows reversed polarity and is just below the Brunhes/Matuyam boundary.

This leaves the possibility that the Pastonian might be correlated with Interglacial II or III, both within the Brunhes Epoch. The presence of *Carpinus* in the Pastonian seems to exclude a correlation with Interglacial II, so that a correlation with Interglacial III seems possible. The pollen diagrams have similarities in the frequencies of *Carpinus* and *Ulmus* and in the near-absence of *Abies*. If, however, we regard the presence of *Eucommia*, *Pterocarya*, *Tsuga*, Cupressaceae (b) and Monocolpate X (*Sabal*?) in the Pastonian as indicating an age older than 'Cromerian complex', from the palaeomagnetic point of view, a correlation with the Waalian (c) would be possible, but not with the intra-Menapian temperate biozone mentioned by Zagwijn (1975) since the Menapian shows reversed polarity except for the basal part (Montfrans, 1971).

The normal polarity found in the silty clays (Pre-Pa a) in section SSB at Sidestrand (Montfrans, 1971) indicates that the Pre-Pa a could be Glacial A or B of the 'Cromerian complex', or Menapian (of which the basal part might be of normal polarity). The Beestonian cold stage could then be of Glacial A, B or C age. (Or, if the Baventian and Pre-Pa a cold episodes are part of the same cold stage, Pre-Pa a could be Glacial A or B, basal Menapian or basal Eburonian, and the Beestonian Glacial A, B or C, or basal Menapian. The former set of possibilities is shown in Fig. 54.)

This brings us to the unsatisfactory conclusion that the Pre-Pa a cold stage might be correlated with the (Eburonian,) Menapian, Glacial A or Glacial B, while the Pastonian might be of Waalian or Interglacial III age. These possibilities are indicated in Fig. 54. The matter of correlation might be improved if the time of arrival of *Macoma balthica* in relation to the known stages of the Netherlands marine Lower Pleistocene could be discovered in offshore boreholes. The absence of marine faunas in Tiglian TC 4c to late 'Cromerian complex' times and the paucity of the faunal record in the Netherlands at this time hinders correlation. On the other hand in East Anglia the post-TC 4c marine and fauna records are substantial and may eventually offer, together with the palaeobotanical record, a more effective means of correlation. The palaeomagnetic results from East Anglia (Montfrans, 1971) are also rather sparse and further studies in this field will undoubtedly also assist correlation.

Appendix 1. Stratigraphy of certain sections

The stratigraphy of certain important sections from which pollen diagrams were obtained is given below. The sections are grouped into the stretches of coast already considered and are arranged alphabetically. The positions of the sections are shown in the figure references. All measurements are in centimetres.

West Runton

WRF (Fig. 4)	0–10	Grey clay, transition at base
	10–18	Brown detritus mud, brecciated around 15 cm
	18–21	Grey clay with marine shells
		Grey sand with marine shells
WRU (Fig. 6)	0–5	Reddish silty mud
	5–120	Brown coarse detritus mud with wood fragments, basal 30 cm with shells
	120–140	Grey calcareous mud with breccia
		Yellow sand with freshwater shell fragments
WRAQ (Fig. 6)	0–+15	Brown coarse detritus mud with breccia at base
	0	Unconformity with sharp contact
	0–65	Brown-grey marl with shells, browner towards base
	65–90	Dark grey and black clay-mud
	90–130	Red ferruginous gravel with marine shell fragments
	130–170	Red sand with marine shells
WRAR (Fig. 6)	0–10	Brown detritus mud
	10–69	Grey clayey sand, contorted with red and yellow sand, with freshwater shells
	60–75	Brown stony detritus mud
	75–120	Rusty ferruginous gravel with marine shell fragments, grading down into red sand
	120–160	Pale laminated silt and clay, with rusty small leaf impressions, caddis fly larval cases and freshwater shells
	160–200	Rusty and yellow sand
WRAYa (Fig. 10)		Till
	0–30	Brown sand
	30–65	Grey silty clay with freshwater shell fragments
	65–75	Yellow sandy clay and sand with shells
	75–120	Grey laminated clayey silt
	120–190	As above, but with thicker sand laminations
	190–230	Grey sand with silty clay laminae
		Gravel
WRCGB (Fig. 6)	215–220	Reddish silty mud
	140–215	Brown coarse detritus mud
	114–140	Brown sandy detritus mud
	40–114	Light brown sandy detritus mud with horizons of brecciated mud, sharp junction at base
	0 (base)–40	Light brown stiff sandy mud with freshwater shells and *Donacia*
		Orange sand

Beeston

BAM (Fig. 16)		Till
	0–+40	Yellow sand with stones
	0	Sharp uneven contact
	0–20	Grey-brown clay, mottled rusty, with small rootlets
	20–65	Darker muddy silt
	65–120	Laminated pale clay and red sand, with caddis fly larval cases
	120–135	Darker muddy silt
	135–138	Yellow sand
	138–160	Brown silty mud
	160–270	Yellow and red sand with laminations of grey silty clay
BKA (Fig. 18)		Gravel
	0–40	Grey laminated silty clay
	40–85	Grey sand with laminated grey silty clay and marine shell fragments
	85–140	Gravel and sand

	140–160	Brown sandy and stony mud
	160–190	Stratified clayey sand
	190–210	Gravelly sand
	210–250	Iron pan, large flints, marine shells
		Brecciated Chalk
BQ (Fig. 18)		Ferrugineous clay and sand
	0–+30	Grey silty clay, brecciated in places
	0–15	Grey laminated silty clay and sand with a pebble horizon at 13 cm
	15–17	Transition
	17–28	Hard brown mud
BR–S (Fig. 18)	0–17	Grey silt laminae in sand
	17–36	Brown compressed peat with wood
	36–60	Grey laminated silty clay
	60–120	Hard brown mud, mottled
	120–125	Grey clayey silt
	125–135	As above, but disturbed
	135–145	Grey laminated sand and clay, disturbed
	145–165	Grey silty sand
	165–200	Dark grey mud
	200–230	Red and grey laminated sand and clay
	230–240	Dark grey silt
	240–245	Red and grey laminated sand and clay
	245–270	Yellow sand
	270–295	Iron pan, large flints, marine shells
		Chalk
Cromer		
CA (Fig. 22)	0–30	Grey laminated silty clay with rusty streaks
	30–80	As above, but with brecciated grey clay
	80–110	Grey laminated silty clay with marine shell fragments
Overstrand		
OB (Fig. 22)		Grey-brown till
	0–9	Grey-brown silty mud
	9–25	Brown compressed mud with large bivalves
	25–105	Red-yellow sand grading down into brown-grey silt
	105–120	Sand and gravel, uneven contact below
	120–150	Blue-grey laminated silty clay
	150–175	Ferruginous gravel
	175–260	Pale stratified sand with thin gravel strings
	260–295	Iron pan, gravel and sand
	295–358	Massive blue-grey silty clay grading down into laminated silty clay; iron-stained rootlets in top 25 cm
Sidestrand		
SSB (Fig. 22)	0–90	Rusty gravelly sand
	90–152	Yellow sand with silty clay lamina (0 cm in pollen diagram)
	152–482	Grey and rusty shelly sand with silty clay laminae
	482–495	Large flints
		Chalk
SSE (Fig. 22)		Till
	0–1000	Yellow and red cross-bedded sand with drift-mud laminae
	1000–1450	Grey cross-bedded sand with thin silty clay laminae
	1450–1550	Grey sand with many thin silty clay laminae
	1550–1600	Grey sand, shelly (marine) at base, with a 5 cm-thick grey silty clay, more organic at base, near base
	1600–1750	Chalk
SSR (Fig. 22)		Brown till
	0–20	Yellow sand with brown streaks
	20–340	Grey and brown silty clay with laminae of sand and burrow structures (0 cm in pollen diagram at 190 cm)
	340–380	Pale gravel
	380–515	Yellow sand with ferrugineous and grey silty clay at 430 cm
SST (Fig. 22)		
SSTA		Yellow sand with drift-mud streaks
	0–10	Grey silty clay laminae in sand
	10–50	Grey-blue stiff stony clay with shell fragments
	50–90	Dark grey sand with silty clay pods
		Erosion surface, with gravel pavement
	90–130	Cross-bedded sand, with marine shell fragments, and drift-mud and silty clay streaks
SSTC	70–75	Grey clay and silt, with clay balls in sand hollows
	75–160	Dark grey-brown clay-mud
	160–200	Brown mud
Paston–Mundesley		
MA (Fig. 33)		Grey till
	0–30	Grey laminated silty clay with organic streaks
	30–70	Yellow sand
	70–80	Yellow sand and gravel
	80–340	Grey-blue laminated silty clay with silt and sand partings, with a horizon of pebbles and drift-mud at 90 cm (0 cm in pollen diagram at 80 cm)
MB (Fig. 33)		Grey till
	0–50	Yellow sand
	50–80	Grey laminated silty clay with organic streaks
	80–130	Yellow sand
	130–140	Grey organic silty clay
	140–160	Yellow sand
		Pebble horizon

	160–190	Yellow sand, ice-wedge cast
	190–210	Brown organic silty clay (0 cm in pollen diagram)
	210–260	Brown cross-bedded sand
	260–305	Grey-blue laminated silty clay with silt and sand partings
MF (Fig. 32)	0–20	Brown detritus mud with shells
	20–23	Grey sand with detritus mud erratic
	23–32	Grey brown clay-mud
		Red ferruginous fine gravel and sand
MN (Fig. 31)	MN 2 30	of grey-blue laminated clay with large wood fragments sealing the next lower horizon
	MN 1 10	of clay conglomerate with mud erratics
	MN 3 40	Grey-blue laminated clay with monocotyledonous remains on bedding planes
MU (Fig. 32)	MU T	Grey laminated silty clay
	0–45	Brown detritus mud, more silty at 23–28 cm
	45–48	Ferruginous fine gravel and sand
	48–78	Grey sand with grey silty clay lamina at MU B
MAH (Fig. 32)		Till
	0–110	Pale sand, horizontally bedded, with a few grey clay streaks
	110–150	Grey-blue laminated silty clay with organic streaks
	150–160	Fine gravel
	160–190	Pale brown stratified sand
	190–285	Black organic stony silt (0 cm in pollen diagram at 195 cm)
	285–290	Grey and yellow sand and silty sand
		Yellow sand with streaks of drift-mud
MAM (Fig. 33)		Till
	0–25	Grey-blue laminated silty clay
	25–52	Brown hard mud and peat, silty at base
	52–60	Rusty sand with wood fragments and grey silty clay seams
	60–72	Contorted grey silty clay and red sand
	72–100	Red sand with pebbles, grey silty clay seams
	100–130	Grey-blue laminated silty clay with rusty laminations
MAT (Fig. 33)	0–213	Sand and gravel (recent beach)
	213–228	Grey silty clay
	228–381	Grey sand
	381–442	Grey sand and gravel
	442–671	Grey sand with grey silty clay laminae
	671–1296	Grey laminated silty clay with grey sand horizons
	1296–1341	Grey silty clay and stones (stone bed)
		Chalk
MPA (Fig. 31)	0–20	Hard mud with *Picea* and *Pinus* cones
	20–60	Brown and grey laminated organic silty clay
PB (Fig. 30)		Till
	0–105	Yellow sand
	105–310	Grey-blue laminated silty clay
	310–316	Pale sand
	316–330	Brecciated grey-blue clay, very fine (0 cm in pollen diagram)
	330–380	Yellow stratified sand with clayey laminae
	380–390	Brown drift-mud
	390–420	Red ferruginous gravel with bone fragments and grey clay layers
PG (Fig. 30)	Up to 120	Red ferruginous bedded sandy gravel with clay balls
	Up to 35	Grey-blue laminated silty clay and silt, with some drift-mud seams
	Up to 100	Grey sand with thin silty clay laminae

Bacton

BI (Fig. 39)		Brown-grey clayey silt
	0–30	Grey laminated silt, muddy at top
	30–60	Red and grey stratified sand
	60–90	Pale sand
	90–120	Red and grey laminated silty clay
	120–160	Greyer laminated silty clay
	160–180	Red silty clay
	180–215	Reddish gravel with clay fragments
	215–225	Grey silty clay
	225–385	Grey laminated silty clay and sand
	385–405	Grey silty clay
	405–455	Red and yellow coarse sand with clayey streaks
	455–475	Red ferruginous gravel
	475–505	Grey silty sand

Ostend

OA (Fig. 39)		Till
	0–52	Grey clay-mud, abundant freshwater shells at 12–16 cm and with wood at 50 cm
	52–88	Brown detritus mud, sharp junction below
	88–104	Blue-grey clay, rootlets present
OD (Fig. 39)		Till
	0–60	Grey-red mottled clay with branches of wood
	60–200	Grey clay with race at top
	200–280	Grey sand
	280–300	Grey sandy clay
	300–350	Grey sand

Happisburgh

HC (Fig. 39)	0–91	Recent beach gravel
	91–381	Grey-blue laminated silty clay with silt and sand partings and drift-mud laminae

	381–396	Gravel
	396–470	Grey sand with mud erratic and thin silty clay lamina
	470–580	Grey sand with siltstone erratics
	580–610	Grey silty clay
	762–1880	Grey sand with grey silty clay laminae
	1880–1920	Sand and gravel
	1920–2120	Grey laminated silty clay, gravelly towards base, and with marine shell fragments
	2120–2200	Sand and gravel
	2200–2895	Grey sand with grey silty clay laminae
		Chalk
HF (Fig. 39)	0–12	Grey silty clay
	12–20	Brown detritus mud with wood
	20–30	Grey silty clay, pebbly in upper part

Corton

CC (Fig. 43)		Till (North Sea Drift)
	0–40	Pale sand, with drift-mud streaks, contorted
	40–55	Grey-brown silt
	55–70	Red stratified sand
	70–81	Grey-blue silty clay with a few stones (0 cm of pollen diagram at 73 cm)
	81–141	Grey-blue laminated silty clay and sand, with burrow structures and disturbed in the upper part
	141–171	Brown and grey silty sand with grey silty clay laminae, sharp junction below
	171–198	Grey-blue clay with a brecciated structure and grey-white concretions at base (rootlet bed)
	198–258	Yellow sand with red spots
CF (Fig. 43)	0–2	Yellow sand
	2–19	Brown sandy mud
	19–39	Red-brown sand
	39–54	Pale fine gravel
	54–55	Grey silty clay laminae involved in involutions
	55–80	Pale sand
CK (Fig. 43)	0–26	Red and grey laminated silty clay, more organic at base, transition to next below
	26–44	Black mud with wood fragments
	44–69	Blue-grey clay with rootlets
BH 2/6 (Fig. 43)	0–30	Recent beach sand
	30–152	Grey-blue clay with race and shells at base
	152–685	Brown sand and gravel
	685–725	Dark grey silty clay
	725–752	Brown sand
	752–838	Dark grey silty clay
	838–910	Grey sand with silty clay laminae

Pakefield

PC (Fig. 46)	0–65	Brown-red sandy silt, laminated in lower half
	65–80	As above, but greyer and silt laminations thicker
	80–95	Transition to next
	95–110	Brown silty mud with some grey silty clay laminations
	110–240	Brown silty mud, with thicker muddy and silty laminations, and sandier at base with twigs
	240–265	Red sandy gravel, hard and ferrugineous at top, paler below
PF (Fig. 46)	0–60	Brown stiff silty clay with race and black specks
	60–110	Brown sand and grey clay interbedded
	110–140	Blue stiff clay
	140–220	Brown clayey silt and pale sand
PL (Fig. 46)	0–243	Talus
	243–549	Brown stiff clay, bluer at base
	549–745	Brown-red sand with small flints
	745–775	Grey silty sand
	775–830	Grey laminated silty clay and sand with much flattened wood
	830–1006	Grey sandy silt with grey and pale silty clay laminae, more sandy below

Kessingland

K (Fig. 46)		Sand (Corton Beds)
	0–4	Brown clay, transition to next below
	4–25	Brown clay-mud
	25–60	Grey and brown clay with sand laminae, with race at base
	60–300	Brown sand with grey silty clay laminae

Appendix 2. Marine mollusc faunas

P. E. P. Norton

The following account describes the assemblages analysed and/or stratigraphically placed in the succession, starting at West Runton and proceeding round the coast to Happisburgh.

1. West Runton

(i) Site WRCO. Bed i of the pre-West Runton Freshwater Bed succession, west of Woman Hithe. Beestonian. The following Mollusca were counted (4375 g material):

Littorina sp. 1 and operculum
Scala cf. *groenlandica* (Chemnitz) or cf. *clathratulum* (Kanmacher) fragments
Mytilus edulis L. fragments
Chlamys opercularis L. fragments
Cardium edule L. 1
Venus cf. *ovata* Pennant ½
Macoma sp. 4
Mya sp. ½
Bivalve (unid.) 1

Also recorded in bed i, at 1 m west of datum by R. A. D. Markham, were a number of complete *Macoma balthica* valves.

This is interpreted as an assemblage of boreal climatic affinity showing no evidence for considerable cold. Although *Venus ovata*, *Scala* aud *Chlamys opercularis* at the present do not live intertidally, they and the other (intertidal type) shells are eroded, pitted and broken in the sample. They had probably been transported from their life locality and incorporated in beach deposits in bed i. The sample contained fewer species than those from WRAT and WRF. The species point to a habitat with rocks (*Mytilus*, *Littorina*) and fine or silty sand (*Macoma*, *Cardium*) in the intertidal region. Sublittorally, sand and perhaps silty (*Chlamys*, *Venus*) with weeds (*Scala*) seem to have occurred.

(ii) Site WRF. Bed c of the pre-West Runton Freshwater Bed succession, east of Woman Hithe.

A silty clay seam above the shelly sand is Pa II in age. Silty clay seams in shelly sands of neighbouring sites WRQ and WRE have a Pre-Pa a type of pollen assemblage. The age of this assemblage may then be Pre-Pa a to Pa II. The following Mollusca were counted (in 603 g material):

Solariella sp. 2
Gibbula sp. 1
Littorina littorea (L.) fragments
Nucella lapillus (L.) fragment
Yoldia sp. 8
Acila cobboldiae (Sowerby) fragments
Mytilus edulis L. fragments
Astarte montagui (Dillwyn) 6
Arctica islandica (L.) 1
Cardium edule L. fragments
Cardium sp. 4
Spisula sp. 1
Donax vittatus (Da Costa) fragments
Donax sp. 4
Macoma balthica (L.) 5
M. obliqua (J. Sowerby) 1
Macoma sp.
Corbula gibba (Olivi) 1
Mya sp. 1

The littoral forms *Gibbula*, *Mytilus*, *Nucella* and *Littorina* indicate intertidal conditions with rocks. *Cardium*, *Donax* and *Macoma* indicate fine or silty intertidal sands. Sublittoral forms (*Astarte*, *Cyprina*, *Corbula*) associated with muddy habitats complete the assemblage. Apart from the presence of *Macoma balthica*, this assemblage recalls those of the classic Norwich Crag, e.g. the shelly crag of the Norwich region (Norton, 1977 and *personal communication*) and the Suffolk region (West & Norton, 1974), though it is comparatively poor in species and individuals.

(iii) Site WRAT. Bed c of the pre-West Runton Freshwater Bed succession west of Woman Hithe. The age of this assemblage is not clear. The pollen assemblage is poorly preserved and may be Pre-Pa a (*see p.* 32). It certainly appears to pre-date the tidal sediments of Pa II seen near Woman Hithe. The following Mollusca were counted:

Rissoa parva (Da Costa) 4
Natica sp. 1
Nucella lapillus (L.) fragment
Gastropod sp. 1
Yoldia sp. 2
Nuculana sp. ? 1

Mytilus edulis L. 3½
Astarte sp. ½
Lepton sp. ½
Cardium edule L. 3½
Cardium sp. ½
Venus ovata Pennant ½
Spisula subtruncata (Da Costa) ½
S. cf. *subtruncata* 2
Spisula sp. 2
Donax vittatus (Da Costa) 5
Macoma balthica (L.) 4½
M. obliqua (Sowerby) 3½
Macoma sp. 6
Tellinid 13½
Hiatella arctica (L.)? ½
Hiatella sp. ½
Corbula gibba (Olivi)? ½
Mya sp. 1
Pholadid ½

Mostly a littoral assemblage (especially the *Macoma*, *Mytilus* and *Cardium*) gathered from rocks and fine or silty intertidal sand. There are some sublittoral shells. All are worn and possibly reworked. If *Macoma balthica* were omitted, the assemblage would be similar to those of the classic Norwich Crag (see comments on West Runton WRF).

(iv) Bed a of the post-West Runton Freshwater Bed succession. This is the *Mya* Bed of Trimmer (1845) containing *Mya truncata* in the position of life. The fauna of this horizon and of the associated sands (*Leda-myalis* Bed of Reid) is listed by Reid (1890). The age of the fauna in the gravel of bed a is Cr IIIb. Reid's list for the *Leda-myalis* Bed is as follows:

Buccinum undatum (L.)
Littorina littorea (L.)
Littorina rudis (Maton)
Nucella lapillus (L.)
Natica catena (Da Costa)?
Trophon antiquus (L.)
T. antiquus (L.) reversed variety
Leda myalis Couthouy
Mya truncata L.
Mytilus edulis L.
Ostrea edulis L.
Astarte borealis Chemnitz
Cyprina islandica L.
Cardium edule L.
Tellina balthica L.

In addition, Trimmer found *Natica helicoides*, *Macoma obliqua* and *Mya arenaria*, and there are also freshwater forms which Reid notes may have been derived into this bed from the sands of the West Runton Freshwater Bed, which are lower in the section.

Reid notes *Trophon antiquus* (reversed) and *Macoma obliqua* as possible derivatives from another bed and West considers that the deposit sampled as WRCO about 30 cm below is a possible source.

Reid reported three species as found in the life position with valves united; these are *Astarte borealis* Chemnitz (which is probably *A. sulcata* (Da Costa), *Leda myalis* Couthouy (now called *Yoldia myalis*) and *Mya truncata* L. All of these occur in muddy or gravelly seabeds in north boreal sublittoral situations today, in which case this may be interpreted as an inshore deposit with other forms rolled down from shallower water. There is some confusion on the correct determination of *Yoldia myalis*. It has been considered (Castell, 1959, and *personal communications*) that *Yoldia myalis* occurred only in the *Leda-myalis* bed and that other Crag records should be amended to take account of their similarity to *Leda oblongoides* Wood 1840. Examination of the original material in the British Museum (Natural History), however, convinces me that it should all have been named *Yoldia myalis* in the first place and that Wood's *oblongoides* is not a valid species. In fact it arises from a misunderstanding of the nomenclatural priority by Wood and appears to be corrected in his 'Crag Mollusca'. It is therefore considered that all Crag shells referred to *Yoldia oblongoides* (Wood) are really *Y. myalis* (Couthouy).

2. Overstrand

An assemblage was described by Markham (1973) from shelly sands in the boreholes at Overstrand (bed b of the Overstrand succession). The assemblage is likely to be Pa II in age, but since the boundary position between Pre-Pa a (stone bed) and Pa II is unknown, the age may better be stated as Pre-Pa a to Pa II. The following Mollusca were listed:

Littorina sp.
Nucella sp.
Nucula cobboldiae
Mytilus sp.
Astarte montagui
Astarte sulcata
Arctica islandica
Mya arenaria
Cardium sp.
Macoma balthica
Macoma obliqua

This assemblage lists both littoral and sublittoral types, interpreted as deposited from rocks and muddy sand in the littoral zone (*Littorina*, *Nucella*, *Mytilus* and *Macoma*, *Mya*, *Cardium*) and from muddy sublittoral sand (*Astarte*, *Arctica*). The presence of *Macoma balthica* distinguishes it from lists of species from classic Norwich Crag assemblages.

3. Sidestrand

Norton (1967) described assemblages from the basal 20 cm of site SSE (Pa II), from shelly sands of Pre-Pa a age (bed b) at site SSB, 3 m above the stone bed, and at site SSK, near site SSB, just above the stone bed (sample K1 and 80 cms above this sample, K6). The fauna is dominated by *Macoma balthica* and is predominantly intertidal. If *Macoma balthica* were omitted from the assemblages, the remainder would resemble those described from the classic Norwich Crag. Tellinids other than *Macoma balthica*, however, are far more infrequent than would be expected: *M. balthica* appears to have replaced them. Boreal marine temperatures are indicated. The almost exclusive occurrence of intertidal species suggests an extensive area of tidal water with deposition at only a few metres depth. The faunal list of the Pre-Pa a assemblages is much less rich than that given by Reid (1882*a*) for the Weybourne Crag, but is based on only a few kilogram material whereas Reid's list was the result of extensive collecting. Possibly, also, it is a cold-stage fauna in comparison to more temperate (Pa) faunas at some of Reid's sites. However, at this locality, the Pa II fauna of SSE is less rich than those of SSK, as may be noted from the Table in Norton (1967).

4. Trimingham

Shelly sand at 7 to 8 m depth in borehole BHT 16. Pollen analysis of the sample indicated a Pa Ib age for the assemblage. A sample of shells contained:

Nucula nucleus (L.) or *Acila cobboldiae* (Sowerby) ½
Arctica islandica (L.) 2
Cardium edule L. 1 and many fragments
Macoma balthica (L.) 12½
Macoma obliqua (Sowerby) 6½
Macoma sp. 3
Mya arenaria L. 1

All the shells were abraded giving the appearance of having been washed around a great deal before final deposition. The *Cardium*, *Mya* and *Macoma balthica* shells suggest an intertidal region of sand or muddy sand with *Arctica* washed up. *Macoma obliqua* is now extinct. It was difficult to decide, in several cases, whether shells were *Macoma balthica* or juvenile *M. obliqua*.

5. Mundesley

A number of samples of Pre-Pa a age in borehole MAT were analysed. All the samples except the last are in the laminated silty clays of bed b.

cm	
842–852	Gastropod sp. 2
	Mytilus edulis L. fragment
	Cardium edule L. ½
	Macoma balthica (L.) 1
850	*Cardium edule* L. 1
	Macoma balthica (L.) 1
884–894	*Littorina littorea* (L.) fragments
	Mytilus edulis L. ½ and fragments
	Cardium edule L. 1½ and fragments
	Donax vittatus (Da Costa) 2
	Tellinid sp. 2
	Corbula gibba (Olivi) ½
981–985	*Mytilus edulis* (L.) fragment
1060–1130	*Buccinum undatum* (L.)? 1 (juvenile)
	Lymnaea palustris (Müller) 1 (broken)
	Cardium edule L. fragment
1297–1303	*Littorina littorea* (L.) fragment
(stone bed)	*Yoldia* sp. 1
	Acila cobboldiae (Sowerby) fragment
	Modiolus sp. or *Acila cobboldiae* fragments
	Mytilus edulis L. ½ and fragments
	Astarte cf *montagui* (Dillwyn) 1
	Arctica islandica (L.) fragments
	Tellinid sp. ½ and fragments
	Mya sp.? ½

These faunas indicate shallow marine conditions, with derivation partly from beach habitats. All the species listed are found intertidally except, typically speaking, *Arctica islandica*, *Corbula gibba*, *Buccinum undatum*, *Astarte montagui* and *Yoldia* (from sublittoral and muddy habitats) and *Lymnaea palustris* (from freshwater paludal situations). The occurrence of *Macoma balthica* at 850 cm is, together with the records from West Runton WRAT and Sidestrand SSK1, the earliest recorded find apart from the Bure Valley Beds, whose age is however not yet clear.

6. Happisburgh

A number of samples from borehole HC were analysed

cm		
2006–2009	bed c, biozone c	*Mytilus edulis* L. fragments
2009–2014	bed c, biozone c	*M. edulis* L. fragments
2014–2017	bed c, biozone c	*M. edulis* L. fragments
2380	bed a, biozone a	*Balanus* plates
		Mytilus edulis L. fragment
		? Pectinid fragment
2560	bed a, biozone a	*Balanus* plates
		Mytilus edulis L. fragments

No ecological conclusions can be drawn from these shells. They are marine, transported and have suffered extreme degradation. Laboratory notes for HC 2006–2009 cm record 'a gravelly deposit in which there happen to be some very battered bits of shell' and this appears true for the rest.

7. Palaeontological note on *Macoma balthica* (L.)

These deposits and the Bure Valley Beds contain some of the oldest verified specimens of *Macoma balthica*, as discussed above. Juveniles especially are hard to separate from *Macoma praetenuis* (Leathes) and *Macoma obliqua* (Sowerby), as discussed by Spaink & Norton (1967). Comparison of specimens from the Bure Valley Beds, the Cromer Forest-bed Formation and more recent deposits shows that in the oldest specimens the difficulty of separating them from other *Macoma*, especially *M. praetenuis*, almost increases with age of the deposit. Thus the Wroxham material from the Bure Valley has proved very difficult; the Trimingham material (Pa Ib), the West Runton WRAT material (Pre-Pa a) and WRF material (Pre-Pa to Pa II) is also difficult. On the other hand the Sidestrand material at all levels was easily determined. Possibly a morphological study of the early evolution of *Macoma balthica* could be undertaken using the well-dated sequence elucidated in this volume. However, dating of the Bure Valley Beds is also necessary.

REFERENCES

Andersen, S. T. (1961). Vegetation and its environment in Denmark in the Early Weichselian Glacial (Last Glacial). *Danm. geol. Unders.* II Række, no. 75.

Andersen, S. T. (1965). Interglacialer og interstadialer i Danmarks Kvartær. *Medd. dansk. geol. Foren.* 15, 486–506.

Andersen, S. T. (1966). Interglacial succession and lake development in Denmark. *Palaeobotanist*, 15, 117–27.

Andersen, S. T. (1973). On the occurrence of pollen similar to *Bruckenthalia spiculifolia* (Salisb.) Reichenb. in Danish Quaternary deposits. *Danm. geol. Unders.*, Årbog 1972, pp. 5–6.

Andersen, S. T. (1974). The Eemian freshwater deposit at Egernsund, South Jylland and the Eemian landscape development in Denmark. *Danm. geol. Unders.*, Årbog 1974, pp. 49–70.

Andrew, R. (1970). The Cambridge pollen reference collection. In *Studies in the Vegetational History of the British Isles* (ed. by D. Walker and R. G. West), pp. 225–31. Cambridge University Press.

Andrew, R. & West, R. G. (1977). Pollen spectra from Pliocene crag at Orford, Suffolk. *New Phytol.* 78, 709–14.

Azzaroli, A. (1953). The deer of the Weybourne Crag and Forest Bed of Norfolk. *Bull. British Museum (Nat. Hist.), Geology*, 2, 3–96.

Baden-Powell, D. F. W. (1950). Field meeting in the Lowestoft district. *Proc. geol. Assoc. Lond.*, 61, 191–7.

Banham, P. H. (1968). A preliminary note on the Pleistocene stratigraphy of north-east Norfolk. Appendix, pp. 507–11. In *Proc. geol. Assoc. Lond.*, 79, 493–512.

Banham, P. H. (1975). Glacitectonic structures: a general discussion with particular reference to the contorted drift of Norfolk. In *Ice Ages: Ancient and Modern* (ed. by A. E. Wright & F. Moseley), pp. 69–94. Liverpool: Seel House Press.

Bell, F. G. (1969). The occurrence of southern, steppe and halophyte elements in Weichselian floras of southern Britain. *New Phytol.* 68, 913–22.

Bell, F. G. (1970). Late Pleistocene floras from Earith, Hunts. *Phil. Trans. R. Soc.* B, 258, 347–78.

Birks, H. H. (1973). Modern macrofossil assemblages in lake sediments in Minnesota. In *Quaternary Plant Ecology* (ed. by H. J. B. Birks & R. G. West), pp. 173–89. Oxford: Blackwell Scientific Publications.

Bishop, M. J. (1974). A preliminary report on the Middle Pleistocene mammal bearing deposits of Westbury-Sub-Mendip, Somerset. *Proc. Univ. Bristol spelaeol. Soc.* 13, 301–18.

Bishop, M. J. (1975). Earliest record of man's presence in Britain. *Nature, Lond.* 253, 95–7.

Blake, J. H. (1877). On the age of the mammalian Rootlet-bed at Kessingland. *Geol. Mag.* dec. ii, 4, 298–300.

Blake, J. H. (1884). Horizontal section 128. Geological Survey E. and W.

Blake, J. H. (1890). *The Geology of the Country near Yarmouth and Lowestoft.* Mem. Geol. Surv. E. and W.

Briggs, D. J., Gilbertson, D. D., Goudie, A. S., Osborne, P. J., Osmaston, H. A., Pettit, M. E., Shotton, F. W. & Stuart, A. J. (1975). New interglacial site at Sugworth. *Nature, Lond.* 257, 477–9.

Brush, G. S. & Brush, L. M. (1972). Transport of pollen in a sediment laden channel: a laboratory study. *Amer. J. Sci.* 272, 359–81.

Castell, C. P. (1959). *British Caenozoic Fossils.* London: British Museum (Natural History).

Cross, A. T., Thompson, G. G. & Zaitzeff, J. B. (1966). Source and distribution of palynomorphs in bottom sediments, southern part of Gulf of California. *Marine Geol.* 4, 467–524.

Dandy, J. E. (1958). *List of British Vascular Plants.* London: British Museum (Natural History) and Botanical Society of the British Isles.

Davis, M. B. & Brubaker, L. B. (1973). Differential sedimentation of pollen grains in lakes. *Limnology and Oceanography* 18, 635–46.

Dickson, J. H. (1973). *Bryophytes of the Pleistocene.* Cambridge University Press.

Duigan, S. L. (1963). Pollen analyses of the Cromer Forest Bed Series in East Anglia. *Phil. Trans. R. Soc.* B, 246, 149–202.

Ellis, A. E. (1962). *Synopses of the British fauna*, 13. *British Freshwater Bivalve Molluscs.* London: Linnaean Society.

Erd, K. (1973). Pollenanalytische gliederung des Pleistozäns der Deutschen Demokratischen Republik. *Z. geol. Wiss. Berlin*, 1, 1087–103.

Funnell, B. M. (1970). The Palaeogene and Early Pleistocene of Norfolk. In *The Geology of Norfolk* (ed. by G. P. Larwood & B. M. Funnell), pp. 340–64. Norwich: Geological Society of Norfolk.

Funnell, B. M. & West, R. G. (1962). The Early Pleistocene of Easton Bavents, Suffolk. *Quart. J. geol. Soc. Lond.* 118, 125–41.

Funnell, B. M. & West, R. G. (1977). Preglacial Pleistocene deposits of East Anglia. In *Recent Advances in British Quaternary Studies* (ed. by F. W. Shotton), pp. 247–65. Oxford: Oxford University Press.

Gardner, K. & West, R. G. (1975). Fossil ice-wedge polygons at Corton, Suffolk. *Bull. geol. Soc. Norfolk*, 27, 47–53.

Godwin, H. (1975). *The History of the British Flora*, 2nd edition. Cambridge University Press.

Goudie, A. S. & Hart, M. G. (1975). Pleistocene events and forms in the Oxford region. In *Oxford and Its Region: Geographical Essays* (ed. by C. G. Smith & D. I. Scargill), pp. 3–13. Oxford University Press.

Groot, J. J. (1966). Some observations on pollen grains in suspension in the estuary of the Delaware River. *Marine Geol.* 4, 409–16.

Groot, J. J. & Groot, C. R. (1966). Marine palynology: possibilities, limitations, problems. *Marine Geol.* 4, 387–95.

Grüger, E. (1968). Vegetationgeschichtliche Untersuchungen an cromerzeitlichen Ablagerungen im nördlichen Randgebiet der deutschen Mittelgebirge. *Eiszeitalter u. Gegenwart*, 18, 204–35.

Gunn, J. (1891). The Cromer Forest Bed and its Mammalia. In *Memorials of John Gunn* ed. by H. B. Woodward & E. T. Newton), pp. 45–66. Norwich: W. A. Nudd.

Guppy, H. B. (1906). *Observations of a Naturalist in the Pacific.* Vol. II. *Plant Dispersal.* London: Macmillan.

Harrison, W. J. (1897). A bibliography of Norfolk glaciology including the Cromer Cliffs, with the Forest-bed Series. *Glacialists' Magazine.*

Hartman, A. A. (1968). A study of pollen dispersal and sedimentation in the western part of the Netherlands. *Acta Bot. Neerlandica*, 17, 506–49.

Harvey, B. I., Langston, M. J., Hughes, M. D. A., Whalley, H. A. and others. (1974). *Records of Wells in North Norfolk.* Well Inventory Series (Metric Units), Institute of Geological Sciences. London: H.M.S.O.

Hey, R. W. (1967). The Westleton Beds reconsidered. *Proc. geol. Assoc. Lond.* 78, 427–45.

Hey, R. W. (1976). Provenance of far-travelled pebbles in the pre-Anglian Pleistocene of East Anglia. *Proc. geol. Assoc. Lond.* 87, 69–81.

Hicks, S. (1977). Modern pollen rain in Finnish Lapland investigated by analysis of surface moss samples. *New Phytol.* 78, 715–34.

Hopkins, J. S. (1950). Differential flotation and deposition of coniferous and deciduous tree pollen. *Ecology*, 31, 633–41.

Huckerby, E. & Oldfield, F. (1976). The Quaternary vegetational history of the French Pays Basque. II. Plant macrofossils and additional pollen-analytical data. *New Phytol.*, 77, 499–526.

Hutchinson, J. N. (1976). Coastal landslides in cliffs of Pleistocene deposits between Cromer and Overstrand, Norfolk, England. In *Contributions to soil mechanics* (ed. by N. Janbu, F. Jørstad & B. Kjærnsli), pp. 155–82. Oslo: Norwegian Geotechnical Institute.

Iversen, J. (1936). Sekundäres pollen als Fehlerquelle. *Danm. geol. Unders.* IV Række, 2, no. 15.

Iversen, J. (1958). The bearing of glacial and interglacial epochs on the formation and extinction of plant taxa. *Uppsala Universitets Årsskrift*, 1958, 210–15.

Kerney, M. P. (1976). A list of the fresh and brackish-water Mollusca of the British Isles. *J. Conch.* 29, 26–8.

Koreneva, E. V. (1957). Spore-pollen analysis of bottom sediments from the Sea of Okhotsk. *Tr. Inst. Oceanol. Akad. Nauk SSSR*, 22, 221–51 (in Russian).

Koreneva, E. V. (1966). Marine palynological researches in the U.S.S.R. *Marine Geol.* 4, 565–74.

Krinsley, D. H. & Funnell, B. M. (1965). Environmental history of quartz sand grains from the Lower and Middle Pleistocene of Norfolk, England. *Quart. J. geol. Soc. Lond.* 121, 36–43.

Lüttig, G. (1965). The Bilshausen type section, West Germany. *Geol. Soc. America Special Paper*, 84, 159–78.

Markham, R. A. D. (1973). Crag from a borehole at Overstrand, Norfolk. *Bull. Ipswich geol. Group*, 12, 2.

Martin, W. E. (1959). The vegetation of Island Beach State Park, New Jersey. *Ecological Monographs*, 29, 1–45.

McQueen, D. R. (1969). Macroscopic plant remains in recent lake sediments. *Tuatara*, 17, 13–19.

Menke, B. (1975). Vegetationgeschichte und Florenstratigraphie Nordwestdeutschlands in Pliozän und Frühquartär. Mit einem Beitrag zur Stratigraphie des Weichselfrühglazials. *Geol. Jb.* A, 26, 3–151.

Montfrans, H. M. van (1971). *Palaeomagnetic Dating in the North Sea Basin.* Rotterdam: Princo N.V.

Müller, H. (1965). Eine pollenanalytische Neuarbeitung des Interglazial-Profils von Bilshausen (Unter-Eichsfeld). *Geol. Jb.* 83, 327–52.

Muller, J. (1959). Palynology of Recent Orinoco delta and shelf sediments. *Micropalaeontology*, 5, 1–32.

Newton, E. T. (1882). *The Vertebrata of the Forest Bed Series of Norfolk and Suffolk.* Mem. Geol. Surv. E. and W.

Newton, E. T. (1891). *The Vertebrata of the Pliocene Deposits of Britain.* Mem. Geol. Surv. U.K.

Norton, P. E. P. (1967). Marine molluscan assemblages in the early Pleistocene of Sidestrand, Bramerton and the Royal Society borehole at Ludham, Norfolk. *Phil. Trans. R. Soc.* B, 253, 161–200.

Norton, P. E. P. (1977). Marine mollusca in the East Anglian Preglacial Pleistocene. In *Recent Advances in British Quaternary Studies* (ed. F. W. Shotton). Oxford: Oxford University Press.

Norton, P. E. P. & Beck, R. B. (1972). Lower Pleistocene molluscan assemblages and pollen from the Crag of Aldeby (Norfolk) and Easton Bavents, Suffolk. *Bull. geol. Soc. Norfolk*, 22, 11–31.

Norton, P. E. P. & Spaink, G. (1973). The earliest occurrence of *Macoma balthica* (L.) as a fossil in the North Sea deposits. *Malacologia*, 14, 33–7.

Oakley, K. P. & Baden-Powell, D. F. W. (1963). *Lexique Stratigraphique International.* I. *Europe.* Fasc. 3a, III. Paris: C.N.R.S.

Peake, N. B. & Hancock, J. M. (1970). The Upper Cretaceous of Norfolk. In *The Geology of Norfolk* (ed. by G. P. Larwood & B. M. Funnell), pp. 293–339. Norwich: Geological Society of Norfolk.

Phillips, L. (1974). Vegetational history of the Ipswichian/Eemian interglacial in Britain and continental Europe. *New Phytol.* 73, 589–604.

Phillips, L. (1976). Pleistocene vegetational history and geology in Norfolk. *Phil. Trans. R. Soc.* B, 275, 215–86.

Polak, B., Maarleveld, G. C. & Nota, D. J. G. (1962). Palynological and sedimentary petrological data of a section in ice-pushed deposits (southern Veluwe, Netherlands). *Geol. en Mijnbouw*, 41, 333–50.

Postma, H. (1967). Sediment transport and sedimentation in the estuarine environment. In *Estuaries* (ed. by G. H. Lauff), pp. 158–179. American Association for the Advancement of Science Publication No. 83.

Praeger, R. L. (1913). On the buoyancy of seeds of some Britannic plants. *Scient. Proc. R. Dublin Soc.* 14, 13–62.

Prestwich, J. (1890). On the relation of the Westleton Beds or Pebbly Sands of Suffolk to those of Norfolk and on their extension inland. Part I. *Quart. J. geol. Soc. Lond.* 46, 84–119.

Reid, C. (1882*a*). *The Geology of the Country around Cromer.* Mem. Geol. Survey E. and W.

Reid, C. (1882*b*). Horizontal Section, Sheet 127, and explanation. Geological Survey E. and W.

Reid, C. (1890). *The Pliocene Deposits of Britain.* Mem. Geol. Surv. U.K.

Reid, P. C. & Downie, C. (1973). The age of the Bridlington Crag. *Proc. Yorks. Geol. Soc.* 39, 315–18.

Reineck, H. E. (1967). Layered sediments of tidal flats, beaches and shelf bottoms of the North Sea. In *Estuaries* (ed. by G. H. Lauff), pp. 191–206. American Association for the Advancement of Science Publication No. 83.

Ridder, N. A. de & Zagwijn, W. H. (1962). A mixed Rhine–Meuse deposit of Holsteinian age from the south-eastern part of the Netherlands. *Geol. en Mijnbouw*, 41, 125–30.

Rose, J. & Allen, P. (1977). Middle Pleistocene stratigraphy in south-east Suffolk. *J. geol. Soc. Lond.*, 133, 83–102.

Rose, J., Allen, P. & Hey, R. W. (1975). Middle Pleistocene stratigraphy in southern East Anglia. *Nature, Lond.* 263, 492–4.

Rudolph, K. (1935). Mikrofloristische Untersuchungen Tertiärer Ablagerungen im nördlichen Böhmen. *Beih. Bot. Zbl.* 54, 244–378.

Salisbury, E. J. (1942). *The Reproductive Capacity of Plants.* London: G. Bell.

Sandberger, F. (1880). Ein Beitrag zur Kenntnis der unterpleistocänen Schichten Englands. *Palaeontographica*, 27, 81.

Spaink, G. & Norton, P. E. P. (1967). The stratigraphical range of *Macoma balthica* (L.) [Bivalvia, Tellinacea] in the Pleistocene of the Netherlands and eastern England. *Meded. Geol. Sticht.*, N.S. 18, 39–44.

Sparks, B. W. (1964). Non-marine Mollusca and Quaternary ecology. *J. Anim. Ecol.* 33 (Suppl.), 87–98.

Stanley, E. A. (1969). Marine palynology. *Oceanogr. Mar. Biol. Ann. Rev.* 7, 277–92.

Stuart, A. J. (1974). Pleistocene history of the British vertebrate fauna. *Biol. Rev.* 49, 225–66.

Stuart, A. J. (1975). The vertebrate fauna of the type Cromerian. *Boreas*, 4, 63–76.

Stuart, A. J. & West, R. G. (1976). Late Cromerian fauna and flora at Ostend, Norfolk. *Geol. Mag.* 113, 469–73.

Stuchlik, L. (1964). Pollen analysis of the Miocene deposits at Rypin. *Acta Palaeobotanica*, 5, No. 3.

Szafer, W. (1954). Pliocene flora from the vicinity of Czorsztyn (West Carpathians) and its relationship to the Pleistocene. *Instytut Geologiczny, Prace*, 11.

Terwindt, J. H. J. (1971). Litho-facies of inshore estuarine and tidal-inlet deposits. *Geol. en Mijnbouw*, 50, 515–26.

Tralau, H. (1963). Asiatic dicotyledonous affinities in the Cainozoic flora of Europe. *Kungl. Svenska Vetensk. Handlingar*, 4th series, 9, no. 3.

Traverse, A. & Ginsburg, N. G. (1966). Palynology of the surface sediments of Great Bahama Bank, as related to water movement and sedimentation. *Marine Geol.* 4, 417–59.

Trimmer, J. (1845). On the cliffs of Northern Drift on the coast of Norfolk. *Quart. J. Geol. Soc. Lond.* 1, 218–19.

Turner, C. (1966). *Middle Pleistocene Vegetational History and Geology in East Anglia.* Ph.D. dissertation, University of Cambridge.

Turner, C. (1970). The Middle Pleistocene deposits at Marks Tey, Essex. *Phil. Trans. R. Soc.* B, 257, 373–440.

Turner, C. (1973). East Anglia. In *A Correlation of Quaternary Deposits in the British Isles* (by Mitchell, G. F., Penny, L. F., Shotton, F. W. & West, R. G.). Geol. Soc. Lond., Special Report No. 4.

Turner, C. (1975). The correlation and duration of Middle Pleistocene interglacial periods in northwest Europe. In *After the Australopithecines* (ed. by K. Butzer & G. L. Isaac), 259–308. The Hague: Mouton Publishers.

Turner, C. & West, R. G. (1968). The subdivision and zonation of interglacial periods. *Eiszeitalter und Gegenwart*, 19, 93–101.

Tutin, T. G. *et al.* (eds) (1964–76). *Flora Europaea*, vols. 1–4. Cambridge University Press.

Valentine, K. H. G. & Dalrymple, J. B. (1975). The identification, lateral variation and chronology of two buried paleocatenas at Woodhall Spa and West Runton, England. *Quaternary Research*, 5, 551–90.

Waldén, H. W. (1976). A nomenclatural list of the land Mollusca of the British Isles. *J. Conch.* 29, 21–5.

Wall, D. & Dale, B. (1968). Early Pleistocene dinoflagellates from the Royal Society borehole at Ludham, Norfolk. *New Phytol.* 67, 315–26.

Warren, S. H. (1940). Geological and Prehistoric traps. *Essex Nat.* 27, 2–19.

West, R. G. (1956). The Quaternary deposits at Hoxne, Suffolk. *Phil. Trans. R. Soc.* B, 239, 265–356.

West, R. G. (1961). Vegetational history of the Early Pleistocene of the Royal Society borehole at Ludham, Norfolk. *Proc. R. Soc.* B, 155, 437–53.

West, R. G. (1968). Evidence for pre-Cromerian permafrost in East Anglia. *Biuletyn Peryglacjalny*, 17, 303–4.

West, R. G. (1972). Relative land/sea-level changes in south-eastern England during the Pleistocene. *Phil. Trans. R. Soc.* A, 272, 87–98.

West, R. G. (1973). Introduction. In *Quaternary Plant Ecology* (ed. by H. J. B. Birks & R. G. West), pp. 1–3. Oxford: Blackwell Scientific Publications.

West, R. G. (1977). *Pleistocene Geology and Biology*, 2nd edition. London: Longman.

West, R. G. & Banham, P. H. (1968). Short field meeting on the north Norfolk coast. *Proc. geol. Assoc. London*, 79, 493–512.

West, R. G. & Donner, J. J. (1958). A note on Pleistocene frost structures in the cliff section at Bacton, Norfolk. *Trans. Norfolk and Norwich Nat. Soc.* 18, 8–9.

West, R. G. & Norton, P. E. P. (1974). The Icenian Crag of southeast Suffolk. *Phil. Trans. R. Soc.* B, 269, 1–28.

West, R. G. & Wilson, D. G. (1966). Cromer Forest Bed Series. *Nature, Lond.* 209, 497–8.

Wilson, D. G. (1973). Notable plant records from the Cromer Forest Bed Series. *New Phytol.* 72, 1207–34.

Wilson, D. G., Marchant, R. & Oldfield, F. (1973). Fossil seeds of *Erica* from the Cromer Forest Bed Series. *New Phytol.* 72, 1235–7.

Woodward, H. B. (1891). Memoir of John Gunn. In *Memorials of John Gunn* (ed. by H. B. Woodward and E. T. Newton), pp. 3–42. Norwich: W. A. Nudd.

Zagwijn, W. H. (1960). Aspects of the Pliocene and Early Pleistocene vegetation in the Netherlands. *Med. Geol. Stichting*. C. III, no. 5.

Zagwijn, W. H. (1974). The palaeogeographic evolution of the Netherlands during the Quaternary. *Geol. en Mijnbouw*, 53, 369–85.

Zagwijn, W. H. (1975). Variations in climate as shown by pollen analysis, especially in the Lower Pleistocene of Europe. In *Ice Ages: Ancient and Modern* (ed. by A. E. Wright & F. Moseley), pp. 137–52. Liverpool: Seel House Press.

Zagwijn, W. H. (1977). Lower and Middle Pleistocene coastlines in the southern North Sea basin. In *North Sea Shorelines*. INQUA Shoreline Commission.

Zagwijn, W. H., Montfrans, H. M. van & Zandstra, J. G. (1971). Subdivision of the 'Cromerian' in the Netherlands; pollen analysis, palaeomagnetism and sedimentary petrology. *Geol. en Mijnbouw*, 50, 41–58.

Zagwijn, W. & Staalduinen, C. J. van (1975). *Toelichting bij geologisches overzichtskaarten van Nederland*. Rijks Geologische Dienst, Haarlem.

Zagwijn, W. H. & Veenstra, H. J. (1966). A pollen analytical study of cores from the Outer Silver Pit, North Sea. *Marine Geol.* 4, 539–51.

Zagwijn, W. H. & Zonneveld, J. I. S. (1956). The interglacial of Westerhoven. *Geol. en Mijnbouw*, 18, 37–46.

Table 1. *Pollen assemblage biozones of substage Pre-Pastonian a*

Site		Location (Fig. no.)	Pollen diagram (Fig. no.)	p.a.b.	Comments
West Runton	WRQ	4	12	*Pinus–Betula*–Gramineae–Ericales	closely associated with crag
	WRE	4	12	*Pinus–Betula*–Gramineae–Ericales	and stone bed
	WRBN	11	12	*Pinus–Betula*–Gramineae–Ericales	
Beeston	BE	16	19	*Pinus–Betula*–Gramineae–Ericales	associated with stone bed
	BKC	18	19	*Pinus–Betula*–Gramineae–Ericales	
Overstrand	BH O 1B	22	25	*Pinus*–Gramineae–Ericales	associated with crag and stone bed
Sidestrand	SSB 50–300	22	27	*Pinus*–Gramineae–Ericales	
	SSD 0–460	22	27	*Pinus*–Gramineae–Ericales	
	SSF 1	22	26	*Pinus*–Gramineae–Ericales	associated with stone bed
	SSH 5	22	38	*Pinus*–Gramineae–Ericales	
	SSH 20	22	38	Gramineae–*Alnus–Pinus*	in lamina of drift mud
	SSF 2	22	26	Gramineae	erratics of coarse
	SSJ	22	38	Gramineae	detritus mud
	SSV 14	23	26	*Pinus*–Gramineae–Ericales	
Paston–Mundesley	BHC 7	33	35	*Pinus*–Gramineae–Ericales	
	BHC 3	33	35	*Pinus*–Gramineae–Ericales	clay laminae in sand
	MAT 680–1340	33	35	*Pinus*–Gramineae–Ericales	massive laminated clay
	BHA1 640	33	35	*Pinus*–Gramineae–Ericales	clay laminae in sand
Happisburgh	HC 712–1859	39	42	*Pinus*–Gramineae–Ericales	clay laminae in sand
Pakefield	PL 854–1006	46	47	*Pinus*–Ericales	clay laminae in sand
	PK 884	46	47	*Pinus*–Ericales	

Table 2. *Macroscopic plant remains from depth in the Happisburgh borehole (HC)*

Taxon	Type of re- mains*	No. of specimens at different depths (cm) 1967–1971	2006–2009	2009–2014	2560
Montane					
Arabis alpina	s	.	.	.	1
Maritime					
cf. *Salicornia* sp.	s	.	.	.	1
Suaeda maritima	s	.	2	1	.
Heath					
Erica sp.	l			1	
Erica sp.	s	.	.	3	3
Dry-ground herbs					
Campanula glomerata	s	.	.	2	.
Damp-ground herbs					
Montia fontana ssp. *fontana*	s	.	.	4	.
Selaginella selaginoides	m	2	2	1	.
Fen and reedswamp halophytes					
Osmunda sp.	ann	.	.	.	2
Aquatics					
Najas cf. *flexilis*	fr	.	.	1	.
Potamogeton sp.	en	.	.	1	.
Stratiotes aloides	ls	.	.	.	2
Typha sp.	s	.	.	.	5
Water ferns					
Azolla filiculoides	m+ma	.	.	.	12
Unclassified					
Carex spp.	n	.	3	.	.
cf. *Cyperus* sp.	n	.	.	.	2
Dryopteris-type	spo	.	.	.	1
cf. *Juncus*	s	.	.	.	1
cf. *Potentilla* sp.	a	.	.	1	.
Umbelliferae	fr	.	.	.	1
Total (excluding water ferns)		2	7	15	19

* Abbreviations: a, achene; ann, annulus of sporangium; b, bud; bar, barb; br, bract; b.sc, bud scale; c, cone; c.sc, cone scale; ca, caryopsis; cal, calyx; c.v, capsule valve; cy, cypsela; e, embryo; en, endocarp; fr, fruit; fr.st, fruit-stone; l, leaf; l.sc, leaf scar; ls, leaf spine; li, lid; m, megaspore; ma, massula; n, nut; oos, oospore; p, plug; per, perianth; r, rhizome; s, seed; spo, sporangium; s.s, sclerenchyma spindle; sh, sheath; st, stem; t, thallus; te, testa; th, thorn; tub, tubercle; u, utricle; w, wood.

Table 3. *Plant macroscopic remains from substage Pre-Pa a at Mundesley MAT*

Taxon	Type of remains*	No. of specimens at different depths (cm) 697–703	884–894	981–985	1185–1189
Trees					
Betula pendula/pubescens	fr	1	.	.	.
Maritime					
Juncus gerardii	s	.	1	.	.
cf. *Salicornia* sp.	s	.	1	.	.
Sphacelaria cf. *plumigera*	t	.	3	.	.
Heath					
Ericaceae	l	1	1	.	.
Ericaceae	s	.	.	.	1
Dry-ground herbs					
Centaurium erythraea	s	.	2	.	.
Damp-ground herbs					
Juncus cf. *conglomeratus*	s	.	3	.	.
Juncus effusus	s	.	1	.	.
Selaginella selaginoides	m	1	4	.	.
Fen and reedswamp					
Osmunda	ann	.	3	.	.
Urtica dioica	n	.	.	.	1

Taxon	Type of remains*	No. of specimens at different depths (cm) 697–703	884–894	981–885	1185–1189
Aquatics					
Callitriche sp.	fr	.	1	.	.
Stratiotes aloides	ls	.	2	1	.
Typha latifolia	s	.	18	.	.
Typha cf. *latifolia*	s	2	.	.	.
Water ferns					
Azolla filiculoides	m	2	8	.	.
Unclassified					
Carex sp.	n	.	1	.	.
Cruciferae	s	.	1	.	.
Dryopteris-type	spo	.	11	4	.
Gramineae	ca	.	4	.	.
Juncus sp.	s	.	4	.	1
Saxifraga sp.	s	.	2	.	.
cf. *Veronica* sp.	s	.	3	.	.
Total (excluding water ferns)		7	66	5	3
Derived megaspores		13	1	.	.

* Abbreviations as in Table 2.

Table 4. *West Runton: pollen assemblage biozones*

Site	Depth (cm)	Location	Bed	Designation of p.a.b.	Stage, substage
WRAYa	55	161 m; Fig. 10	post-WRFB c	*Betula*	Cr IVc
WRAY	57, 107, 127	162 m; Fig. 10	post-WRFB c	*Betula*	Cr IVc
WRAYa	80	161 m; Fig. 10	post-WRFB c	*Pinus–Alnus–Picea–Betula*	Cr IVa
WRAY	95	162 m; Fig. 10	post-WRFB c	*Pinus–Alnus–Picea–Betula*	Cr IVa
WRAW	50–130	43 m; Fig. 7	post-WRFB b		
WRAYb	153–180	159 m; Fig. 10	post-WRFB b		
WRAYa	130–170	161 m; Fig. 10	post-WRFB b	*Abies–Carpinus*	Cr IIIb
WRAZ		295 m; Fig. 11	post-WRFB b		
WRBQ		127 m; Fig. 6	post-WRFB b		
WRBZb		5 m; Fig. 7	post-WRFB b		
WRBCa		406 m; Fig. 11	post-UFB b	*Carpinus*	Cr IIIa
WRU	0–115	334 m; Fig. 6	WRFB f		
WRCGB	140–110	315 m; Fig. 6	WRFB f, d	*Quercus–Ulmus–Tilia*	Cr IIb
WRAR	5	345 m; Fig. 6	WRFB f		
WRBXb	0–5	16 m; Fig. 7	WRFB f		
WRU	115–125	334 m; Fig. 6	WRFB f, d		
WRAQ	+5	286 m; Fig. 6	WRFB d	*Pinus–Quercus–Ulmus*	Cr IIa
WRCGB	110–80	315 m; Fig. 6	WRFB d		
WRT	40	524 m; Fig. 4	WRFB z		
WRU	125–140	334 m; Fig. 6	WRFB d		
WRAQ	0–15	286 m; Fig. 6	WRFB a		
WRCGB	80–40	315 m; Fig. 6	WRFB d	*Pinus–Ulmus*	Cr Ib
WRAR	75	345 m; Fig. 6	WRFB b		
WRT	25–5	524 m; Fig. 4	WRFB z		
WRAQ	15–35	286 m; Fig. 6	WRFB a		
WRCGB	40–0	315 m; Fig. 6	WRFB a	*Betula–Pinus*	Cr Ia
WRCGC	3–17	308 m; Fig. 6	WRFB a		
WRAQ	35–65	286 m; Fig. 6	WRFB a	Gramineae–Cyperaceae–*Betula*	l Be b
WRAQ	65–90	286 m; Fig. 6	WRFB a	Gramineae–Cyperaceae–*Artemisia*	l Be a
WRCNA		450 m; Fig. 4	pre-WRFB g	*Alnus–Betula–Pinus–Quercus*	Pa III
WRAA		inset; Fig. 4	pre-WRFB e	*Carpinus*	Pa III
WRAX	0–160	36 m; Fig. 7	pre-WRFB d i		
WRM	15–35	70 m; Fig. 8	pre-WRFB d i		
WRBD		32 m; Fig. 7	pre-WRFB d i	*Pinus–Quercus–Carpinus–Ulmus*	Pa II
WRCMa		inset; Fig. 4	pre-WRFB e		
WRF	0–20	inset; Fig. 4	pre-WRFB d ii		
WRAC	5–0	inset; Fig. 4	pre-WRFB d ii		
WRAC	15–10	inset; Fig. 4	pre-WRFB dii	Gramineae–*Pinus–Quercus–Ulmus–Carpinus*	(l) Pa II
WRAS		191 m; Fig. 10	pre-WRFB c	*Pinus–Picea*	?Pa I
WRAT		205 m; Fig. 10	pre-WRFB c	*Pinus–Picea*–Ericales	?Pre-Pa a
WRQ		inset; Fig. 4	pre-WRFB c	*Pinus–Betula*–Gramineae–Ericales	Pre-Pa a
WRE		inset; Fig. 4	pre-WRFB c	*Pinus–Betula*–Gramineae–Ericales	Pre-Pa a
WRBN	0, tooth	705 m; Fig. 11	pre-WRFB b	*Pinus–Betula*–Gramineae–Ericales	Pre-Pa a

Table 5. *West Runton: macroscopic plant remains*

			No. of specimens at different depths										
			Pa				I Be			Cr			
			II			III	a	a/b	b	b/Ia	Ia		
	Taxon	Type of remains*	WRF 18–25	WRF 13–18	WRF 0–13	WR CNA	WRAQ 80–90	WRAQ 60–80	WRAQ 50–60	WRAQ 30–40	WRCGB 25–40	WRCGB 0–20	WRAQ 10–25
	Trees												
1	*Alnus* sp.	c	.	.	.	.	.	.	.	.	.	.	.
2	*Alnus* sp.	c.sc	.	.	.	4	.	.	.	.	.	.	.
3	cf. *Alnus* sp.	c	.	.	.	.	.	1	.	.	.	.	.
4	*Alnus glutinosa*	fr	.	.	.	27	.	.	.	.	.	.	.
5	*Betula* sp.	c.sc	.	.	.	.	.	.	.	.	22	.	.
6	*Betula pendula*	fr	.	.	.	.	.	.	.	.	.	.	.
7	*B. pendula*	c.sc	.	.	.	.	.	.	.	.	.	2	.
8	*B. pendula/pubescens*	fr	.	.	.	.	.	.	.	.	.	36	1
9	*B. pendula/pubescens*	c.sc	.	.	.	.	.	.	.	.	.	10	.
10	*B. pubescens*	c.sc	.	.	.	.	.	.	.	.	.	2	.
11	*B.* tree sp. × *B. nana*	fr	.	.	.	11	.	.	.	.	.	.	.
12	cf. Coniferae	l	.	.	.	1	.	.	.	.	.	.	.
13	*Ilex aquifolium*	s	.	.	.	.	.	.	.	.	.	.	.
14	cf. *Ilex* sp.	s	.	.	.	.	.	.	.	.	.	.	.
15	*Prunus domestica*	fr.st	.	.	.	.	.	.	.	.	.	.	.
	Shrubs												
16	*Betula nana*	fr	.	.	.	.	.	4	.	.	.	.	.
17	*Betula* cf. *nana*	fr	.	.	.	.	.	.	.	2	.	.	.
18	*Rosa* sp.	th	.	.	.	.	.	.	.	.	.	.	.
19	*Rubus idaeus*	a	.	.	.	.	.	.	.	.	.	.	.
20	*Rubus* cf. *idaeus*	a	.	.	.	.	.	.	.	.	.	.	.
21	*Salix* sp.	b	.	.	.	.	.	.	.	.	.	.	.
22	*Salix* sp.	b.sc	.	.	.	.	.	6	.	.	.	.	.
23	*Salix* sp.	b + b.sc	.	.	.	13	.	.	.	.	.	.	.
24	*Salix* sp.	l.sc	.	.	.	.	.	.	.	.	.	.	.
	Maritime												
25	*Euphorbia paralias*	s	.	.	.	.	.	.	.	.	.	.	.
26	*Frankenia laevis*	s	.	.	.	.	.	.	.	.	.	.	1
27	*Salicornia* sp.	s	.	.	.	7	.	.	.	.	.	.	.
	Heath												
28	*Carex pauciflora*	n + u	.	.	.	.	.	.	.	.	.	.	.
29	*Empetrum* sp.	s	.	.	.	2	.	.	.	.	.	.	.
30	*Erica* sp.	s	.	.	.	.	.	.	.	.	.	.	.
31	cf. *Erica* sp.	s	.	.	.	1	.	.	.	.	.	.	.
32	*Erica tetralix*	s	.	.	.	.	.	.	.	.	.	.	.
33	*Eriophorum vaginatum*	s.s	.	.	.	9	.	.	.	.	.	.	.
34	*Viola* cf. *lactea*	s	.	.	.	.	.	.	.	.	1	.	.
	Montane												
35	*Ranunculus hyperboreus*	a	.	.	.	.	.	.	.	1	.	.	.
	Dry-ground herbs												
36	*Atriplex* sp.	s	.	.	.	.	1	.	.	9	.	.	.
37	cf. *Atriplex* sp.	s	.	1	.	2	.	.	.	.	.	.	.
38	*Atriplex hastata*	s	.	.	.	.	.	.	2	.	.	.	.

Ia/b	Ib	Ib/IIa	IIa	IIa/b	IIb												II	
WRAQ 5– 20	WRU 130– 140	WRU 120– 130	WRAQ 0– +15	WRU 110– 120	WRU 100– 110	WRU 90– 100	WRU 80– 90	WRU 70– 80	WRU 60– 70	WRU 50– 60	WRU 40– 50	WRU 30– 40	WRU 20– 30	WRU 10– 20	WRU 0– 10	WRBXb	FWB	
.	.	.	2	3	5	2	.	1	.	2	1	.	.	3	2	.	3	1
.	.	.	.	.	2	.	.	.	.	.	.	.	.	.	.	.	.	2
.	.	.	.	.	.	.	.	.	.	.	.	.	.	.	.	.	.	3
.	6	5	32	43	145	151	73	35	12	2	3	.	11	9	.	.	.	4
.	.	.	.	.	.	.	.	.	.	.	.	.	.	.	.	.	.	5
.	.	.	.	.	17	.	.	.	.	.	.	.	.	.	.	.	.	6
.	.	.	.	.	.	.	.	.	.	.	.	.	.	.	.	.	.	7
.	5	3	6	8	.	12	10	4	.	.	.	.	1	1	.	.	.	8
.	.	.	.	.	.	.	.	.	.	.	.	.	.	.	.	.	.	9
.	.	.	.	.	.	.	.	.	.	.	.	.	.	.	.	.	.	10
.	.	.	.	.	.	.	.	.	.	.	.	.	.	.	.	.	.	11
.	.	.	.	.	.	.	.	.	.	.	.	.	.	.	.	.	.	12
.	.	.	.	.	.	.	.	.	.	.	.	.	.	.	.	.	1	13
.	.	.	.	.	.	.	.	.	.	.	.	.	.	.	.	.	1	14
.	.	.	.	.	.	.	.	.	.	.	.	.	.	.	.	.	2	15
.	.	.	.	.	.	.	.	.	.	.	.	.	.	.	.	.	.	16
.	.	.	1	.	.	.	.	.	.	.	.	.	.	.	.	.	.	17
.	.	.	.	.	1	.	.	.	.	.	.	.	.	.	.	.	.	18
.	.	.	.	.	1	1	.	.	.	.	.	.	.	.	.	.	.	19
.	.	1	.	.	.	.	.	.	.	.	.	.	.	.	.	.	.	20
.	.	.	.	2	.	.	.	.	.	.	.	.	.	.	.	.	.	21
.	.	7	.	.	2	.	.	.	.	.	.	.	.	.	.	.	.	22
.	.	.	.	.	.	.	.	.	.	.	.	.	.	.	.	.	.	23
.	.	.	.	1	.	.	.	.	.	.	.	.	.	.	.	.	.	24
.	.	2	.	.	.	.	.	.	.	.	.	.	.	.	.	.	.	25
.	.	.	.	.	.	.	.	.	.	.	.	.	.	.	.	.	.	26
.	.	.	.	.	.	.	.	.	.	.	.	.	.	.	.	.	.	27
.	.	.	1	.	.	.	.	.	.	.	.	.	.	.	.	.	.	28
.	.	.	.	.	.	.	.	.	.	.	.	.	.	.	.	.	.	29
.	.	.	.	.	.	.	.	.	.	.	.	.	.	.	.	1	.	30
.	.	.	.	.	.	.	.	.	.	.	.	.	.	.	.	.	.	31
.	.	.	.	.	.	.	.	.	.	.	.	.	.	.	.	1	.	32
.	.	.	.	.	.	.	.	.	.	.	.	.	.	.	.	.	.	33
.	.	.	.	.	.	.	.	.	.	.	.	.	.	.	.	.	.	34
.	.	.	.	.	.	.	.	.	.	.	.	.	.	.	.	.	.	35
.																		
.	.	98	.	.	.	.	.	15	.	.	18	.	.	.	.	.	.	36
.	.	.	.	.	.	.	.	.	.	.	.	21	4	.	.	.	.	37
.	.	17	.	.	.	.	.	.	.	.	.	.	.	.	.	.	.	38

Table 5 (*cont.*)

			No. of specimens at different depths										
			Pa				1 Be			Cr			
			II			III	a	a/b	b	b/Ia	Ia		
	Taxon	Type of remains*	WRF 18–25	WRF 13–18	WRF 0–13	WR CNA	WRAQ 80–90	WRAQ 60–80	WRAQ 50–60	WRAQ 30–40	WRCGB 25–40	WRCGB 0–20	WRAQ 10–25
	Dry-ground herbs (*cont.*)												
39	*A. hastata/patula*	s	.	.	.	.	.	.	.	.	.	12	.
40	*A. littoralis/hastata*	s	.	.	.	.	.	.	.	.	.	.	.
41	*A. patula*	s	.	.	.	.	.	.	.	.	1	.	.
42	*Atriplex* cf. *patula*	s	.	.	.	.	.	.	.	.	.	.	.
43	*Ballota nigra*	n	.	.	.	.	.	.	.	.	.	.	.
44	*Calamintha* sp.	s	.	.	.	.	.	.	.	.	.	.	.
45	*Campanula rotundifolia*	s	.	.	.	7	.	.	.	.	.	.	.
46	*Cerastium holosteoides*	s	.	.	.	.	.	.	.	.	.	.	.
47	*Chaenorhinum minus*	s	.	.	.	1	.	.	.	.	.	.	.
48	Chenopodiaceae	s	10	.	4	.	.	.	.	.	.	.	.
49	*Chenopodium* sp.	s	.	.	.	1	.	.	.	.	.	.	.
50	cf. *Chenopodium* sp.	s	.	5	.	.	.	.	.	.	.	.	.
51	*Chenopodium album*	s	.	1	.	.	.	.	.	.	.	.	.
52	cf. *Chenopodium album*	s	.	.	.	.	.	.	.	.	.	.	.
53	*Chenopodium glaucum*	s	.	.	.	.	.	.	.	.	6	7	.
54	*C. murale*	s	.	.	.	.	.	1	.	.	.	.	.
55	*C. rubrum/botryodes*	s	.	.	.	.	.	.	.	.	.	.	.
56	*Hypericum humifusum*	s	.	.	.	.	.	.	.	.	.	2	.
57	*Pastinaca sativa*	fr	.	.	.	.	.	.	.	.	.	.	.
58	*Peplis portula*	s	.	.	.	1	.	.	.	.	.	.	.
59	cf. *Picris* sp.	cy	.	.	.	1	.	.	.	.	.	.	.
60	*Plantago major*	s	.	.	.	1	.	.	.	.	.	.	.
61	*Polygonum lapathifolium*	n	.	.	.	1	.	.	.	.	.	.	.
62	*P. lapathifolium* ssp. *nodosum*	n	.	.	.	.	.	.	.	.	.	.	.
63	*Ranunculus* cf. *bulbosus*	a	.	.	.	1	.	.	.	.	.	.	.
64	*Rumex acetosella* agg.	n	.	.	.	.	.	.	.	.	.	.	.
65	*R. crispus*	n+per.	.	.	.	.	.	.	.	.	.	.	.
66	*Sagina* sp.	s	.	.	.	.	.	.	.	.	.	1	.
67	cf. *Sagina* sp.	s	.	.	.	.	.	.	.	.	.	.	.
68	*Sagina ciliata*	s	.	.	.	.	.	.	.	.	.	.	.
69	*Sagina* cf. *procumbens*	s	.	.	.	.	.	.	.	.	.	.	.
70	*Silene dioica*	s	.	.	.	.	.	.	.	.	.	.	.
71	*Solanum nigrum*	s	.	.	.	.	.	.	.	.	.	.	.
72	*Sonchus oleraceus*	cy	.	.	.	.	.	8	.	.	1	.	2
73	*Stellaria media*	s	.	.	.	.	.	.	.	.	.	.	.
74	*Thalictrum minus*	a	.	.	.	1	.	.	.	2	.	.	.
75	*Urtica urens*	n	.	.	.	.	.	.	.	.	.	.	.
76	*Viola arvensis*	s	.	.	.	.	.	.	.	.	.	.	.
	Damp-ground herbs												
77	*Ajuga reptans*	n	.	.	.	1	.	.	.	.	.	.	.
78	*Barbarea vulgaris*	s	.	.	.	.	.	.	.	.	.	.	.
79	*B. vulgaris/stricta*	s	.	.	.	.	.	.	.	.	.	.	.
80	*Carduus* cf. *acanthoides*	cy	.	.	.	.	.	.	.	.	.	.	.
81	*Conium maculatum*	fr	.	.	.	.	.	.	.	.	.	.	.
82	*Juncus compressus*	s	.	.	.	.	.	.	.	.	.	.	.
83	*Ranunculus acris*	a	.	.	.	.	.	.	.	.	.	.	.
84	*R. acris/repens*	a	.	.	.	.	.	.	.	.	.	.	.

Ia/b	Ib	Ib/IIa	IIa	IIa/b	IIb												II	
WRAQ 5–20	WRU 130–140	WRU 120–130	WRAQ 0–+15	WRU 110–120	WRU 100–110	WRU 90–100	WRU 80–90	WRU 70–80	WRU 60–70	WRU 50–60	WRU 40–50	WRU 30–40	WRU 20–30	WRU 10–20	WRU 0–10	WRBXb	FWB	
.	124	.	21	.	.	.	.	.	.	.	.	.	.	.	.	.	.	39
.	.	.	.	.	.	.	.	.	.	.	.	.	.	.	.	.	5	40
.	.	27	.	.	8	.	19	.	.	.	.	.	.	.	.	.	1	41
.	.	.	.	52	.	20	.	.	.	.	.	.	.	.	.	.	.	42
.	.	.	.	.	.	.	.	.	.	.	.	.	.	1	.	.	.	43
.	.	.	.	.	.	7	.	.	.	.	.	.	.	.	.	.	.	44
.	.	.	.	.	.	.	.	.	.	.	.	.	.	.	.	.	.	45
.	1	.	.	.	.	.	.	.	.	.	.	.	.	.	.	.	.	46
.	.	.	.	.	.	.	.	.	.	.	.	.	.	.	.	.	.	47
.	.	.	.	.	.	.	.	.	14	14	.	.	.	.	.	.	.	48
.	.	.	.	.	.	.	.	.	.	.	.	.	.	.	.	.	.	49
.	.	.	.	.	.	.	.	.	.	.	.	.	.	.	.	1	.	50
.	.	.	.	.	16	6	3	.	.	.	.	.	.	.	.	.	.	51
.	.	1	.	.	.	.	.	.	.	.	.	.	.	.	.	.	.	52
.	.	3	.	11	.	.	.	.	.	.	.	.	.	.	.	.	.	53
.	.	.	.	.	.	.	.	.	.	.	.	.	.	.	.	.	.	54
.	.	.	.	.	.	.	.	1	.	.	.	.	.	.	.	.	.	55
.	.	.	.	.	.	.	1	.	.	.	.	.	.	.	.	.	.	56
.	.	1	.	.	.	.	.	.	.	.	.	.	.	.	.	.	.	57
.	.	.	.	.	.	.	5	1	.	.	.	.	.	.	.	.	.	58
.	.	.	.	.	.	.	.	.	.	.	.	.	.	.	.	.	.	59
.	2	.	.	.	.	.	.	.	.	.	.	.	.	.	.	.	.	60
.	.	.	.	1	.	.	.	.	.	.	.	.	.	.	.	.	.	61
.	3	7	1	.	.	.	.	.	.	.	.	.	.	.	.	.	.	62
.	.	.	.	.	.	.	.	.	.	.	.	.	.	.	.	.	.	63
.	3	.	.	.	1	.	.	.	.	.	.	.	.	.	.	.	.	64
.	.	.	.	.	.	.	.	.	.	.	.	.	.	.	.	.	1	65
.	.	.	.	.	.	.	.	.	.	.	.	.	.	.	.	.	.	66
.	1	.	.	.	.	.	.	.	.	.	.	.	.	.	.	.	.	67
.	.	1	.	.	.	.	.	.	.	.	.	.	.	.	.	.	.	68
.	1	.	.	.	.	.	.	.	.	.	.	.	.	.	.	.	.	69
.	.	.	.	.	1	.	.	.	.	.	.	.	.	.	.	.	.	70
.	.	.	.	.	.	.	3	.	.	.	.	.	.	.	.	.	.	71
.	.	.	.	.	.	.	.	.	.	.	.	.	.	.	.	.	.	72
.	.	2	.	.	.	.	.	.	.	.	.	.	.	.	.	.	.	73
.	.	.	.	.	.	.	.	.	.	.	.	.	.	.	.	.	.	74
.	.	2	.	.	.	3	1	.	.	.	.	.	.	.	.	.	.	75
.	1	.	.	.	.	.	.	.	.	.	.	.	.	.	.	.	.	76
.	.	.	.	1	.	.	.	.	.	.	.	.	.	.	.	.	.	77
.	.	.	.	.	.	1	.	.	.	.	.	.	.	.	.	.	.	78
.	.	1	.	.	.	.	.	.	.	.	.	.	.	.	.	.	.	79
.	.	.	.	.	1	.	.	.	.	.	.	.	.	.	.	.	.	80
.	4	4	.	.	.	.	.	.	.	.	.	.	.	.	.	.	.	81
.	.	.	.	.	1	.	.	.	.	.	.	.	.	.	.	.	.	82
.	.	.	.	.	.	3	.	.	.	.	.	.	.	.	.	.	.	83
.	.	.	.	.	.	.	.	.	.	.	.	.	1	.	.	.	.	84

Table 5 (*cont.*)

			No. of specimens at different depths										
			Pa				I Be				Cr		
			II			III	a	a/b	b	b/Ia	Ia		
	Taxon	Type of remains*	WRF 18–25	WRF 13–18	WRF 0–13	WR CNA	WRAQ 80–90	WRAQ 60–80	WRAQ 50–60	WRAQ 30–40	WRCGB 25–40	WRCGB 0–20	WRAQ 10–25
	Damp-ground herbs (*cont.*)												
85	*R. repens*	a	.	.	.	.	.	.	.	.	.	.	.
86	*Scirpus setaceus*	n	.	.	.	.	.	.	.	.	.	1	.
87	*Selaginella selaginoides*	m	.	.	.	.	.	1	.	.	1	.	.
88	*Senecio aquaticus*	cy	.	.	.	.	.	.	.	.	.	.	1
89	*Solanum dulcamara*	s	.	.	.	.	.	.	.	.	.	.	.
90	*Stachys sylvatica*	n	.	.	.	.	.	.	.	.	.	.	.
	Muddy substrates, shallow water												
91	*Alisma* sp.	fr	.	.	.	14	.	.	.	.	.	.	
92	cf. *Alisma* sp.	fr	.	.	.	.	.	.	.	.	.	.	.
93	*Alisma plantago-aquatica*	fr	.	.	43	.	.	1	27	23	5	.	.
94	*Eleocharis* sp.	n	.	.	.	.	.	.	.	14	.	.	88
95	cf. *Eleocharis* sp.	n	.	.	.	.	.	.	2	2	.	4	.
96	*Eleocharis palustris* ssp. *palustris*	n	.	.	.	.	100	.	.	.	.	.	55
97	*E. palustris* ssp. *microcarpa*	n	.	.	.	.	.	.	.	.	.	.	.
98	*E. palustris + uniglumis*	n	.	.	.	.	.	474	.	.	.	.	.
99	*E. uniglumis*	n	.	.	.	.	.	.	.	6	.	.	56
100	*Limosella* sp.	s	.	.	.	2	.	.	.	.	.	.	.
101	*Ranunculus* cf. *flammula*	a	.	.	.	.	.	.	.	1	.	.	.
102	*R. flammula/reptans*	a	.	.	.	1	.	.	.	.	.	.	.
103	*Rorippa nasturtium-aquaticum*	s	.	.	.	.	.	.	.	.	.	.	.
104	*Rumex maritimus*	per	.	.	.	.	.	.	.	.	.	.	.
105	*R. maritimus*	n	.	.	.	1	.	.	.	.	.	.	.
106	*R.* cf. *maritimus*	n	.	.	.	.	.	.	.	.	.	.	.
	Fen and reedswamp helophytes												
107	*Apium* sp.	fr	.	.	.	.	.	.	.	.	.	.	.
108	*Bidens* sp.	cy	.	.	.	3	.	.	.	.	.	.	.
109	*Carex* cf. *aquatilis*	n+u	.	.	.	.	.	.	.	.	.	.	.
110	*Carex dioica*	n+u	.	.	.	.	.	.	.	.	.	.	.
111	*C. riparia*	n	.	.	12	.	.	.	5	2	.	.	.
112	*C.* cf. *riparia*	n	.	.	.	.	.	2	.	.	.	.	.
113	*Cicuta virosa*	fr	.	.	.	.	.	5	.	.	.	.	.
114	*Cladium mariscus*	n	.	.	.	.	.	.	.	.	.	.	.
115	*Cyperus longus*	n	.	.	.	.	.	.	.	.	12	1	.
116	*Epilobium parviflorum*	s	.	.	.	.	.	.	.	1	.	.	.
117	*Eupatorium cannabinum*	cy	.	.	.	1	.	.	.	.	.	.	.
118	*Hypericum tetrapterum*	s	.	.	.	.	.	.	.	.	5	.	.
119	*Lycopus europaeus*	n	.	.	.	.	.	.	.	1	5	2	.
120	*Lythrum salicaria*	s	.	.	.	1	.	.	.	.	.	.	.
121	*Myosoton aquaticum*	s	.	.	.	.	.	.	.	.	.	.	.
122	*Osmunda* sp.	spo	.	.	.	1	.	.	.	.	.	.	.
123	*Osmunda* sp.	ann	.	.	.	1	.	.	.	.	.	.	.
124	*Ranunculus lingua*	a	.	.	.	.	.	.	.	.	.	.	1
125	*R. sceleratus*	a	3	.	.	13	.	1	52	45	142	46	.
126	*Sium latifolium*	fr	.	.	.	.	.	.	.	.	.	.	.
127	*Urtica dioica*	n	2	.	32	8	.	.	356	120	387	86	3
128	*Viola palustris*	s	.	.	.	.	.	.	.	.	1	.	.

Ia/b	Ib	Ib/IIa	IIa	IIa/b	IIb												II	
WRAQ 5–20	WRU 130–140	WRU 120–130	WRAQ 0–+15	WRU 110–120	WRU 100–110	WRU 90–100	WRU 80–90	WRU 70–80	WRU 60–70	WRU 50–60	WRU 40–50	WRU 30–40	WRU 20–30	WRU 10–20	WRU 0–10	WRBXb	FWB	
.	6	.	6	.	3	4	.	.	.	2	.	1	.	1	3	.	.	85
.	.	.	.	.	.	.	.	.	.	.	.	.	.	.	.	.	.	86
.	.	.	.	.	.	.	.	.	.	.	.	.	.	.	.	.	.	87
.	.	.	.	.	.	.	.	.	.	.	.	.	.	.	.	.	.	88
.	.	.	.	.	1	.	.	.	.	.	4	9	2	.	.	.	.	89
.	.	.	.	.	.	.	.	.	.	.	.	1	1	.	.	.	.	90
.	.	.	.	.	.	.	.	.	.	.	.	.	.	.	.	.	.	91
1	.	.	.	.	.	.	.	.	.	.	.	.	.	.	.	.	.	92
.	16	24	10	45	68	57	47	23	21	19	6	7	27	40	8	.	.	93
.	.	.	.	.	19	.	.	1	.	.	.	.	.	2	.	.	.	94
.	.	.	.	.	.	.	15	.	.	.	.	.	.	.	.	.	.	95
.	.	.	.	.	.	.	.	.	.	.	.	.	.	.	.	.	.	96
.	.	2	.	.	.	.	.	.	.	.	.	.	.	.	.	.	.	97
.	.	.	.	.	.	.	.	.	.	.	.	.	.	.	.	.	.	98
.	.	.	.	.	.	.	.	.	.	.	.	.	.	.	.	.	.	99
.	.	.	.	.	.	.	.	.	.	.	.	.	.	.	.	.	.	100
.	.	.	.	.	.	.	.	.	.	.	.	.	.	.	.	.	.	101
.	.	.	.	.	.	.	.	.	.	.	.	.	.	.	.	.	.	102
.	1	.	.	.	.	.	.	.	.	.	.	.	.	.	.	.	.	103
.	24	31	19	54	71	79	59	13	2	.	4	.	3	2	.	.	.	104
.	1	.	.	3	.	.	.	.	.	.	.	.	.	.	.	.	.	105
.	.	5	.	.	6	.	.	.	.	.	.	.	.	.	.	.	.	106
.	.	.	.	.	1	1	.	.	.	.	.	.	.	.	.	.	.	107
.	.	.	.	.	.	1	1	.	.	.	.	.	.	.	.	.	.	108
.	.	.	1	.	.	.	.	.	.	.	.	.	.	.	.	.	.	109
.	.	.	.	.	5	.	.	.	.	.	.	.	.	.	.	.	.	110
.	.	.	.	.	4	.	3	8	.	7	.	2	.	.	.	.	.	111
.	.	.	.	.	.	.	.	.	.	.	.	.	.	.	.	.	.	112
.	.	.	.	.	.	1	.	.	.	2	4	.	.	.	.	.	.	113
.	.	.	.	.	1	.	.	.	.	.	.	.	.	.	.	.	.	114
.	.	.	.	.	3	.	6	.	.	.	.	.	.	.	.	.	.	115
.	.	.	.	.	.	.	.	.	.	.	.	.	.	.	.	.	.	116
.	.	.	.	.	.	1	.	.	.	.	.	.	.	.	.	.	.	117
.	.	.	.	.	.	.	.	.	.	.	.	.	.	.	.	.	.	118
.	2	.	.	1	18	14	9	7	4	2	8	10	5	6	2	.	.	119
.	.	.	.	.	.	.	.	.	.	.	.	.	.	.	.	.	.	120
.	3	.	.	.	.	.	.	.	.	.	.	.	.	.	.	.	.	121
.	.	.	.	.	.	.	.	.	.	.	.	.	.	.	.	.	.	122
.	.	.	.	.	.	.	.	.	.	.	.	.	.	.	.	.	.	123
.	.	.	.	.	.	.	.	.	.	.	.	.	.	.	.	.	.	124
6	46	60	30	109	136	131	93	73	26	41	32	28	23	32	8	.	.	125
.	.	1	.	.	.	.	.	.	.	.	.	.	.	.	.	.	.	126
169	508	331	98	297	405	345	336	101	20	19	22	68	134	72	32	.	.	127
.	.	.	.	.	.	.	.	.	.	.	.	.	.	.	.	.	.	128

Table 5 (*cont.*)

			No. of specimens at different depths										
			Pa				I Be				Cr		
			II			III	a	a/b	b	b/Ia	Ia		
	Taxon	Type of remains*	WRF 18–25	WRF 13–18	WRF 0–13	WR CNA	WRAQ 80–90	WRAQ 60–80	WRAQ 50–60	WRAQ 30–40	WRCGB 25–40	WRCGB 0–20	WRAQ 10–25
	Aquatics												
129	*Callitriche* sp.	fr	.	.	.	22	.	.	.	.	.	.	.
130	*Callitriche hermaphroditica*	fr	.	.	.	3	.	.	.	.	.	.	.
131	*C. intermedia*	fr	.	.	.	1	.	.	.	.	.	.	.
132	*C. platycarpa*	fr	.	.	.	1	.	.	.	.	.	.	.
133	*C. stagnalis*	fr	.	.	.	7	.	.	.	.	.	.	.
134	*Ceratophyllum* sp.	fr	.	19	.	.	.	.	.	.	.	.	.
135	*Ceratophyllum demersum*	fr	.	.	.	.	.	.	.	.	.	.	.
136	cf. *C. demersum*	fr	.	.	.	.	.	.	.	.	.	.	.
137	*C.* cf. *submersum*	fr	.	1	.	.	.	.	.	.	.	.	.
138	*Chara* sp.	oos	.	.	.	6	.	13	.	.	1	5	849
139	*Elatine hydropiper*	s	.	.	.	1	.	.	.	.	.	.	.
140	*Hippuris vulgaris*	a	149	.	.	4	1	4	.	.	5	.	1
141	*H. vulgaris*	p	.	.	.	.	.	.	.	.	1	.	1
142	*Hydrocharis morsus-ranae*	s	.	.	3	2	.	.	1	.	.	.	.
143	*Lemna* sp.	s	.	.	.	.	.	.	.	1	.	.	.
144	*Lemna trisulca*	s	4	2	5	.	.	.	.	.	.	.	.
145	*Menyanthes trifoliata*	s	.	.	.	2	.	.	.	.	.	.	.
146	*Myriophyllum spicatum/verticillatum*	n	.	.	.	.	.	.	.	.	.	.	7
147	*Najas minor*	fr	.	.	.	.	.	.	.	.	.	.	.
148	*Nitella/Tolypella*	oos	.	.	.	68	.	.	.	.	.	.	.
149	*Nuphar* sp.	s	.	.	.	.	.	.	.	1	7	2	.
150	*Nuphar* sp.	p	.	.	.	.	.	.	1	1	17	6	.
151	*Nuphar lutea*	s	.	.	.	.	.	.	.	.	1	2	.
152	*Nymphaea* sp.	s	.	.	.	.	.	.	.	.	.	.	.
153	*Potamogeton* cf. *acutifolius*	en	.	.	.	.	.	.	.	.	.	.	.
154	*P. alpinus*	en	.	.	.	.	.	.	.	.	.	.	.
155	*P.* cf. *alpinus*	en	.	.	.	.	.	.	.	.	.	.	.
156	*P. berchtoldii*	en	.	.	.	.	.	.	.	.	.	.	.
157	*P.* cf. *compressus*	en	.	.	.	.	.	.	.	.	.	.	.
158	*P. densus*	en	.	.	.	.	.	.	.	.	.	.	.
159	*P. friesii*	en	.	.	.	.	.	1	.	.	.	.	6
160	*P.* cf. *friesii*	en	.	.	.	.	.	1	.	.	.	.	.
161	*P. friesii foliosus*	en	.	.	.	.	.	.	.	.	.	.	.
162	*P. lucens*	en	.	.	.	.	.	.	.	.	.	.	.
163	*P. lucens*	li	.	.	.	.	.	.	.	.	.	.	.
164	*P.* cf. *lucens*	en	.	.	.	.	.	.	.	.	.	.	.
165	*Potamogeton* cf. *lucens*	li	.	.	.	.	.	.	.	.	.	.	.
166	*P. natans*	en	.	.	.	.	.	2	.	.	1	.	6
167	*P.* cf. *natans*	en	.	.	.	.	.	.	.	.	.	.	.
168	*P.* cf. *panormitanus*	en	.	.	.	.	.	.	.	.	.	.	.
169	*P. pectinatus*	en	.	15	1	.	.	.	.	.	.	2	1
170	*P.* cf. *pectinatus*	en	.	.	.	.	.	.	.	.	.	.	.
171	*P. pectinatus*	li	.	1	.	.	.	.	.	.	.	.	.
172	*P.* cf. *perfoliatus*	en	.	.	.	.	.	.	.	.	.	.	.
173	*P. trichoides*	en	.	.	.	.	.	.	.	.	.	.	.
174	*P.* cf. *trichoides*	en	.	.	.	.	.	.	.	.	.	.	.
175	*P. trichoides*	li	.	.	.	.	.	.	.	.	.	.	.
176	*P.* spp.	en	.	5	.	5	.	1	.	1	.	.	8

Ia/b	Ib	Ib/IIa	IIa	IIa/b	IIb												II	
WRAQ 5–20	WRU 130–140	WRU 120–130	WRAQ 0–+15	WRU 110–120	WRU 100–110	WRU 90–100	WRU 80–90	WRU 70–80	WRU 60–70	WRU 50–60	WRU 40–50	WRU 30–40	WRU 20–30	WRU 10–20	WRU 0–10	WRBXb	FWB	
.	.	.	1	1	.	.	.	.	.	.	.	.	.	1	.	.	.	129
.	.	.	.	.	.	.	.	.	.	.	.	.	.	.	.	.	.	130
.	1	.	.	.	.	.	.	.	.	.	.	.	.	.	.	.	.	131
.	.	.	.	.	3	.	.	.	.	.	.	.	.	.	.	.	.	132
.	.	.	.	.	.	.	3	.	.	.	.	.	.	.	.	.	.	133
.	.	.	.	.	.	.	.	.	.	.	.	.	.	.	.	.	.	134
.	.	1	.	.	.	.	.	.	.	.	.	2	.	4	.	.	1	135
.	.	.	.	.	.	.	1	.	.	.	.	.	.	.	.	.	.	136
.	.	.	.	.	.	.	.	.	.	.	.	.	.	.	.	.	.	137
.	22	42	.	69	113	8	.	3	.	.	.	.	.	.	.	.	.	138
.	.	.	.	.	.	.	.	.	.	.	.	.	.	.	.	.	.	139
.	.	1	.	.	.	.	.	.	.	.	.	.	.	.	.	.	.	140
.	.	.	.	.	.	.	.	.	.	.	.	.	.	.	.	.	.	141
.	.	.	.	1	5	9	5	1	2	.	.	.	.	.	.	.	.	142
.	.	.	.	.	.	.	.	.	.	.	.	.	.	.	.	.	.	143
.	.	.	.	.	3	1	.	.	.	.	.	.	.	.	.	.	.	144
.	.	.	.	.	.	.	.	.	.	.	.	.	.	.	.	.	.	145
.	.	.	.	.	.	.	.	.	.	.	.	.	.	.	.	.	.	146
.	.	.	.	.	.	.	.	.	.	.	.	.	4	8	6	.	.	147
.	.	.	.	.	.	.	.	.	.	.	.	.	.	.	.	.	.	148
.	5	2	1	5	14	7	5	3	5	2	1	1	.	.	.	.	.	149
.	3	.	.	10	15	14	5	.	.	.	.	.	.	.	.	.	.	150
.	1	.	1	2	1	2	.	1	1	1	.	.	4	3	.	.	2	151
.	.	.	.	1	1	.	.	.	.	.	.	.	.	.	.	.	.	152
.	.	.	6	.	.	.	.	.	.	.	.	.	.	.	.	.	.	153
.	.	.	.	.	.	.	.	.	.	.	.	.	1	2	.	.	.	154
.	.	.	.	.	.	.	.	.	.	.	.	.	2	.	.	.	.	155
.	.	.	.	.	.	.	.	.	.	1	.	.	.	.	.	.	.	156
.	.	1	.	.	.	.	.	.	.	.	.	.	.	.	.	.	.	157
.	.	.	.	.	.	.	.	.	.	.	.	.	5	.	2	.	.	158
.	.	.	2	.	.	.	.	.	.	.	1	.	.	.	4	.	.	159
.	.	.	.	.	2	.	.	.	.	.	.	.	.	.	.	.	.	160
.	.	.	.	7	36	13	5	4	2	1	.	.	.	.	.	.	.	161
.	8	7	3	18	2	.	.	.	.	.	.	.	.	.	.	.	.	162
.	.	.	2	1	.	.	.	.	.	.	.	.	.	.	.	.	.	163
.	.	.	.	5	2	.	.	.	.	.	.	.	.	.	.	.	.	164
.	.	.	.	.	3	.	.	.	.	.	.	.	.	.	.	.	.	165
.	.	.	.	5	5	4	4	4	.	.	.	.	.	.	.	.	.	166
.	.	.	.	.	.	.	.	.	4	.	1	.	.	.	.	.	.	167
.	.	.	.	.	.	.	.	.	.	.	.	.	.	2	.	.	.	168
.	.	.	.	1	1	.	.	.	.	.	.	.	.	.	.	.	.	169
.	.	.	.	.	1	.	.	.	.	.	.	.	.	.	.	.	.	170
.	.	.	.	.	.	.	.	.	.	.	.	.	.	.	.	.	.	171
.	.	.	.	.	.	.	.	.	.	.	.	.	.	3	.	.	.	172
.	.	.	.	.	.	.	.	.	.	.	.	.	5	2	.	.	.	173
.	.	.	.	.	.	.	.	1	.	.	.	.	.	.	.	.	.	174
.	.	.	.	.	.	.	.	.	.	.	.	.	4	.	.	.	.	175
.	1	9	.	5	4	6	8	3	3	2	.	.	2	5	.	.	.	176

Table 5 (*cont.*)

	Taxon	Type of remains*	No. of specimens at different depths										
			Pa				l Be			Cr			
			II			III	a	a/b	b	b/Ia	Ia		
			WRF 18–25	WRF 13–18	WRF 0–13	WRCNA	WRAQ 80–90	WRAQ 60–80	WRAQ 50–60	WRAQ 30–40	WRCGB 25–40	WRCGB 0–20	WRAQ 10–25
	Aquatics (*cont.*)												
177	*Potamogeton* spp.	li	.	.	.	.	.	.	.	.	3	.	.
178	*Ranunculus* subg. *Batrachium*	a	7	55	10	72	.	12	.	4	130	44	19
179	*Sagittaria sagittifolia*	fr	.	.	.	4	.	.	5	.	11	2	.
180	*Scirpus lacustris*	n	.	.	.	.	1	2	.	.	10	22	19
181	*S. lacustris + tabernaemontani*	n	.	.	.	.	.	.	.	.	.	.	.
182	*Sparganium* sp.	n	.	.	.	.	.	.	.	.	.	.	.
183	*S. angustifolia*	n	.	.	.	.	.	.	.	.	.	.	.
184	*S.* cf. *angustifolia*	n	.	.	.	.	.	.	.	.	.	.	.
185	*S. emersum*	n	.	.	.	.	.	.	.	.	.	.	.
186	*S.* cf. *emersum*	n	.	.	.	.	.	.	.	.	.	.	.
187	*Stratiotes aloides*	s	.	.	.	.	.	.	.	.	.	.	.
188	*S. aloides*	ls	.	.	.	19	.	.	.	.	.	.	.
189	*Typha* spp.	s	.	10	78	.	.	.	65	.	81	.	5
190	*T. angustifolia*	s	.	.	.	.	.	.	.	6	.	2	.
191	*T. latifolia*	s	.	.	.	.	.	12	.	9	.	13	.
192	*T.* cf. *latifolia*	s	.	.	.	52	.	.	.	.	.	.	.
193	*T. angustifolia + latifolia*	s	24	.	.	.	.	.	.	.	.	.	.
194	*Zannichellia palustris*	a	.	.	.	6	.	.	1	.	33	2	.
	Water ferns												
195	*Azolla filiculoides*	m, m + ma	104	9	496	24	.	.	.	.	.	.	.
196	*Salvinia natans*	m	.	.	.	2	.	.	.	.	.	.	.
197	*S. natans*	ma	.	.	.	16	.	.	.	.	.	.	.
	Unclassified												
198	*Carex* spp.	n	2	.	.	61	.	.	.	5	7	4	.
199	*C.* sp.	u	.	.	.	.	.	.	.	.	1	.	.
200	Caryophyllaceae	s	.	.	.	.	.	.	.	.	.	.	.
201	*Cerastium* sp.	s	.	.	.	.	.	.	.	.	.	1	.
202	*Cirsium* sp.	cy	.	.	.	.	.	.	.	.	.	.	.
203	cf. *Cirsium* sp.	cy	.	.	.	.	.	.	.	.	.	.	.
204	*Cirsium palustre/vulgare/dissectum*	cy	.	.	.	.	.	.	.	.	.	.	1
205	*C. palustre/acaulon*	cy	.	.	.	.	.	.	.	.	.	.	1
206	Compositae	cy	.	.	.	1	.	.	3	.	1	.	.
207	cf. *Crepis* sp.	cy	.	.	.	.	.	.	.	.	.	.	.
208	*Cruciferae*	s	.	.	.	100	.	.	.	.	.	.	.
209	*Cuscuta* cf. *europaea*	s	.	.	.	.	.	.	.	.	.	.	.
210	Cyperaceae	n	.	.	.	.	.	.	.	.	10	.	5
211	cf. *Cyperus*	n	.	.	.	.	.	.	.	.	.	.	.
212	*Dryopteris*-type	spo	.	.	.	18	.	.	.	.	.	.	.
213	*Epilobium* sp.	s	.	.	.	.	.	.	.	.	.	.	.
214	*Equisetum* sp.	sh, st	.	.	.	.	5	11	.	.	.	.	a.
215	Gramineae	ca	.	.	11	12	.	5	3	.	2	3	1
216	*Juncus* sp.	s	.	.	.	5	.	.	.	.	.	.	.
217	Labiatae	n	.	.	.	.	.	.	.	.	1	.	.

Ia/b	Ib	Ib/IIa	IIa	IIa/b	IIb												II	
WRAQ 5–20	WRU 130–140	WRU 120–130	WRAQ 0–+15	WRU 110–120	WRU 100–110	WRU 90–100	WRU 80–90	WRU 70–80	WRU 60–70	WRU 50–60	WRU 40–50	WRU 30–40	WRU 20–30	WRU 10–20	WRU 0–10	WRBXb	FWB	
.	2	5	.	9	6	.	2	.	.	.	.	.	.	5	.	.	.	177
.	47	59	124	132	259	203	127	94	45	37	14	6	90	190	56	.	.	178
.	.	1	2	2	6	.	2	.	9	7	.	.	1	2	.	.	.	179
.	.	14	36	.	113	78	71	167	105	39	20	.	.	.	.	.	.	180
.	.	.	.	118	.	.	.	.	.	.	.	.	.	.	.	.	.	181
.	.	3	.	.	.	1	.	.	1	.	.	.	1	.	.	.	.	182
.	.	.	.	7	3	.	.	5	.	.	.	.	.	.	.	.	.	183
.	.	.	.	.	.	.	1	.	.	.	.	.	.	.	1	.	.	184
.	.	.	.	.	2	.	.	.	.	.	.	.	.	.	.	.	.	185
.	.	.	7	.	.	.	.	.	.	.	.	.	.	.	.	.	.	186
.	.	.	.	.	.	.	1	.	.	.	.	.	3	1	.	.	.	187
.	.	.	.	9	49	20	25	52	6	13	2	3	5	6	5	.	.	188
.	.	.	1	.	.	.	9	.	.	.	1	.	.	.	.	.	.	189
.	.	.	.	2	6	.	.	.	.	.	.	.	.	.	6	.	.	190
.	.	.	.	.	9	6	.	4	.	.	.	.	5	7	15	3	.	191
.	.	.	.	.	.	.	.	.	.	.	.	.	.	.	.	.	.	192
.	.	.	.	.	.	.	.	.	.	.	.	.	.	.	.	.	.	193
.	14	28	31	27	36	14	26	6	1	1	1	.	.	3	2	.	.	194
.	.	.	.	15	177	.	100	.	13	78	16	.	.	v.a.	v.a.	51	.	195
.	.	.	.	.	.	.	.	.	.	.	.	.	.	.	.	.	.	196
.	.	.	.	.	.	.	.	.	.	.	.	.	.	.	.	.	.	197
.	6	2	8	8	16	33	32	72	87	.	.	9	.	35	.	.	.	198
.	.	.	.	.	.	.	.	.	.	.	.	.	.	.	.	.	.	199
.	.	.	1	1	2	.	.	.	.	.	.	.	.	.	.	.	.	200
.	.	.	.	.	.	.	.	.	.	.	.	.	.	.	.	.	.	201
.	1	.	.	.	.	.	.	.	.	.	.	.	.	.	.	.	.	202
.	.	1	.	.	.	.	.	.	.	.	.	.	.	.	.	.	.	203
.	.	.	.	.	.	.	.	.	.	.	.	.	.	.	.	.	.	204
.	.	.	.	.	.	.	.	.	.	.	.	.	.	.	.	.	.	205
.	.	.	.	.	.	.	.	.	.	.	.	.	.	.	.	.	.	206
.	.	.	.	.	.	.	.	.	.	.	.	.	.	1	.	.	.	207
.	1	.	.	.	.	.	.	.	.	.	.	.	.	.	.	.	.	208
.	.	.	.	.	.	.	.	.	.	.	.	.	1	.	.	.	.	209
.	2	.	.	8	10	.	.	.	.	51	27	.	28	.	.	.	.	210
.	.	.	.	.	.	.	.	.	.	2	.	.	.	.	.	.	.	211
.	.	.	.	.	.	.	.	.	.	.	.	.	.	.	.	3	.	212
.	.	2	.	.	.	.	.	.	.	.	.	.	.	.	.	.	.	213
.	.	3	15	a.	v.a.	a.	a.	v.a.	20	8	8	.	.	.	.	.	.	214
1	.	.	.	.	.	.	1	.	.	.	.	.	.	.	.	.	.	215
.	.	.	.	.	.	.	.	.	.	.	.	.	.	.	.	.	.	216
.	23	2	9	18	15	23	5	7	3	.	.	11	9	1	.	.	.	217

Table 5 (*cont.*)

		No. of specimens at different depths										
		Pa				I Be				Cr		
		II			III	a	a/b	b	b/Ia	Ia		
Taxon	Type of remains*	WRF 18–25	WRF 13–18	WRF 0–13	WRCNA	WRAQ 80–90	WRAQ 60–80	WRAQ 50–60	WRAQ 30–40	WRCGB 25–40	WRCGB 0–20	WRAQ 10–25
Unclassified (*cont.*)												
218 *Mentha* sp.	n	.	.	5	.	.	1	7	.	.	.	.
219 *Mentha arvensis/aquatica*	n	.	.	.	.	.	.	12	.	.	35	.
220 *Polygonum* sp.	n	.	.	.	.	.	.	.	1	.	.	.
221 *Potentilla* sp.	a	.	.	.	1	.	.	.	.	.	.	.
222 *Ranunculus* subg. *Ranunculus*	a	.	.	.	.	.	.	.	.	.	.	.
223 *Rumex* sp.	n	.	.	.	.	.	.	.	.	.	1	.
224 cf. *Rumex* sp.	n	.	.	.	.	.	.	.	.	.	.	.
225 *Rumex* subg. *Rumex*	tub + per	.	.	.	.	.	.	.	1	.	.	.
226 *Saxifraga* sp.	s	.	.	.	15	.	.	.	.	.	1	.
227 *Scirpus* sp.	n	.	.	.	.	.	.	.	.	.	.	.
228 *Solanum* sp.	s	.	.	1	.	.	.	.	.	1	.	.
229 Umbelliferae	fr	.	.	.	1	.	.	.	1	.	.	.
230 *Valeriana* sp.	n	.	.	.	.	.	.	.	.	.	.	3
231 *Viola* sp.	s	.	.	.	17	.	.	.	.	.	.	.
232 cf. *Viola* sp.	s	.	.	.	.	.	.	.	.	.	.	.
233 Total (excluding water ferns and *Equisetum*)		201	115	205	545	103	558	542	260	913	359	1141
234 Derived megaspores		1	.	.	222	.	.	.	.	1	.	.

* Abbreviations as in Table 2.

Ia/b	Ib	Ib/IIa	IIa	IIa/b	IIb												II	
WRAQ 5–20	WRU 130–140	WRU 120–130	WRAQ 0–+15	WRU 110–120	WRU 100–110	WRU 90–100	WRU 80–90	WRU 70–80	WRU 60–70	WRU 50–60	WRU 40–50	WRU 30–40	WRU 20–30	WRU 10–20	WRU 0–10	WRBXb	FWB	
.	.	.	.	.	.	.	.	.	.	13	3	.	.	.	.	.	.	218
.	12	12	9	27	103	73	46	21	14	.	.	3	5	16	4	1	.	219
.	.	.	.	.	.	.	.	.	.	.	.	.	.	.	.	.	.	220
.	.	.	.	.	.	.	.	.	.	.	.	.	.	.	.	.	.	221
.	.	5	.	2	5	.	.	2	.	.	.	.	.	.	.	.	.	222
.	.	.	4	.	.	1	.	.	.	.	.	.	.	.	.	.	.	223
.	.	.	.	.	.	.	.	.	.	.	.	.	.	.	.	1	.	224
.	.	.	.	.	.	.	.	.	.	.	.	.	.	.	.	.	.	225
.	.	.	.	.	.	.	.	.	.	.	.	.	.	.	.	.	.	226
.	.	.	.	.	.	.	.	.	.	.	.	7	.	.	.	.	.	227
.	.	1	.	.	.	2	.	.	.	.	.	.	.	.	.	.	.	228
.	1	.	.	.	.	.	.	.	.	.	.	.	1	.	.	.	.	229
.	.	.	.	.	.	.	.	.	.	.	.	.	.	.	.	.	.	230
.	.	.	.	.	.	.	.	.	.	.	.	.	.	.	.	.	.	231
.	.	.	.	.	.	.	.	.	.	.	.	1	.	.	.	.	.	232
177	908	835	476	1132	1793	1358	1068	733	387	280	173	190	387	466	156	.	.	233
.	.	.	.	.	.	.	.	.	.	.	.	.	.	.	.	.	.	234

Table 6. *Freshwater Mollusca from the West Runton Freshwater Bed*

			No. of specimens at different depths (cm)														
			WRU							WRAQ							
	Taxon	Group	140–130	130–120	120–110	110–100	100–90	90–80	80–70	60–50	50–40	40–20	20–5	0–15	WRBZ	WRBZb	
1	*Viviparus viviparus* (Linné)	4	7	43	30	9	8	1	1	—	—	—	—	8	66	7	1
2	*Valvata cristata* Müller	3	13	20	20	26	12	—	—	—	—	—	1	9	5	—	2
3	*V. piscinalis* (Müller)	4	231	571	528	209	106	26	21	—	—	2	4	206	333	21	3
4	*V. goldfussiana* Wüst	4	—	6	14	—	—	—	—	—	—	—	—	—	—	38	4
5	*Marstoniopsis scholtzi* (Schmidt)		1	8	6	1	—	—	—	—	—	—	—	—	5	—	5
6	*Nematurella runtoniana* Sandberger		—	—	—	—	—	—	—	—	—	—	—	—	?1	—	6
7	*Bithynia inflata* (Hansén)	4	83	164	193	63	54	14	20	—	2	—	3	72	115	3	7
8	*Physa fontinalis* (Linné)	4	—	1	2	1	—	—	—	—	—	—	—	—	—	—	8
9	*Lymnaea truncatula* (Müller)	1	3	4	—	—	—	—	—	—	—	—	—	4	—	—	9
10	*L. palustris* (Müller)	2	45	136	109	9	12	3	4	—	—	—	—	35	9	—	10
11	*L. stagnalis* (Linné)	4	—	2	3	4	—	—	—	—	—	—	—	1	4	—	11
12	*L. peregra* (Müller)	2	5	6	7	3	—	—	—	—	—	—	—	5	4	—	12
13	*Lymnaea* sp.		—	—	—	—	—	—	—	—	—	6	9	—	39	—	13
14	*Myxas glutinosa* (Müller)	3	—	1	—	—	?2	—	—	—	—	—	—	—	—	—	14
15	*Planorbis planorbis* (Linné)	3	—	—	—	—	—	—	?1	—	—	—	—	—	—	—	15
16	*P. carinatus* Müller	3	3	7	11	2	4	—	—	—	—	—	—	3	1	—	16
17	*Anisus leucostoma* (Millet)	1	7	18	4	—	—	—	—	—	—	—	—	—	2	—	17
18	*A. vortex* (Linné)	3	—	6	9	2	—	—	—	—	—	—	—	1	2	—	18
19	*A. vorticulus* (Troschel)	3	1	5	6	4	—	—	—	—	—	—	—	3	—	—	19
20	*Bathyomphalus contortus* (Linné)	2	3	3	6	11	—	—	—	—	—	—	—	6	—	—	20
21	*Gyraulus laevis* (Alder)		19	30	3	—	—	—	—	—	—	—	—	8	—	—	21
22	*G. albus* (Müller)	2	12	58	156	112	10	—	1	—	—	—	—	14	22	—	22
23	*Armiger crista* (Linné)	2	29	82	338	202	19	4	—	—	—	—	—	192	26	—	23
24	*Hippeutis complanatus* (Linné)	2	3	14	8	6	1	1	—	—	—	—	—	—	—	—	24
25	*Planorbarius corneus* (Linné)	4	—	3	1	—	—	—	—	—	—	—	—	—	2	—	25
26	*Ancylus fluviatilis* Müller	4	—	—	1	—	—	—	—	—	—	—	—	—	—	—	26
27	*Acroloxus lacustris* (Linné)	3	9	28	77	80	33	22	2	—	—	—	—	41	7	—	27
28	*Unio pictorum* (Linné)	4	—	2	8	—	—	—	—	—	—	—	—	—	3	—	28
29	*Anodonta cygnea* (Linné)		1	—	—	—	—	—	—	—	—	—	—	—	—	—	29
30	*A. anatina* (Linné)		—	—	—	—	—	—	—	—	—	—	—	—	2	—	30
31	*Unio* or *Anodonta* sp.		—	—	1	frag.	—	—	—	—	—	—	—	frag.	—	frag.	31
32	*Sphaerium corneum* (Linné)	2	11	11	6	—	1	—	—	—	—	—	—	—	1	—	32
33	*S. bulleni* Kennard		—	—	—	—	—	—	—	—	—	—	—	—	—	8	33
34	*S. lacustre* (Müller)	1	—	1	—	1	—	—	—	—	—	—	—	—	—	—	34
35	*Pisidium amnicum* (Müller)	4	15	25	19	16	8	4	2	—	—	—	—	11	8	2	35
36	*P. clessini* Neumayr	4	10	25	15	4	2	1	—	—	—	—	—	5	15	15	36
37	*P. casertanum* (Poli)	1	12	18	41	41	12	—	2	—	—	—	—	13	10	12	37
38	*P. casertanum* var. *ponderosa* Stelfox		—	—	—	—	—	—	—	—	—	—	—	—	—	5	38
39	*P. obtusale* (Lamarck)	1	—	4	51	41	2	—	—	—	—	—	—	9	2	—	39
40	*P. milium* Held	2	2	3	7	11	—	—	—	—	—	—	—	6	—	—	40
41	*P. subtruncatum* Malm	2	8	25	36	18	—	—	—	—	—	—	—	11	9	—	41
42	*P. supinum* Schmidt	4	7	6	3	—	—	—	—	—	—	—	—	—	3	32	42
43	*P. henslowanum* (Sheppard)	4	48	127	200	119	10	3	2	—	—	—	—	59	24	46	43
44	*P. hibernicum* Westerlund		—	—	—	1	—	—	—	—	—	—	—	—	—	—	44
45	*P. nitidum* Jenyns	2	60	108	236	197	15	5	1	—	—	—	—	72	18	5	45
46	*P. moitessierianum* Paladilhe	4	62	226	308	144	89	22	3	—	—	—	—	156	61	22	46
47	*Pisidium* spp.		—	—	—	—	—	—	—	—	—	2	—	—	—	—	47
	Total		710	1797	2463	1337	400	106	60	0	2	10	17	950	799	216	
	Land (Table 7)		328	560	309	13	18	1	1	6	6	14	14	80	47	0	
	Combined total		1038	2357	2772	1350	418	107	61	6	8	24	31	1030	846	216	

Table 7. *Land Mollusca from the West Runton Freshwater Bed*

	No. of specimens at different depths (cm)													
	WRU							WRAQ						
Taxon	140–130	130–120	120–110	110–100	100–90	90–80	80–70	60–50	50–40	40–20	20–5	0–15	WRBZ	WRBZb
Carychium minimum Müller	34	63	29	—	—	—	—	—	—	1	1	6	4	—
Succinea putris (Linné)	1	—	—	—	—	—	—	—	—	—	—	—	—	—
S. oblonga Draparnaud	1	4	—	—	—	—	—	—	—	—	—	—	—	—
S. putris (Linné) or *Oxyloma pfeifferi* (Rossmässler)	80	162	97	8	7	—	—	1	3	1	2	18	13	—
Cochlicopa lubrica (Müller)	18	22	13	—	1	—	—	—	—	1	—	5	5	—
Vallonia costata (Müller)	2	—	—	—	—	—	—	—	—	—	—	—	—	—
V. pulchella (Müller)	29	49	15	—	—	—	—	—	—	—	—	—	—	—
V. enniensis Gredler	2	7	—	—	2	—	—	—	—	—	—	1	—	—
Punctum pygmaeum (Draparnaud)	1	2	2	—	—	—	—	—	—	—	—	—	—	—
Discus ruderatus (Studer)	—	1	—	—	—	—	—	—	—	—	—	—	—	—
Vitrina semilimax	11	13	7	—	—	—	—	—	—	—	—	4	1	—
Zonitoides nitidus (Müller)	13	15	8	—	—	—	—	?frag	?frag	6	6	1	4	—
Limax spp.	3	14	5	—	—	—	—	—	—	—	—	1	—	—
Deroceras spp.	118	191	125	5	8	1	1	4	2	5	4	39	19	—
Clausilia bidentata (Ström)	—	1	—	—	—	—	—	—	—	—	—	—	—	—
Clausilia sp.	1	—	—	—	—	—	—	—	—	—	—	—	—	—
Trichia plebeia (Draparnaud)	—	5	—	—	—	—	—	—	—	—	—	—	—	—
T. hispida (Linné)	—	3	7	—	—	—	—	—	—	—	1	—	1	—
Trichia spp.	12	7	—	—	—	—	—	—	—	—	—	5	—	—
Arianta arbustorum (Linné)	—	1	—	—	—	—	—	—	—	—	—	—	—	—
Cepaea sp.	—	—	1	—	—	—	—	—	—	—	—	—	—	—
Cepaea or *Arianta* sp.	2	—	—	—	—	—	—	—	—	—	—	—	—	—
Total	328	560	309	13	18	1	1	6	6	14	14	80	47	0

Table 8. *West Runton: summary of environmental history*

Bed	Sediment	Hiatus	Sedimentary regime	Sea-level	Periglacial: Ice-wedge casts	Periglacial: Involutions	Stage	Pollen assemblage biozone	Regional vegetation	Climate
Post-West Runton Freshwater Bed succession										
e	till		glacial	.	.	.	An	.	.	.
d	sand		freshwater	.	+	.	An	.	.	.
										cold
						+	.	.	.	.
c	massive silt		alluvial, freshwater at top	regression	+	+	Cr IVc	*Betula*	woodland	cool
							Cr IVa	*Pinus–Alnus–Picea–Betula*	woodland	cool
b	laminated silty clay, sandier at base		tidal	.	.	.	Cr IIIb	*Abies–Carpinus*	woodland	temperate
							Cr IIIa	*Carpinus*	woodland	temperate
a	gravel		beach	marine transgression	.	.	.	.	.	.
West Runton Freshwater Bed succession										
g	laminated silt		fluviatile	.	.	.	.	.	.	.
f	detritus mud		slow-running freshwater	.	.	.	. ↑↑Cr IIb	*Quercus–Ulmus–Tilia*	woodland	temperate
e	shelly sand		fluviatile	.	.	.	.	.	.	.
d	breccia-containing sediments		fluviatile	.	.	.	Cr IIa ↓	*Pinus–Quercus–Ulmus*	woodland	temperate
c	shelly sand		fluviatile	.	.	.		.	.	.
b	stony detritus mud		slow-running freshwater	.	.	.	Cr Ib	*Pinus–Ulmus*	woodland	temperate
							↓Cr Ia	*Betula–Pinus*	open woodland	temperate
a	marl, calcareous mud		freshwater, pool	.	.	.	l Be b	Gramineae–Cyperaceae–*Betula*	grassland	.
		+		regression	.	.	l Be a	Gramineae–Cyperaceae–*Artemisia*	.	.
Pre-West Runton Freshwater Bed succession										
i	gravel, sand, with shells		beach, shallow marine	marine transgression	.	+	.	.	.	.
					+ ?	.	.	.	.	cold
							Be			
h	laminated pale silty clay		freshwater	.	.	.	.	.	.	.
					+					
g	silty clay, organic laminations		freshwater/marine	.	.	+	.	.	.	.
				regression	.	.	Pa III	*Alnus–Betula–Pinus–Quercus*	woodland	temperate
f	cross-bedded sand		shallow marine	.	.	.	.	.	.	.
e	gravel, silty clay, clay-conglomerate		shallow marine	.	.	.	Pa III	*Carpinus*	woodland	temperate
							Pa II	*Pinus–Quercus–Carpinus–Ulmus*	woodland	.
d ii	silty clay and detritus mud							*Pinus–Quercus–Carpinus–Ulmus*	woodland	temperate
	(east of Woman Hithe)		brackish	.	.	.	Pa II	Gramineae–*Pinus–Quercus–Ulmus–Carpinus*	fen, woodland	.
d i	laminated silty clay		tidal	.	.	.	Pa II	*Pinus–Quercus–Carpinus–Ulmus*	woodland	temperate
	(west of Woman Hithe)			marine transgression						
		+			+ ?					
c	shelly sand (Weybourne Crag)		marine	.	.	.	.	.	.	.
							Pre-Pa a	*Pinus–Betula*–Gramineae–Ericales	.	
b	stone bed		shallow marine	.	.	.	Pre-Pa a	*Pinus–Betula*–Gramineae–Ericales	grassland (heath ?), some woodland	cold
				marine transgression	.	.	.	.	.	.
a	brecciated Chalk		land surface	.	solifluction					

Table 9. *Beeston: pollen assemblage biozones*

Site	Sample no. *or* depth (cm)	Location	Bed	Designation of p.a.b.	Stage, substage
BP	5, 6	166 m; Fig. 16	u	*Pinus–Betula–Alnus*–Gramineae	Cr IVa
BA-B	4	45 m; Fig. 18	u	*Pinus–Betula–Alnus*–Gramineae–Cyperaceae	Cr IVa
BAM	55	281 m; Fig. 16	r	Cyperaceae–Gramineae–*Pinus*	Be
BAM	130–220	281 m; Fig. 16	r	Gramineae–Cyperaceae–*Pinus–Alnus–Betula*	Be
BM	2	170 m; Fig. 16	r	Gramineae–Cyperaceae–*Pinus*	Be
BK		136 m; Fig. 17	r	Gramineae–Cyperaceae–*Pinus–Betula–Alnus*	Be
BA-B	3	44 m; Fig. 18	r	Gramineae–*Betula–Pinus–Alnus*	Be
BF	4	112 m; Fig. 17	q	Gramineae–*Betula–Pinus–Alnus*	Be
BA-B	2	43 m; Fig. 18	q	Gramineae–*Betula–Pinus*	Be
BV	12, 13	343 m; Fig. 16	p	Gramineae–Cyperaceae	Be
BU	3	290 m; Fig. 16	p	Gramineae–Cyperaceae–*Betula–Alnus*	Be
BF	3	114 m; Fig. 17	p	Gramineae–Cyperaceae–*Betula–Pinus–Alnus*	Be
BM	1	170 m; Fig. 16	o	Gramineae–*Betula*	Be
BAV		200 m; Fig. 16	o	Gramineae–*Betula–Pinus*	Be
BF	2	114 m; Fig. 17	n	Gramineae–*Betula*	Be
BF	1	113 m; Fig. 17	n	Gramineae–*Betula–Pinus*	Be
BR-S	5	23 m; Fig. 18	m	*Betula–Pinus*	Pa IVc
BG	40	17 m; Fig. 18	k	*Pinus–Alnus–Picea–Betula*–Ericales	Pa IVa
BR-S	18–33	23 m; Fig. 18	l	*Pinus–Alnus–Picea–Betula*–Ericales	Pa IVa
BKA	0–40	65 m; Fig. 18	k	*Pinus–Alnus–Picea–Betula*–Ericales	Pa IVa
BKA	60–78	65 m; Fig. 18	k	*Pinus–Betula–Quercus*	Pa III
BR-S	40	23 m; Fig. 18	k	*Pinus–Betula–Quercus*	Pa III
BV	16	342 m; Fig. 16	k		Pa II
BU	2	290 m; Fig. 16	k		Pa II
BAU		223 m; Fig. 16	k		Pa II
BA-B	1	43 m; Fig. 18	k	*Pinus–Quercus–Carpinus–Ulmus*	Pa II
BG	70	17 m; Fig. 18	k		Pa II
BQ	0–15	21 m; Fig. 18	k		Pa II
BR-S	45–50	23 m; Fig. 18	k		Pa II
BQ	15–25	21 m; Fig. 18	i		Pa Ib
BR-S	60–80	23 m; Fig. 18	i		Pa Ib
BG	75	17 m; Fig. 18	i		Pa Ib
BTB	3	184 m; Fig. 16	i	*Pinus*	Pa Ib
BC	2, 1	178 m; Fig. 16	i		Pa Ib
BAX	100–150	127 m; Fig. 17	i		Pa Ib
BW	8, 10, 12	305 m; Fig. 16	g, i		Pa Ib
BU	1	290 m; Fig. 16	g		Pa Ib
BKA	138–158	65 m; Fig. 18	i	*Betula–Pinus*–Gramineae	Pa Ia
BTB	1	184 m; Fig. 16	g	*Betula–Pinus*–Gramineae	Pa Ia
BTA	9, 8	137 m; Fig. 17	g	*Betula–Pinus*–Gramineae	Pa Ia
BAX	190	127 m; Fig. 17	i	Gramineae–Cyperaceae–*Betula–Pinus*	Pre-Pa d
BR-S	100–120	23 m; Fig. 18	i	Gramineae–Cyperaceae–*Betula–Pinus*	Pre-Pa d
BR-S	140–165	23 m; Fig. 18	g	Gramineae–*Pinus–Betula–Alnus*	Pre-Pa c
BKF	1, 2	71 m; Fig. 18	f	Gramineae–*Pinus–Betula–Alnus*	Pre-Pa c
BR-S	170–235	23 m; Fig. 18	d, e	Gramineae–*Artemisia*	Pre-Pa b
BG	140	17 m; Fig. 18	e	Gramineae–*Artemisia*	Pre-Pa b
BA-B	5	43 m; Fig. 18	d	Gramineae–*Artemisia*	Pre-Pa b
BE		300 m; Fig. 16	b	*Pinus–Betula*–Gramineae–Ericales	Pre-Pa a
BKC		68 m; Fig. 18	b	*Pinus–Betula*–Gramineae–Ericales	Pre-Pa a

Table 10. *Beeston: macroscopic plant remains*

Taxon	Type of re-mains*	No. of specimens at different depths														
		Pre-Pa				Pa						Be				Cr
		b	c		d/I	Ia		Ib	II	II/III	IVa					IVa
		BR–S 180–200	BR–S 140–155	BKF 2	BAX 170–190	BKA	*Stratiotes* sample	BR–S 80–90	BQ 5 BG 70	BR–S 40–50	BR–S 20–30	BV 13	BV 12	BAM 40–50	BA–B 3	BA–B 4
Trees																
cf. *Alnus* sp.	fr	.	.	.	.	.	.	1	.	.	.	.	.	.	.	.
Betula sp.	fr	.	.	.	1	.	.	.	.	.	.	.	.	.	.	.
cf. *Betula* sp.	fr	.	.	.	.	.	.	.	.	.	.	.	.	.	1	.
Betula sp.	c.sc	.	.	.	.	.	.	.	.	.	4	.	1	.	.	.
Betula pendula/pubescens	fr	.	.	.	.	.	.	1	.	.	2	.	.	.	.	.
B. pendula/pubescens	c.sc	.	.	.	.	.	.	.	.	.	.	.	1	.	.	.
Betula cf. *pubescens*	fr	.	.	.	.	.	.	.	.	13	.	.	.	.	.	.
Betula pubescens	c.sc	.	.	.	.	.	.	.	.	4	.	.	.	.	.	.
Coniferae	l	.	.	.	.	.	.	.	.	9	2	.	.	.	.	.
Picea sp.	b	.	.		.	.	.	.	.	.	1	.	.	.	.	.
Picea sp.	b.sc	.	.	.	.	.	.	.	.	.	1	.	.	.	.	.
Picea cf. *omorika*	l	.	.	.	.	.	.	.	.	1	.	.	.	.	.	.
Pinus sp.	w	.	.	.	.	.	.	.	.	.	1	.	.	.	.	.
Pinus sp.	s	.	.	.	.	.	.	.	.	.	10	.	.	.	.	.
Shrubs																
cf. *Betula nana*	l	.	.	.	.	.	.	.	.	.	.	.	.	1	.	.
Betula cf. *nana*	fr	.	.	.	.	.	.	.	.	.	.	.	.	.	5	.
Rubus idaeus	a	.	.	.	.	.	.	.	.	1	.	.	.	.	.	.
Salix sp.	b+b.sc	.	.	.	1	.	.	2	.	.	.	.	2	2	.	.
Salix sp.	l	.	.	.	.	.	.	.	.	.	.	.	.	2	.	.
Salix cf. *polaris*	l	.	.	.	.	.	.	.	.	.	.	.	.	2	.	.
Salix cf. *repens*	l	.	.	.	.	.	.	.	.	.	.	.	.	2	.	.
Salix cf. *reticulata*	l	.	.	.	.	.	.	.	.	.	.	.	.	4	.	.
Maritime																
Armeria maritima	cal	.	.	.	.	.	.	.	.	.	.	.	.	1	1	.
Aster tripolium	cy	.	.	.	.	.	.	1	.	1	.	.	.	.	.	.
Juncus gerardii	s	.	.	.	.	.	.	.	.	.	.	.	.	.	.	1
Ruppa sp.	fr.st	.	.	.	.	.	.	.	.	1	.	.	.	.	.	.
Salicornia sp.	s	.	.	.	.	.	.	1	1	46	.	.	.	.	.	.
Gramineae cf. *Spartina* sp.	ca	.	.	.	.	.	.	.	.	24	.	.	.	.	.	.

Table 10 (*cont.*)

Taxon	Type of re-mains*	No. of specimens at different depths: Pre-Pa b BR–S 180–200	Pre-Pa c BR–S 140–155	Pre-Pa c BKF 2	Pre-Pa d/I BAX 170–190	Pa Ia BKA	Pa Ia *Stratiotes* sample	Pa Ib BR–S 80–90	Pa II BQ 5 BG 70	Pa II/III BR–S 40–50	Pa IVa BR–S 20–30	Be BV 13	Be BV 12	Be BAM 40–50	Be BA–B 3	Cr IVa BA–B 4
Heath																
Andromeda polifolia	s	.	.	.	.	.	.	.	.	1	.	.	.	.	.	.
Calluna vulgaris	fr+s	.	.	.	.	.	.	.	.	.	13	.	.	.	.	.
Chamaedaphne calyculata	s	.	.	.	.	.	.	.	.	2	5	.	.	.	.	.
cf. *Chamaedaphne calyculata*	l	.	.	.	.	.	.	.	.	1	.	.	.	.	.	.
cf. Ericaceae	l	.	.	.	.	.	.	.	.	.	.	.	.	.	1	.
Erica spp.	s	.	.	.	.	.	.	.	.	25	156	.	.	.	.	.
Erica lusitanica	s	.	.	.	.	.	.	.	.	.	4	.	.	.	.	.
Erica cf. *lusitanica*	s	.	.	.	.	.	.	.	.	.	1	.	.	.	.	.
Erica umbellata	s	.	.	.	.	.	.	.	.	.	1	.	.	.	.	.
Erica cf. *tetralix*	s	.	.	.	.	.	.	.	.	.	1	.	.	.	.	.
Erica cf. *tetralix*	l	.	.	.	.	.	.	.	.	2	.	.	.	.	.	.
Eriophorum vaginatum	s.s	.	.	.	.	.	.	.	.	28	34	.	.	.	.	.
Vaccinium myrtillus	fr	.	.	.	.	.	.	.	.	.	14	.	.	.	.	.
Montane																
Cherleria sedoides	s	.	.	.	.	.	.	.	.	.	.	.	4	.	.	.
Thalictrum alpinum	a	.	1	.	.	.	.	.	.	.	.	.	.	1	.	.
Saxifraga oppositifolia	l	.	.	.	.	.	.	.	.	.	.	.	.	15	.	.
Dry-ground herbs																
Atriplex patula	s	.	.	.	.	.	.	.	.	.	.	.	.	.	4	.
Campanula rotundifolia	s	.	.	.	.	.	.	.	.	.	.	.	.	.	.	1
Cerastium arvense	s	.	.	.	.	.	.	.	.	.	.	.	.	.	2	.
Chenopodium rubrum/botryodes	s	.	.	.	.	1	.	.	.	.	.	.	.	.	1	.
Corispermum sp.	s	.	.	.	.	.	.	3	.	.	.	.	.	.	.	.
Potentilla anserina	a	.	.	.	.	.	.	.	.	.	.	.	.	.	2	.
Potentilla nivea/tabernaemontani/Sibbaldia procumbens	a	.	.	.	.	.	.	.	.	.	.	.	1	.	.	.
Ranunculus bulbosus	a	.	.	.	6	.	.	.	.	.	.	.	.	.	.	.
Rumex acetosella agg.	a	.	.	.	.	.	.	2	.	.	.	.	.	.	.	.
Thalictrum minus	a	.	.	.	.	.	.	.	.	.	.	.	.	.	1	.
Urtica urens	n	.	.	.	4	.	.	.	.	.	.	.	.	.	.	.

Table 10 (*cont.*)

Taxon	Type of re-mains*	No. of specimens at different depths														
		Pre-Pa				Pa						Be				Cr
		b	c		d/I	Ia		Ib	II	II/III	IVa					IVa
		BR–S 180–200	BR–S 140–155	BKF 2	BAX 170–190	BKA	*Stratiotes* sample	BR–S 80–90	BQ 5 BG 70	BR–S 40–50	BR–S 20–30	BV 13	BV 12	BAM 40–50	BA–B 3	BA–B 4
Damp-ground herbs																
Carex sect. *Acutae*	n+u	.	.	.	.	.	.	.	.	.	.	.	11	.	.	.
Juncus cf. *effusus*	s	.	.	.	.	.	.	.	.	.	.	.	1	.	.	.
Juncus subnodulosus	s	.	.	.	.	.	.	.	.	.	.	1	.	.	.	.
Selaginella selaginoides	m	.	3	.	.	.	.	1	1	.	.	.	.	.	.	4
Muddy substrates/shallow water																
Alisma sp.	fr	.	8	.	35	5	.	3	.	.	.	.	.	.	.	.
Alisma plantago-aquatica	fr	.	.	.	.	.	.	.	.	5	.	.	.	.	.	.
Eleocharis sp.	n	.	3	.	.	2	.	.	.	2	.	.	1	.	.	.
Eleocharis carniolica	n	.	.	.	.	.	.	.	.	.	.	1	.	.	.	.
Eleocharis cf. *palustris* ssp. *palustris*	n	.	.	.	.	8	.	.	.	.	.	.	.	.	.	.
Isoetes cf. *hystrix*	m	.	.	.	.	.	.	.	.	.	.	.	.	.	.	1
Ranunculus flammula	a	.	.	.	.	1	.	.	.	1	.	1	.	.	.	.
Rorippa microphylla	s	.	.	.	.	.	.	.	.	.	.	1	.	.	.	.
Rumex maritimus	per	.	.	.	6	1	.	.	.	.	.	.	.	.	.	.
Fen and reedswamp helophytes																
Bidens tripartita	cy	.	.	.	4	.	.	.	.	.	.	.	.	.	.	.
Carex aquatilis/bigelowii	n+u, n	.	.	.	.	.	.	.	.	.	.	.	.	9	9	.
Cladium mariscus	n	.	.	.	.	2	.	.	.	.	.	.	.	.	.	.
Eupatorium cannabinum	cy	.	.	.	.	.	.	.	.	.	.	.	.	.	.	3
cf. *Glyceria* sp.	ca	.	.	.	.	.	.	.	.	1	.	.	.	.	.	.
Hydrocotyle vulgaris	s	.	.	.	.	6	.	.	.	.	.	.	.	.	.	.
Lycopus europaeus	n	.	.	.	1	.	.	1	.	1	.	.	.	.	.	.
Oenanthe aquatica	fr	.	.	.	.	.	.	.	.	1	.	.	.	.	.	.
Phragmites communis	r	.	.	.	.	.	.	.	.	.	1	.	.	.	.	.
Ranunculus lingua	a	.	.	.	5	.	.	.	.	.	.	.	.	.	.	.
R. sceleratus	a	.	.	.	.	.	.	1	.	6	.	1	.	.	3	.
Urtica dioica	n	1	1	.	6	.	.	1	.	6	.	.	1	.	.	.
Aquatics																
Callitriche sp.	fr	.	.	.	.	.	.	2	.	.	.	.	.	.	.	.
Callitriche hermaphroditica	fr	.	.	.	.	.	.	.	.	.	.	.	.	.	16	.

Table 10 (*cont.*)

Taxon	Type of re-mains*	No. of specimens at different depths: Pre-Pa, b, BR-S 180–200	c, BR-S 140–155	BKF 2	d/I, BAX 170–190	Pa, Ia, BKA	*Stratiotes* sample	Ib, BR-S 80–90	II, BQ 5 BG 70	II/III, BR-S 40–50	IVa, BR-S 20–30	Be, BV 13	BV 12	BAM 40–50	BA-B 3	Cr, IVa, BA-B 4
Aquatics (*cont.*)																
Callitriche intermedia/stagnalis	fr	.	.	.	3	.	.	.	.	.	.	.	.	.	.	.
Chara sp.	oos	.	100	.	.	.	.	86	.	.	.	.	.	.	.	32
cf. *Chara* sp.	oos	.	.	.	.	.	.	.	.	.	.	1	.	.	.	.
Elatine hydropiper	s	.	.	.	.	.	.	.	.	1	.	.	.	.	.	.
Hippuris vulgaris	a	.	.	.	3	1	.	.	.	2	.	2	.	.	95	.
Menyanthes trifoliata	s	.	.	.	.	.	.	.	.	3	.	.	.	.	.	.
Myriophyllum sp.	n	.	.	.	.	.	.	.	.	.	.	.	.	.	1	.
Nitella/Tolypella sp.	oos	.	.	.	.	.	.	.	.	.	.	4	50	.	.	.
Nuphar sp.	s	.	.	.	4	.	.	.	.	.	.	.	.	.	.	.
Nuphar lutea	s	.	.	.	.	2	.	.	.	.	.	.	.	.	.	.
Potamogeton cf. *epihydrus*	en	.	.	.	4	.	.	.	.	.	.	.	.	.	.	.
Potamogeton gramineus	en	.	.	.	1	.	.	.	.	.	.	.	.	.	.	.
Potamogeton cf. *gramineus*	en	.	.	.	1	.	.	.	.	.	.	.	.	.	.	.
Potamogeton lucens	en	.	.	.	2	.	.	.	.	.	.	.	.	.	.	.
P. lucens	li	.	.	.	2	.	.	.	.	.	.	.	.	.	.	.
P. natans	en	.	.	.	.	2	.	.	.	.	.	.	.	.	.	.
Potamogeton cf. *natans*	en	.	.	.	7	.	.	.	.	.	.	.	.	.	.	.
Potamogeton cf. *natans*	li	.	.	.	1	.	.	.	.	.	.	.	.	.	.	.
Potamogeton cf. *nodosus*	en	.	.	.	.	.	.	.	.	1	.	.	1	.	.	.
Potamogeton cf. *nodosus*	li	.	.	.	.	.	.	.	.	2	.	.	.	.	.	.
Potamogeton pectinatus	en	.	.	.	.	.	.	1	.	.	.	.	.	.	.	.
Potamogeton cf. *perfoliatus*	en	.	.	.	.	1	.	.	.	.	.	.	.	.	.	.
P. trichoides	en	.	.	.	13	1	.	.	.	.	.	.	.	.	.	.
P. trichoides	li	.	.	.	3	.	.	.	.	.	.	.	.	.	.	.
P. rutilus	en	.	.	.	.	.	.	.	.	1	.	.	.	.	.	.
Potamogeton spp.	en	.	1	2	6	.	.	1	.	1	.	.	.	.	.	.
Potamogeton spp.	li	.	.	.	5	.	.	.	.	5	.	.	.	.	.	.
Ranunculus subg. *Batrachium* spp.	a	.	.	3	212	30	.	2	.	2	.	.	31	.	4	.
Sagittaria sp.	fr	.	.	.	15	14	.	1	.	.	.	.	.	.	.	.
Sagittaria sagittifolia	fr	.	.	.	.	.	.	.	.	.	.	1	.	.	.	.
Scirpus lacustris	n	.	.	.	.	7	3	.	.	.	.	.	.	.	.	.
S. tabernaemontani	n	.	.	.	.	10	.	.	.	.	.	.	.	.	.	.

Table 10 (*cont.*)

Taxon	Type of re-mains*	No. of specimens at different depths Pre–Pa b BR–S 180–200	c BR–S 140–155	BKF 2	Pa d/I BAX 170–190	Ia BKA	*Stratiotes* sample	Ib BR–S 80–90	II BQ 5 BG 70	II/III BR–S 40–50	IVa BR–S 20–30	Be BV 13	BV 12	BAM 40–50	BA–B 3	Cr IVa BA–B 4
Aquatics (*cont.*)																
Sparganium sp.	n	.	.	.	1	.	.	.	.	.	.	.	.	.	.	.
Stratiotes aloides	s	.	.	.	15	9	1	.	.	.	.	.	.	.	.	.
S. aloides	ls	.	.	.	800	150	22	1	.	2	.	.	.	.	.	.
Typha spp.	s	.	7	.	.	.	.	.	.	.	.	1	5	.	.	22
Typha angustifolia	s	.	.	.	.	.	.	.	.	15	.	.	.	.	.	.
T. latifolia	s	.	.	.	.	3	.	.	.	.	.	.	.	.	.	.
T. angustifolia + latifolia	s	.	.	.	20	.	.	18	.	.	.	.	.	.	.	.
Zannichellia palustris	a	.	.	.	.	.	.	24	.	1	.	.	.	.	5	.
Water ferns																
Azolla filiculoides	m	.	9	.	2	.	.	7	.	.	.	1	2	.	.	.
A. filiculoides	m + ma	.	.	.	.	.	.	.	.	7	.	1	.	.	.	25
Salvinia natans	m	.	.	.	.	.	.	.	.	.	.	.	.	.	.	3
Unclassified																
Carex spp.	n	.	.	.	2	28	1	.	.	17	2	1	3	.	6	1
cf. *Cerastium* sp.	s	.	.	.	.	.	.	.	.	.	.	1	.	.	.	.
cf. *Cirsium* sp.	cy	.	.	.	.	.	.	.	.	.	.	.	1	.	.	.
Cruciferae	s	.	.	.	.	.	.	.	.	.	.	1	.	.	.	.
Gramineae	ca	.	3	.	.	.	.	1	.	.	.	2	3	.	.	.
Isoetes sp.	m	.	1	.	.	.	.	.	.	.	.	.	.	.	.	.
Juncus sp.	s	.	.	.	.	.	.	1	.	3	.	.	1	.	.	.
Mentha sp.	n	.	.	.	.	.	.	1	.	.	.	.	1	.	.	.
cf. *Ranunculus* subg. *Ranunculus*	a	.	.	.	.	.	.	.	.	.	.	.	.	.	1	.
Rumex sp.	n	.	.	.	.	.	.	.	.	1	.	.	.	.	.	.
Saxifraga sp.	s	.	.	.	.	.	.	.	.	.	.	1	.	.	.	.
Scirpus sp.	n	.	.	.	.	5	.	.	.	.	1	.	.	.	.	.
Stellaria sp.	s	.	.	.	1	.	.	.	.	.	.	.	.	.	.	.
Taraxacum sp.	cy	.	.	.	.	.	.	.	.	.	.	1	.	.	.	.
Thalictrum sp.	a	.	1	.	.	.	.	.	.	.	.	.	.	.	.	.
Umbelliferae sp.	fr	.	.	.	.	.	.	.	.	.	.	.	1	.	.	.
Total (excluding water ferns)		1	128	5	1190	289	27	157	2	239	254	20	119	39	158	67
Derived megaspores		.	50	.	4	3	.	7	.	8	.	.	.	.	1	1

* Abbreviations as in Table 2.

Table 11. *Beeston: summary of environmental history*

Bed	Sediment	Hiatus	Sedimentary regime	Sea-level	Periglacial group no.: Ice-wedge casts	Periglacial group no.: Involutions	Stage	Pollen assemblage biozones	Regional vegetation	Climate
w	till		glacial	.	.	.	An			
v	sand		fluviatile							
u	silt		alluvium/tidal?	?	.	.	Cr IVa?	*Pinus–Betula–Alnus*–Gramineae	open woodland	cool
					6	3				
t	sand		fluviatile							
s	gravel		fluviatile channel cutting	.	.	.	.	Cyperaceae–Gramineae–*Pinus*		
		+	land surface, weathering	.	5	.	.	Gramineae–Cyperaceae–*Pinus–Alnus–Betula*		
r	laminated silt		alluvium	.	.	.	.	Gramineae–Cyperaceae–*Pinus*		
						2	.	Gramineae–Cyperaceae–*Pinus–Betula–Alnus*		
								Gramineae–*Betula–Pinus–Alnus*		
					4	.	.	Gramineae–*Betula–Pinus–Alnus*		
q	sand, gravel		fluviatile	.	.	.	.	Gramineae–*Betula–Pinus*		
							Be			
								Gramineae–Cyperaceae–*Betula–Alnus*		
p	laminated muddy silt, mud		freshwater, pool	.	.	.	.	Gramineae–Cyperaceae	grassland	cold
					3	.	.	Gramineae–Cyperaceae–*Betula–Pinus–Alnus*		
				regression						
o	gravel, sand		beach	marine transgression	.	.	.	Gramineae–*Betula*		
								Gramineae–*Betula–Pinus*		
					2					
n	laminated sandy silt, sand, gravel		fluviatile channel cutting	.	1	1	.	Gramineae–*Betula*		
								Gramineae–*Betula–Pinus*		
		+								
m	sand with silt laminae		fluviatile	.	.	.	Pa IVc	*Betula–Pinus*	woodland	cool
l	wood peat		freshwater/telmatic	.	.	.	Pa IVa	*Pinus–Alnus–Picea*–Ericales	woodland, heath	cool
				regression						
k	laminated silty clay		tidal	.	.	.	Pa III	*Pinus–Betula–Quercus*		
							Pa II	*Pinus–Quercus–Carpinus–Ulmus*	woodland	temperate
j	gravel		beach							
				marine transgression						
i	silty mud		freshwater, pool	.	.	. .	Pa I b	*Pinus*	woodland	
							Pa I a	*Betula–Pinus*–Gramineae	open woodland	temperate
							↓ Pre-Pa d	Gramineae–Cyperaceae–*Betula–Pinus*	grassland	cold
h	gravel		fluviatile channel cutting							
g	sand with silt laminae		fluviatile							
							Pre-Pa c	Gramineae–*Pinus–Betula–Alnus*	grassland, some woodland	cool
f	gravel		fluviatile							
e	silty mud		freshwater, pool							
							Pre-Pa b	Gramineae–*Artemisia*	grassland	cold
d	laminated silt and sand		fluviatile							
c	sand		fluviatile or marine?	regression						
		+								
b	stone bed		shallow marine	marine transgression	.	.	Pre-Pa a	*Pinus–Betula*–Gramineae–Ericales	grassland (heath?), some woodland	cold
a	brecciated Chalk		land surface	.	solifluction					

Table 12. *East Runton, Cromer and Overstrand: pollen assemblage biozones*

Site	Depth (cm)	Location	Bed	Designation of p.a.b.	Bio-zone	Substage
East Runton						
ER A	10–100	Fig. 22	.	Gramineae–Cyperaceae	a	Pre-Pa d
Cromer						
CA	5	Fig. 22	g	*Pinus–Picea*	b	Pa IVb
CA	110	Fig. 22	g	*Pinus–Quercus–Carpinus–Ulmus*	a	Pa II
Overstrand						
OB	0–25	250 m NW; Fig. 22	m	*Pinus–Betula*	f	Cr IVb
OB	120–150	250 m NW; Fig. 22	k	*Abies–Carpinus*	e	Cr IIIb
OB	295–358	250 m NW; Fig. 22	g	*Pinus–Picea*	d	Pa IVb
BH O 3B	+110	Figs. 2, 22	g	*Pinus–Picea*	d	Pa IVb
BH O 3B	+70	Figs. 2, 22	g	*Pinus–Picea–Alnus–Betula*	c	Pa IVa
BH O 4S	−210	Figs. 2, 22	d	*Pinus–Quercus–Ulmus–Carpinus*	b	Pa II
BH O 1B	−410	Figs. 2, 22	a	*Pinus*–Gramineae–Ericales	a	Pre-Pa a

Table 13. *Plant macroscopic remains from East Runton* (*ERA 0–25 cm*)

Taxon	Type of remains*	No. of specimens	Taxon	Type of remains*	No. of specimens
Shrubs			Aquatics		
Salix sp.	b. sc	2	*Hippuris vulgaris*	s	1
Montane			*H. vulgaris*	p	1
Ranunculus hyperboreus	a	5	*Nitella/Tolypella* sp.	oos	18
Dry-ground herbs			*Potamogeton* spp.	en	3
Potentilla anserina	a	4	*Ranunculus* subg. *Batrachium* spp.	en	14
Damp-ground herbs			Unclassified		
Carex cf. *bigelowii*	n+u	3	*Carex* spp.	n	2
Carex bigelowii/nigra	n+u	2	Caryophyllaceae	s	2
Selaginella selaginoides	m	3	*Equisetum* sp.	sh	5
Muddy substrates			Gramineae	ca	8
Alisma plantago-aquatica	fr	2	*Ranunculus* sp.	a	2
Eleocharis palustris ssp. *microcarpa*	n	1	cf. *Saxifraga* sp.	s	1
E. uniglumis	n	1	*Stellaria* sp.	s	2
Fen and reedswamp			Total		84
Ranunculus sceleratus	a	2	Derived megaspores		4

* Abbreviations as in Table 2.

Table 14. *Cromer and Overstrand: summary of environmental history*

Bed	Sediment	Sedimentary regime	Sea-level	Stage	Pollen assemblage biozone	Regional vegetation	Climate
n	till	glacial	.	.	.	.	.
m	detritus mud, silty at top	freshwater	.	Cr IVb	*Pinus–Betula*	woodland	cool
			regression				
l	silty sand and gravel	tidal, beach?	.	.	.	.	.
k	laminated silty clay	tidal	.	Cr IIIb	*Abies–Carpinus*	woodland	temperate
			marine transgression				
j	ferruginous gravel	.	.	.	.	.	.
i	cross-bedded sand	.	.	.	.	.	.
h	ferruginous gravel, sand, clay conglomerate	.	.	.	.	.	.
	.	land surface, rootlets	.	.	.	.	.
			regression				
g	laminated silty clay	tidal	.	Pa IVb Pa IVa	*Pinus–Picea* *Pinus–Picea–Alnus–Betula*	woodland	cool
f	sand	marine	.	.	.	.	.
e	gravel and sand	marine	.	.	.	.	.
d	silty clay	tidal	.	Pa II	*Pinus–Quercus–Ulmus–Carpinus*	woodland	temperate
c	gravel and sand	marine	.	.	.	.	.
b	shelly sand ('Weybourne Crag')	marine	.	.	.	.	.
a	stone bed	shallow marine	.	.	.	.	.
			marine transgression	Pre-Pa a	*Pinus*–Gramineae–Ericales	grassland (heath?) some woodland	cold

Table 15. *Sidestrand: pollen assemblage biozones*

Site	Sample no. *or* depth (cm)	Location	Bed	Designation of p.a.b.	Biozone	Substage
SS TA	0–30	389 m SE; Fig. 22	e	*Pinus–Betula*	n	Cr IVb
SS V	8	40 m; Fig. 23	k	*Pinus–Alnus–Picea–Betula*	m	Cr IVa
SS R	5–150	0 m; Fig. 22	k	*Abies–Carpinus*	l	Cr IIIb
SS V	6	37 m; Fig. 23	k	*Abies–Carpinus*	l	Cr IIIb
SS TA	55	389 m SE; Fig. 22	k	*Abies–Carpinus*	l	Cr IIIb
SS TC	6, 70–200	375 m SE; Fig. 22	i, k	*Pinus–Ulmus–Betula*	k	Cr Ib
SS E	1	660 m SE; Fig. 22	g	*Pinus–Picea–Alnus–Betula*	i	Pa IVa
SS E	2	660 m SE; Fig. 22	g	*Alnus–Pinus*	(l) j	Pa IVa
SS E	3, 4	660 m SE; Fig. 22	g	*Pinus–Picea–Alnus–Betula*	i	Pa IVa
SS TA	110	389 m SE; Fig. 22	g	*Pinus–Picea–Alnus–Betula*	i	Pa IVa
SS TB	0–225	383 m SE; Fig. 22	f, g	*Pinus–Picea–Alnus–Betula*	i	Pa IVa
SS U	0	400 m SE; Fig. 22	f, g	*Pinus–Picea–Alnus–Betula*	i	Pa IVa
SS P	50	35 m SE; Fig. 22	f?	*Pinus–Picea–Alnus–Betula*	i	Pa IVa
SS P	30–0	35 m SE; Fig. 22	f?	*Pinus–Picea*–Cupressaceae–*Alnus–Betula*	(l) h	Pa IVa
SS V	11, 2	15 m; Fig. 23	g	*Alnus–Pinus*	(l) g	Pa II
SS F	4	300 m SE; Fig. 22	g	*Alnus–Pinus*	(l) g	Pa II
SS E	5, 6, 7	660 m SE; Fig. 22	f			
SS F	3, 5	300 m SE; Fig. 22	g			
SS R	235	0 m; Fig. 22	g			
SS V	3a, 3b, 4, 13	27 m; Fig. 23	g	*Pinus–Quercus–Ulmus–Carpinus*	f	Pa II
SS TB	265–400	383 m SE; Fig. 22	f, g			
SS U	50–130	400 m SE; Fig. 22	f,			
SS N	10–30	35 m SE; Fig. 22	f?			
SS V	10, 12	SE end; Fig. 23	d	*Pinus–Picea*	e	Pa Ib
SS B	0	215 m SE; Fig. 22	b	*Alnus–Pinus*–Gramineae–*Artemisia*	d	Pre-Pa c?
SS J		130 m SE; Fig. 22	b	Gramineae	c	Pre-Pa a
SS F	2	300 m SE; Fig. 22	b	Gramineae	c	Pre-Pa a
SS H	20	180 m SE; Fig. 22	b	Gramineae–*Alnus–Pinus*	(l) b	Pre-Pa a
SS H	5	180 m SE; Fig. 22	b			
SS D	0–460	140 m SE; Fig. 22	b			
SS B	50–300	215 m SE; Fig. 22	b	*Pinus*–Gramineae–Ericales	a	Pre-Pa a
SS F	1	300 m SE; Fig. 22	a			
SS V	14	SE end; Fig. 23	a			

Table 16. *Sidestrand and Trimingham: macroscopic plant remains*

Taxon	Type of remains*	No. of specimens at different depths: Sidestrand, Pa II, SSTB 260–270	Sidestrand, Cr Ib, SSTC 180–200	Sidestrand, Cr Ib, SSTC 130–150	Sidestrand, Cr Ib, SSTC 90	Trimingham, Cr IIa, TB	Trimingham, e An, TL
Trees							
Betula pendula/pubescens	fr	.	5	.	.	.	.
Quercus sp.	fr	1	.	.	.	.	.
Shrubs							
Rosa sp.+*Rubus* sp.	th	.	16	.	.	.	.
Rubus idaeus	a	1	.	.	.	.	.
Salix sp.	b.sc	.	.	.	.	9	.
cf. *Salix* sp.	l.sc	.	.	.	.	3	.
Thelycrania cf. *sanguinea*	fr.st	.	.	.	.	2	.
Maritime							
Salicornia sp.	s	122	.	.	.	.	.
Heath							
Erica cf. *tetralix*	l	3	.	.	.	.	.
Eriophorum vaginatum	s.s	.	2	.	.	.	6
cf. *Eriophorum* sp.	n	.	2	.	.	.	.
Myrica gale	fr	.	4	.	.	.	.
cf. *Myrica gale*	fr	.	1	.	.	.	.
Dry-ground herbs							
Aphanes microcarpa	a	.	.	.	.	4	.
Arenaria serpyllifolia	s	.	1	.	.	.	.
Atriplex sp.	s	.	3	4	.	.	.
cf. *Atriplex* sp.	e	6	.	.	.	.	.
Cerastium holosteoides	s	.	29	.	.	.	.
Chenopodium sp.	s	.	.	4	.	.	.
Chenopodium sect. *Pseudoblitum*	s	.	1	.	.	.	.
Chenopodium album	s	1	2	.	.	.	.
C. glaucum	s	.	.	.	.	1	.
C. polyspermum	s	.	.	.	.	1	.
Cirsium vulgare	cy	.	1	.	.	.	.
Diplotaxis tenuifolia	s	1	.	.	.	.	.
Galeopsis angustifolia	n	.	2	1	.	.	.
Helianthemum chamaecistus	s	.	.	.	.	1	.
Lamium purpureum	n	.	.	1	.	.	.
Marrubium vulgare	n	.	.	.	.	1	.
Origanum sp.	n	.	9	3	.	.	.
Plantago major	s	.	7	10	.	5	.
Polygonum aviculare	n	.	1	.	.	.	.
Rumex cf. *acetosa*	n	.	4	.	.	.	.
Scleranthus sp.	cal	.	.	.	.	2	.
Stellaria media	s	.	1	.	.	.	.
cf. *Thlaspi* sp.	s	.	.	1	.	.	.
Thymus sp.	n	.	6	.	.	.	.
Urtica urens	n	.	3	.	.	.	.
Damp-ground herbs							
Carduus cf. *acanthoides*	cy	.	.	.	.	1	.
Cirsium palustre	cy	.	1	.	.	.	.

Table 16 (*cont.*)

Taxon	Type of remains*	No. of specimens at different depths: Sidestrand: Pa II SSTB 260–270	Sidestrand: Cr Ib SSTC 180–200	Sidestrand: Cr Ib SSTC 130–150	Sidestrand: Cr Ib SSTC 90	Trimingham: Cr IIa TB	Trimingham: e An TL
Damp-ground herbs (*cont.*)							
Conium maculatum	fr	.	4	.	.	.	.
Myosurus minimus	a	.	.	.	.	.	1
Selaginella selaginoides	m	4	.	.	.	1	.
Solanum dulcamara	s	1	1	.	.	.	.
Stellaria neglecta	s	.	1	.	.	.	.
Muddy substrates, shallow water							
Alisma sp.	fr	.	.	.	.	2	.
Alisma plantago-aquatica	fr	6	17	2	.	.	.
Eleocharis sp.	n	1	.	.	.	.	.
Eleocharis carniolica	n	.	39	.	.	.	.
E. palustris subsp. *palustris*	n	.	.	.	.	.	14
Rorippa sp.	s	.	1	.	.	.	.
Rorippa cf. *sylvestris*	s	.	1	.	.	.	.
Rumex maritimus	n	.	.	1	.	.	.
R. maritimus	per	.	41	10	.	3	.
R. palustris	n+per	.	11	2	.	.	.
Fen and reedswamp helophytes							
cf. *Angelica sylvestris*	fr	.	6	.	.	.	.
Bidens sp.	cy	.	.	2	.	.	.
Bidens tripartita	cy	.	10	.	.	.	.
Carex riparia	n	.	4	.	.	.	.
Cicuta virosa	fr	.	11	.	.	.	.
Cyperus longus	n	.	234	25	.	.	.
Eriophorum latifolium	n	.	1	.	.	.	.
Hypericum tetrapterum	s	.	2	.	.	1	.
Lycopus europaeus	n	.	25	2	.	6	.
Myosoton aquaticum	s	.	.	.	.	3	.
Oenanthe aquatica	fr	.	2	1	.	.	.
Oenanthe cf. *aquatica*	fr	.	.	.	.	2	.
Osmunda sp.	ann	1	.	.	.	.	.
Ranunculus lingua	a	1	.	.	.	.	.
Ranunculus sceleratus	a	5	75	12	1	3	.
Urtica dioica	n	.	180	31	.	55	.
Aquatics							
Ceratophyllum demersum	fr	.	.	2	.	.	.
Chara sp.	oos	.	1	.	.	4	6
Elatine hydropiper	s	1	.	.	.	.	.
Hippuris vulgaris	a	5	.	.	.	.	69
cf. *Hippuris vulgaris*	a	2	.	.	.	.	.
Hottonia palustris	s	.	.	.	.	1	.
Hydrocharis morsus-ranae	s	.	.	.	.	11	.
cf. *Hydrocharis morsus-ranae*	s	.	.	.	.	6	.
Lemna sp.	s	2	.	.	.	.	.
Menyanthes trifoliata	s	10	.	.	.	.	.
cf. *Menyanthes trifoliata*	s	.	.	1	.	.	.
Myriophyllum verticillatum	n	.	.	.	.	.	2

Table 16 (*cont.*)

Taxon	Type of remains*	No. of specimens at different depths					
		Sidestrand				Trimingham	
		Pa II	Cr Ib			Cr IIa	e An
		SSTB 260–270	SSTC 180–200	SSTC 130–150	SSTC 90	TB	TL
Aquatics (*cont.*)							
Najas marina	fr	3	.	.	.	.	.
N. minor	fr	.	.	.	.	1	.
N. tenuissima	fr	.	.	.	.	1	.
Nitella/Tolypella	oos	.	.	.	.	.	45
Nuphar lutea	s	1	.	.	.	.	.
Potamogeton berchtoldii	en	2	.	.	.	.	.
P. distinctus	en	.	.	2	.	.	.
P. filiformis	en	.	.	.	.	.	2
P. filiformis/vaginatus	en	1	.	.	.	.	.
P. gramineus	en	.	.	.	.	.	11
P. cf. *gramineus*	en	1	.	.	.	.	.
P. natans	en	.	.	1	.	.	.
P. pectinatus	en	2	.	.	.	.	.
P. praelongus	en	.	.	.	.	.	1
Potamogeton spp.	en	5	2	1	.	.	2
Ranunculus sub. *Batrachium* spp.	a	4	.	1	.	2	60
Sagittaria sagittifolia	fr	1	15	.	.	.	.
Scirpus lacustris	n	2	2	16	.	.	.
Sparganium erectum ssp. *erectum*	n	.	.	1	.	.	.
Stratiotes aloides	s	.	.	2	.	.	.
S. aloides	ls	.	19	2	1	1	.
Typha spp.	s	.	.	.	1	.	.
Typha angustifolia	s	11	38	6	.	.	.
T. latifolia	s	.	223	27	.	7	.
Zannichellia palustris	a	1	.	.	.	48	3

Table 16 concludes overleaf

Table 16 (*cont.*)

		No. of specimens at different depths					
		Sidestrand				Trimingham	
		Pa II	Cr Ib				
Taxon	Type of remains*	SSTB 260–270	SSTC 180–200	SSTC 130–150	SSTC 90	Cr IIa TB	e An TL
Water ferns							
Azolla filiculoides	m+ma	v.a.	.	.	.	.	.
Salvinia natans	m	.	430	37	a	.	.
S. natans	ma	.	.	.	a	.	.
Unclassified							
Carex spp.	n	11	14	.	.	4	.
Carex spp.	n+u	.	.	3	.	.	.
Caryophyllaceae	s	.	8	.	.	.	.
Cirsium sp.	cy	.	2	.	.	.	.
Compositae	cy	.	3	.	.	.	.
Cruciferae	s	2	18	10	.	.	.
Cyperaceae	n	6	.	.	.	.	.
Dryopteris-type	spo	.	.	.	1	.	.
Gramineae	ca	1	22	8	.	1	.
Hypericum sp.	s	1	.	.	.	.	.
Mentha sp.	n	.	50	13	.	.	.
Mentha arvensis/aquatica	n	.	.	.	.	13	.
Potentilla sp.	a	.	10	6	.	8	.
Ranunculus subg. *Ranunculus*	a	.	2	1	.	.	.
cf. *Ranunculus* subg. *Ranunculus*	a	4	.	.	.	.	.
Rumex sp.	n	.	1	4	.	.	.
Scirpus sp.	n	.	.	19	.	.	.
Solanum sp.	s	.	.	2	.	.	.
Stellaria sp.	s	.	.	.	.	1	.
Umbelliferae	fr	.	.	1	.	.	.
Veronica sp.	s	.	71	.	.	.	.
Total (excluding water ferns)		233	1268	241	4	215	228
Derived megaspores		46	.	5	.	.	.

* Abbreviations as in Table 2.

Table 17. *Sidestrand: summary of environmental history*

Bed	Sediment	Hiatus	Sedimentary regime	Sea-level	Stage	Pollen assemblage biozone	Regional vegetation	Climate
n	till	.	glacial	.	An	.	.	.
m	bedded sand	.	.	.	.	.	.	.
l	stony silty clay	.	freshwater ?	.	Cr IVb	*Pinus–Betula*	woodland	cool
				regression				
k	laminated silty clay and sand		tidal					
					Cr IVa	*Pinus–Alnus–Picea–Betula*	woodland	cool
					Cr IIIb	*Abies–Carpinus*	woodland	temperate
j	ferruginous gravel	.	.	marine transgression				
		+						
i	mud and silty mud	.	freshwater (brackish ?)	.	Cr Ib	*Pinus–Ulmus–Betula*	woodland	temperate
h	sands, gravels, clays in channel fillings	.	.	.	.	.	.	.
				regression				
		+						
g	sand, cross-bedded, with drift mud and clay-conglomerate	.	marine	.	Pa IVa ↑	*Pinus–Picea–Alnus–Betula*	woodland	cool
f	laminated silty clay and sand	.	tidal	.	↓ Pa II	*Pinus–Quercus–Ulmus–Carpinus* and local p.a.b.s	woodland	temperate
e	ferruginous gravel	.	.	.	.	.	.	.
d	sand	.	marine	.	Pa Ib	*Pinus–Picea*	woodland	temperate
c	gravel	.	marine		.	.	.	.
b	shelly sand, silty clay laminae, clay-conglomerate ('Weybourne Crag')	.	marine	.	. Pre-Pa a	. *Pinus*–Gramineae–Ericales and local p.a.b.s	. grassland (heath ?), some woodland	. cold
a	stone bed	.	shallow marine	.	Pre-Pa a	*Pinus*–Gramineae–Ericales		
				marine transgression				

Table 18. *Trimingham: pollen assemblage biozones*

Site	Depth (cm)	Location (m) (Fig. 29)	Designation of p.a.b.	Biozone	Substage
TL		2555	Cyperaceae	j	e An
TC	0–150	270	*Abies–Carpinus*	i	Cr IIIb
TH	60	1000	*Abies–Carpinus*	i	Cr IIIb
BH 14A	326	2575	Gramineae–*Quercus–Ulmus*	h	Cr IIa
TB	20–140	175	*Ulmus–Quercus–Pinus*	g	Cr IIa
TC	300	270	Gramineae–*Pinus–Betula*	f	Pa IV?
TJ		1090	Gramineae–*Pinus*	e	Pa IV?
TM		1000	*Pinus–Alnus–Picea–Betula*	d	Cr or Pa IVa
TF	10–125	2000	*Pinus–Alnus–Picea–Betula*	d	Cr or Pa IVa
TA		130	Gramineae–*Pinus–Quercus–Ulmus–Carpinus*	c	Pa II
BH 10	180	3275	*Pinus*	b	Pa Ib
BH 10	625	3275	*Pinus–Picea*	a	Pa Ib
BH 16	142, 536, 750	2200	*Pinus–Picea*	a	Pa Ib

Table 19. *Trimingham, Mundesley, Ostend and Corton: non-marine molluscs*

Taxon	Cr IIa Trimingham TB	Cr IVa/b Mundesley MB	Cr IVc MAU	Cr IVa Ostend OA	Cr II Corton CAB
Valvata naticina Menke	3	—	—	—	—
V. piscinalis (Müller)	—	4	16	15	—
Bithynia tentaculata (Linné)	23	—	—	—	—
Bithynia ? inflata (Hansén)	—	—	—	307	—
Bithynia sp.	—	1	85	—	—
Lymnaea truncatula (Müller)	—	—	1	—	1
L. peregra (Müller)	—	—	2	—	—
Lymnaea sp.	—	15	—	1	—
Planorbarius corneus (Linné)	1	—	—	14	—
Anisus leucostoma (Millet)	—	3	1	—	—
Armiger crista (Linné)	—	—	1	—	—
Succineidae	—	8	2	—	1
Cochlicopa sp.	1	—	1	—	—
Discus sp.	frag	—	—	—	—
Deroceras ? agreste (Linné)	—	—	1	—	3
Deroceras spp.	10	—	—	—	—
Clausilia sp.	2	—	—	—	—
Trichia hispida (Linné)	—	—	—	—	1
Cepaea sp.	2	—	—	—	—
?*Unio*	—	—	frags	—	—
Sphaerium sp.	2	—	?frag	—	—
Pisidium amnicum (Müller)	2	—	1	9	—
P. clessini Neumayr	—	—	—	frag	—
P. casertanum (Poli)	—	1	—	—	—
P. ? supinum Schmidt	—	1	—	—	—
P. henslowanum (Sheppard)	39	?1	3	27	—
P. moitessieranum Paladilhe	15	—	61	20	—
Derived marine molluscs	—	—	frag	—	—
Totals	100	34	177	393	6

Table 20. *Trimingham: summary of environmental history*

Sediment	Sedimentary regime	Stage	Pollen assemblage biozone	Regional vegetation	Climate
Till	glacial	An	.	.	.
Silty mud, sand	fluviatile, freshwater pools	e An	Cyperaceae	herb	cold
Laminated clay	tidal	Cr IIIb	*Abies–Carpinus*	woodland	temperate
Silty mud	freshwater	Cr IIa	Gramineae–*Quercus–Ulmus* *Ulmus–Quercus–Pinus*	maritime, woodland woodland	temperate
Silty clay laminae in sand	marine?	Pa IV	Gramineae–*Pinus–Betula* Gramineae–*Pinus*	maritime, woodland	cool
Laminated clay	tidal	Cr/Pa IVa	*Pinus–Alnus–Picea–Betula*	woodland	cool
Silty clay laminae in cross-bedded sand	marine	Pa II	Gramineae–*Pinus–Quercus–Ulmus–Carpinus*	maritime, woodland	temperate
Clay and shelly sand	marine	Pa Ib	*Pinus–Picea* *Pinus*	woodland	temperate

Table 21. *Paston–Mundesley: pollen assemblage biozones*

Site	Sample no. or depth (cm)	Location	Bed	Designation of p.a.b.	Biozone	Substage
ME	2	48 m SE; Fig. 33	m	Gramineae–Cyperaceae	w	e An
MAH	0–90	460 m SE; Fig. 32	l	Gramineae–Cyperaceae–*Betula*–*Pinus*	v	e An
MAU	0–35	33 m NW; Fig. 34	k	*Betula*	u	Cr IVc
MB	0	297 m NW; Fig. 33	j	*Pinus–Alnus–Betula*	t	Cr IVb
MAD		392 m SE; Fig. 32	j			
MAH	1	457 m SE; Fig. 32	j			
PB	0	1688 m SE; Fig. 30	i (s)			
PA	20–230	1660 m SE; Fig. 30	i (lc)	*Pinus–Alnus–Picea–Betula*	s	Cr IVa
MAR	127	181 m NW; Fig. 33	i (lc)			
MAM	0–23	50 m NW; Fig. 33	i (lc)			
MC		134 m NW; Fig. 33	i (lc)			
MA	0	333 m NW; Fig. 33	i (lc)			
MAM	25–52	50 m NW; Fig. 33	i (o)	Gramineae–*Pinus–Betula*	(l) s	Cr IVa
MAU	40–140	33 m NW; Fig. 34	i (lc)			
MB	55, 60	297 m NW; Fig. 33	i (lc)			
MAB		34 m SE; Fig. 33	i (b)			
MAR	147–207	181 m NW; Fig. 33	i (lc)			
MAL		343 m NW; Fig. 33	i (lc)	*Abies–Carpinus*	r	Cr IIIb
MAM	75	50 m NW; Fig. 33	i (lc)			
MAK	0	1 m SE; Fig. 33	i (lc)			
MA	60–260	33 m NW; Fig. 33	i (lc)			
MO		63 m SE; Fig. 32	i (b)			
MR		123 m SE; Fig. 32	i (b)			
MW	1	93 m SE; Fig. 32	i (b, s)	*Carpinus*	q	Cr IIIa
MU	T	74 m SE; Fig. 32	i (lc)	*Carpinus*	q	Cr IIIa
ME	1	48 m SE; Fig. 33	i (b)	*Quercus–Ulmus–Tilia*	p	Cr IIb
MAK	10	1 m SE; Fig. 33	h	*Pinus–Betula*–Gramineae	(l) n	Cr II

Table 21 (*cont.*)

Site	Sample no. or depth (cm)	Location	Bed	Designation of p.a.b.	Biozone	Substage
MP	0	146 m SE; Fig. 32	h	Gramineae–*Betula–Pinus*	(l) m	Cr II
MF	0–20	141 m SE; Fig. 32	h	Gramineae–*Betula–Pinus*	(l) m	Cr II
MU	0–25	75 m SE; Fig. 32	h	Gramineae–*Betula–Pinus*	(l) m	Cr II
MG	5	139 m SE; Fig. 32	h	*Quercus–Ulmus–Pinus*	l	Cr IIa
MF	22–32	141 m SE; Fig. 32	h	*Quercus–Ulmus–Pinus*	l	Cr IIa
MU	30–45	75 m SE; Fig. 32	h	*Quercus–Ulmus–Pinus*	l	Cr IIa
PE	0	1322 m SE; Fig. 31	g	*Myrica–Pinus–Betula*	k	Pa IVc
MAP		530 m SE; Fig. 31	e	*Pinus–Betula*	j	Pa IVc
BHB1	230–450	490 m SE; Fig. 31	e	*Pinus–Picea*	i	Pa IVb
BHB2	240	957 m SE; Fig. 31	e	*Pinus–Picea*	i	Pa IVb
MPA	0	662 m SE; Fig. 31	g	*Betula–Alnus*	(l) h	Pa IVa
MPA	10	662 m SE; Fig. 31	e	*Pinus–Betula–Alnus*	g	Pa IVa
BHB1	600	490 m SE; Fig. 31	e	*Pinus–Alnus–Picea–Betula*	f	Pa IVa
ML		1375 m SE; Fig. 30	e	*Pinus–Alnus–Picea–Betula*	f	Pa IVa
MAQ	20–60	894 m SE; Fig. 31	e	*Pinus–Alnus–Picea–Betula*	f	Pa IVa
PE	10–20	1322 m SE; Fig. 31	e	*Pinus–Alnus–Picea–Betula*	f	Pa IVa
PF	0–70	1110 m SE; Fig. 31	e	*Pinus–Alnus–Picea–Betula*	f	Pa IVa
MP	10	146 m SE; Fig. 32	e	*Pinus–Betula–Quercus*	e	Pa III
PH	2–65	1295 m SE; Fig. 31	e	*Alnus–Betula–Quercus*–Ericales	d	Pa III
PG	4–1	1695 m SE; Fig. 30	e	*Pinus–Quercus–Carpinus–Ulmus*	c	Pa II
PB	80	1688 m SE; Fig. 30	f	*Pinus–Quercus–Carpinus–Ulmus*	c	Pa II
BHB3		1395 m SE; Fig. 30	e	*Pinus–Quercus–Carpinus–Ulmus*	c	Pa II
BHB2	460	957 m SE; Fig. 31	e	*Pinus–Quercus–Carpinus–Ulmus*	c	Pa II
BHA1	330	33 m SE; Fig. 33	e	*Pinus–Quercus–Carpinus–Ulmus*	c	Pa II
MAT	220–480	15 m SE; Fig. 33	c	*Pinus–Quercus–Carpinus–Ulmus*	c	Pa II
BHA2	260–570	1518 m SE; Fig. 30	e	*Pinus–Quercus–Carpinus–Ulmus*	c	Pa II
MW	2	98 m SE; Fig. 32	f	*Pinus–Quercus–Carpinus–Ulmus*	c	Pa II
MU	B	75 m SE; Fig. 32	e	*Pinus–Quercus–Carpinus–Ulmus*	c	Pa II
MQ		146 m SE; Fig. 32	e	*Pinus–Quercus–Carpinus–Ulmus*	c	Pa II
MN	2, 3	1295 m SE; Fig. 31	e	*Pinus–Quercus–Carpinus–Ulmus*	c	Pa II
MK		1450 m SE; Fig. 31	e	*Pinus–Quercus–Carpinus–Ulmus*	c	Pa II
PH	2	1295 m SE; Fig. 30	e	*Pinus–Quercus–Carpinus–Ulmus*	c	Pa II
MN	1	1295 m SE; Fig. 31	e	*Alnus–Betula–Quercus*	(l) b	Pa II
BHC7		600 m NW; Fig. 33	c?	*Pinus*–Gramineae–Ericales	a	Pre-Pa a
BHC3		270 m NW; Fig. 33	c	*Pinus*–Gramineae–Ericales	a	Pre-Pa a
MAT	680–1340	15 m SE; Fig. 33	a, b	*Pinus*–Gramineae–Ericales	a	Pre-Pa a
BHA1	640	33 m SE; Fig. 33	b	*Pinus*–Gramineae–Ericales	a	Pre-Pa a

Table 22. *Paston–Mundesley: macroscopic plant remains*

			No. of specimens at different depths												
			Pa							Cr					
			II	III	IVa				IVc	IIa		II			
	Taxon	Type of remains*	PG	PH 2	ML	MAQ	MPA 2	MPA 1	PE	MG 0–10	MU 25–45	MU 20–30	MU 0–15	MF 0–20	MAK 6–11
	Trees														
1	*Alnus* sp.	c	.	.	.	.	.	2	.	1	.	.	2	.	.
2	*Alnus* sp.	fr	1	1	.	.	.	.	.	.	.	.	.	.	.
3	cf. *Alnus* sp.	fr	2	1	.	.	.	.	.	.	.	.	.	.	.
4	*Alnus glutinosa*	fr	.	.	.	.	6	13	.	.	.	.	.	.	.
5	cf. *Alnus glutinosa*	fr	.	.	.	.	.	2	.	.	.	.	.	.	.
6	*Betula* sp.	fr	.	7	.	.	.	.	.	.	.	.	.	3	.
7	cf. *Betula* sp.	fr	.	1	.	.	2	.	.	.	.	.	.	.	.
8	*Betula pendula*	c.sc	.	.	.	.	.	4	.	.	.	.	.	.	.
9	*B. pendula/pubescens*	fr	.	.	.	2	1	34	.	.	.	.	.	.	10
10	*B.* cf. *pendula/pubescens*	fr	.	.	.	.	.	.	.	.	.	.	.	.	.
11	cf. *Betula* tree sp. × *Betula nana*	fr	.	.	.	.	.	.	.	.	.	.	.	.	.
12	Coniferae	l	.	.	.	.	2	.	.	.	.	.	.	.	.
13	*Fraxinus* sp.	w	.	.	.	.	.	.	.	1	.	.	.	.	.
14	*Picea abies*	l	.	.	.	.	.	54	.	.	.	.	.	.	.
15	*Pinus* sp.	w	.	.	1	.	.	.	.	.	.	.	.	.	.
16	*Prunus* sp.	fr.st	.	.	.	.	.	.	.	.	.	.	.	1	.
17	*P. avium*	fr.st	.	.	.	.	.	.	.	.	.	.	1	.	.
18	*Quercus* sp.	w	.	.	.	.	.	.	.	.	.	.	1	.	.
19	*Quercus* cf. *robur*	fr	.	.	.	.	.	.	.	.	.	.	1	.	.
20	*Taxus baccata*	w	.	.	.	.	.	.	.	.	.	.	.	.	.
	Shrubs														
21	*Betula nana*	fr	.	.	.	.	.	.	.	.	.	.	.	.	.
22	*Betula* cf. *nana*	fr	2	.	.	.	.	.	.	.	.	.	.	.	.
23	*Rubus idaeus*	a	.	.	.	.	1	9	.	.	.	.	.	.	.
24	cf. *Rubus* sp.	a	.	.	.	.	.	.	.	.	.	.	.	.	1
25	*Salix* cf. *herbacea*	l	.	.	.	.	.	.	.	.	.	.	.	.	.
26	*Salix* cf. *polaris*	l	.	.	.	.	.	.	.	.	.	.	.	.	.
27	*Salix* cf. *viminalis/phylicifolia*	l	.	.	.	.	.	.	.	.	.	.	.	.	.
28	cf. *Salix* sp.	l	.	.	.	.	.	.	.	.	.	.	.	.	.
29	*Salix* sp.	b	2	30	.	.	.	.	.	.	.	.	.	.	.
30	*Salix* sp.	b.sc	.	30	.	.	24	.	.	.	2	.	.	.	.
31	cf. *Salix* sp.	b.sc	.	.	.	.	.	.	.	.	.	.	.	.	.
32	cf. *Salix* sp.	l.sc	.	.	.	.	.	15	.	.	.	.	.	.	.
33	*Viburnum opulus*	fr.st	.	2	.	.	.	.	.	.	.	.	.	.	.
	Maritime														
34	*Armeria maritima*	cal	.	.	.	.	.	.	.	.	.	.	.	.	.
35	*Glaux maritima*	s	.	4	.	.	.	.	.	.	.	.	.	.	.
36	*Juncus* cf. *gerardii*	s	.	.	.	.	.	.	.	.	.	.	.	.	.
37	*Salicornia* sp.	s	36	.	.	22	.	.	.	.	.	.	.	.	.
38	cf. *Salicornia* sp.	s	.	3	.	.	.	.	.	.	.	.	.	.	.
39	*Triglochin maritima*	s	.	.	.	.	.	.	.	.	.	.	.	.	.

										An								
IIIb		IVa							IVc	e An								
MAK +9– –15	MAR 180– 190	MAR 130– 145	MAR 100– 120	MAM 40– 50	MAM 30– 40	MAM 23	PA 200	PA 30	MAU 0– 35	PH 1	MAH 10– 30	MAJ	MAA	ME 5	ME 2	ME 4	PC	
.	.	.	.	.	.	.	.	.	.	.	.	.	.	.	.	.	.	1
.	.	.	.	.	.	.	.	.	.	.	.	.	.	.	.	.	.	2
.	.	.	.	.	.	.	.	.	.	.	.	.	.	.	.	.	2	3
.	.	8	.	.	.	.	.	.	.	.	.	.	.	.	.	.	.	4
.	.	.	.	.	.	.	.	.	.	.	.	.	.	.	.	.	.	5
.	.	.	.	.	.	.	.	.	.	.	.	.	.	.	.	.	.	6
.	.	.	.	.	.	.	.	.	.	.	.	1	.	1	.	.	.	7
.	.	.	.	.	.	.	.	.	.	.	.	.	.	.	.	.	.	8
5	.	5	8	.	.	.	.	.	4	1	2	.	.	.	.	.	.	9
.	.	.	.	.	1	.	.	.	.	.	.	.	.	.	.	.	.	10
.	.	.	.	.	.	.	.	.	.	.	1	.	.	.	.	.	.	11
.	.	6	.	1	.	.	.	.	.	.	1	.	.	.	.	.	.	12
.	.	.	.	.	.	.	.	.	.	.	.	.	.	.	.	.	.	13
.	.	.	.	.	.	.	.	.	.	.	.	.	.	.	.	.	.	14
.	.	.	.	.	.	.	.	.	.	.	.	.	.	.	.	.	.	15
.	.	.	.	.	.	.	.	.	.	.	.	.	.	.	.	.	.	16
.	.	.	.	.	.	.	.	.	.	.	.	.	.	.	.	.	.	17
.	.	.	.	.	.	.	.	.	.	.	.	.	.	.	.	.	.	18
.	.	.	.	.	.	.	.	.	.	.	.	.	.	.	.	.	.	19
.	.	.	.	.	.	.	.	.	.	.	.	.	1	.	.	.	.	20
.	.	.	.	.	.	.	.	.	.	1	1	.	.	.	1	.	.	21
.	.	.	.	.	.	.	.	.	.	.	.	.	.	.	.	.	.	22
.	.	1	.	.	.	.	.	.	.	.	.	.	.	.	.	.	.	23
.	.	.	.	.	.	.	.	.	.	.	.	.	.	.	.	.	.	24
.	.	.	.	.	.	.	.	.	.	.	.	.	.	.	1	.	.	25
.	.	.	.	.	.	.	.	.	.	.	.	.	.	.	1	.	.	26
.	.	.	.	.	.	.	.	.	.	.	.	.	.	.	1	.	.	27
.	.	.	.	.	.	.	.	.	.	.	.	.	.	1	.	.	.	28
.	.	.	.	.	.	.	.	.	.	1	7	.	.	.	2	1	4	29
.	.	.	.	.	.	.	.	.	.	.	3	.	.	.	1	.	.	30
.	.	.	.	.	.	.	.	.	2	.	.	.	.	.	.	.	.	31
.	.	.	.	.	.	.	.	.	.	.	.	.	.	.	.	.	.	32
.	.	.	.	.	.	.	.	.	.	.	.	.	.	.	.	.	.	33
.	.	.	.	.	.	.	.	.	.	.	.	.	.	.	1	2	.	34
.	.	.	.	.	.	.	.	.	.	.	.	.	.	.	.	.	.	35
.	.	2	.	.	.	.	.	.	.	.	.	.	.	.	.	.	.	36
2	6	5	8	.	.	.	.	8	.	1	.	.	.	.	.	.	.	37
.	.	.	.	.	.	.	.	.	.	.	.	.	.	.	.	.	.	38
1	.	.	.	.	.	.	.	.	.	.	.	.	.	.	.	.	.	39

Table 22 (*cont.*)

	Taxon	Type of re-mains*	No. of specimens at different depths												
			Pa							Cr					
			II	III	IVa				IVc	IIa		II			
			PG	PH 2	ML	MAQ	MPA 2	MPA 1	PE	MG 0–10	MU 25–45	MU 20–30	MU 0–15	MF 0–20	MAK 6–11
	Heath														
40	*Carex* cf. *binervis*	n+u	.	.	.	.	.	.	.	.	.	2	.	.	.
41	*Eriophorum vaginatum*	s.s	.	.	.	15	1	.	.	.	.	.	.	.	.
42	*Juncus squarrosus*	s	.	1	.	.	.	.	.	.	.	.	.	.	.
	Montane														
43	*Luzula* cf. *spicata*	s	.	.	.	.	.	.	.	.	.	.	.	.	.
44	*Oxyria digyna*	fr	.	.	.	.	2	.	.	.	.	.	.	.	.
45	*Saxifraga oppositifolia*	l	.	.	.	.	.	.	.	.	.	.	.	.	.
46	*Thalictrum alpinum*	a	.	.	.	.	1	.	.	.	.	.	.	.	.
	Dry-ground herbs														
47	*Aphanes arvensis*	s	.	1	.	.	.	.	.	.	.	.	.	.	.
48	*Arenaria* sp.	s	.	1	.	.	.	.	.	.	.	.	.	.	.
49	*Atriplex patula*	s	.	.	.	.	.	.	.	.	.	.	.	.	.
50	cf. *Atriplex* sp.	e	.	.	.	.	.	.	.	.	.	.	.	.	.
51	*Cerastium arvense*	s	.	.	.	.	.	.	.	.	.	.	.	.	.
52	*C. holosteoides*	s	.	.	.	.	.	.	.	.	.	.	.	.	.
53	cf. Chenopodiaceae	s	.	.	.	1	.	.	.	.	.	.	.	.	.
54	*Chenopodium* cf. *album*	s	.	.	.	.	.	.	.	.	.	.	.	.	.
55	*C. rubrum*	s	.	.	.	.	.	.	.	.	.	.	.	.	.
56	*C. rubrum/botryodes*	s	4	.	.	.	.	.	.	.	.	.	.	9	.
57	*Chenopodium* sp.	s	.	.	.	.	.	.	.	.	.	.	.	.	.
58	*Cirsium arvense*	cy	.	.	.	.	.	.	.	.	.	.	.	.	.
59	*Diplotaxis muralis*	s	.	.	.	.	6	.	.	.	.	.	.	.	.
60	*Draba* cf. *muralis*	s	.	1	.	.	.	.	.	.	.	.	.	.	.
61	cf. *Euphorbia cyparissias*	s	.	.	.	.	.	.	.	.	.	.	.	.	1
62	*Linum anglicum*	s	.	.	.	.	.	.	.	.	.	.	.	.	.
63	*Minuartia* sp.	s	.	.	.	.	.	.	.	.	.	.	.	.	.
64	*Plantago major*	s	.	.	.	.	.	.	.	.	.	.	.	.	.
65	*Polygonum aviculare*	n	.	1	.	.	.	.	.	.	.	.	.	.	.
66	*P. lapathifolium*	n	.	.	.	.	.	.	.	.	1	.	.	.	.
67	*Prunella vulgaris*	n	.	.	.	.	.	.	.	.	.	.	.	.	.
68	*Silene alba*	s	.	1	.	.	.	.	.	.	.	.	.	.	.
69	*Sonchus oleraceus*	cy	.	.	.	.	.	.	.	.	1	.	.	.	.
70	*Thalictum minus* agg.	a	.	.	.	.	.	.	.	.	.	.	.	.	.
71	cf. *T. minus* agg.	a	.	.	.	.	.	.	.	.	.	.	.	.	.
72	*Urtica urens*	n	3	.	.	.	.	.	.	.	.	.	.	.	.
73	cf. *U. urens*	n	.	1	.	.	.	.	.	.	.	.	.	.	.
74	*Valerianella locusta/carinata*	fr	.	.	.	.	.	.	.	.	.	.	.	.	.
	Damp-ground herbs														
75	*Carex pendula*	n+u	.	.	.	.	.	.	.	.	.	.	.	4	.
76	*Hypericum* cf. *hirsutum*	s	3	.	.	.	.	.	.	.	.	.	.	.	.
77	*Ranunculus* cf. *acris*	a	1	.	.	.	.	.	.	.	.	.	.	.	.
78	*R. repens*	a	.	.	.	.	.	.	.	.	.	.	.	.	.

										An								
IIIb		IVa							IVc	e An								
MAK +9–15	MAR 180–190	MAR 130–145	MAR 100–120	MAM 40–50	MAM 30–40	MAM 23	PA 200	PA 30	MAU 0–35	PH 1	MAH 10–30	MAJ	MAA	ME 5	ME 2	ME 4	PC	
.	.	.	.	.	.	.	.	.	.	.	.	.	.	.	.	.	.	40
.	.	.	1	.	.	.	.	.	.	.	.	.	.	.	.	.	3	41
.	.	.	.	.	.	.	.	.	.	.	.	.	.	.	.	.	.	42
.	.	.	.	.	.	.	.	.	.	.	1	.	.	.	.	.	.	43
.	.	.	.	.	.	.	.	.	.	.	.	.	.	.	8	7	.	44
.	.	.	.	.	.	.	.	.	.	.	.	.	.	.	.	3	.	45
.	.	.	.	.	.	.	.	.	.	.	.	.	.	.	.	.	.	46
.	.	.	.	.	.	.	.	.	.	.	.	.	.	.	.	.	.	47
.	.	.	.	.	.	.	.	.	.	.	.	.	.	.	.	.	.	48
.	.	.	.	.	.	.	.	.	1	.	.	.	.	.	.	.	.	49
.	.	.	.	.	.	.	.	.	1	.	.	.	.	.	1	.	.	50
.	.	.	.	.	.	.	.	.	.	.	.	.	.	1	.	.	.	51
.	.	.	.	.	.	.	.	.	.	.	.	.	.	.	2	.	.	52
.	.	.	.	.	.	.	.	.	.	.	.	.	.	.	.	.	.	53
.	.	.	.	.	.	.	.	.	1	.	.	.	.	.	.	.	.	54
.	.	.	.	.	.	.	.	.	.	.	.	.	.	.	.	.	.	55
.	.	.	.	.	.	.	.	.	.	.	.	.	.	.	.	.	.	56
.	.	.	.	.	.	.	.	.	1	.	.	.	.	.	.	.	.	57
.	.	.	.	.	.	.	.	.	.	.	.	.	.	.	.	.	.	58
.	.	.	.	.	.	.	.	.	.	.	.	.	.	.	.	.	.	59
.	.	.	.	.	.	.	.	.	.	.	.	.	.	.	.	.	.	60
.	.	.	.	.	.	.	.	.	.	.	.	.	.	.	.	.	.	61
.	.	.	.	.	.	.	.	.	.	.	.	.	.	.	1	3	.	62
.	.	.	.	.	.	.	.	.	.	.	.	.	.	.	1	.	.	63
.	.	.	1	.	.	.	.	.	.	.	.	.	.	.	.	.	.	64
.	.	.	.	.	.	.	.	.	.	.	.	.	.	.	1	.	.	65
.	.	.	.	.	.	.	.	.	.	.	.	.	.	.	.	.	.	66
.	.	.	.	.	.	.	.	.	1	.	.	.	.	.	.	.	.	67
.	.	.	.	.	.	.	.	.	.	.	.	.	.	.	.	.	.	68
.	.	.	.	.	.	.	.	.	.	.	.	.	.	.	.	.	.	69
.	.	.	.	.	.	.	.	.	2	.	.	.	.	.	.	.	.	70
.	.	.	.	.	.	.	.	.	1	.	.	.	.	.	.	.	.	71
.	.	.	.	.	.	.	.	.	.	.	.	.	.	.	.	.	.	72
.	.	.	.	.	.	.	.	.	.	.	.	.	.	.	.	.	.	73
.	.	.	.	.	.	.	.	.	1	.	.	.	.	.	.	.	.	74
.	.	.	.	.	.	.	.	.	.	.	.	.	.	.	.	.	.	75
.	.	.	.	.	.	.	.	.	.	.	.	.	.	.	.	.	.	76
.	.	.	.	.	.	.	.	.	.	.	.	.	.	.	.	.	.	77
.	.	.	.	.	.	.	.	.	2	.	.	.	.	.	.	.	.	78

Table 22 (*cont.*)

	Taxon	Type of re-mains*	No. of specimens at different depths												
			Pa							Cr					
			II	III	IVa				IVc	IIa		II			
			PG	PH 2	ML	MAQ	MPA 2	MPA 1	PE	MG 0–10	MU 25–45	MU 20–30	MU 0–15	MF 0–20	MAK 6–11
	Damp-ground herbs (*cont.*)														
79	*Ranunculus* cf. *repens*	a	.	.	.	.	.	.	.	.	.	.	.	.	.
80	*Selaginella selaginoides*	m	.	4	.	.	.	.	.	.	.	.	.	.	.
81	*Senecio aquaticus*	cy	.	.	.	.	.	.	.	.	.	.	.	.	.
	Muddy substrates/shallow water														
82	*Alisma* sp.	fr	12	7	.	.	.	.	.	.	.	.	.		.
83	cf. *Alisma* sp.	fr	.	.	.	.	2	.	.	.	.	.	.	.	.
84	*Alisma plantago-aquatica*	fr	.	.	.	2	.	.	.	.	.	.	.	.	37
85	*Eleocharis* sp.	n	7	.	.	.	2	1	.	.	.	.	.	.	.
86	*E. mamillata*	n	.	.	.	.	.	.	.	.	.	.	.	.	.
87	*E. palustris* subsp. *palustris*	n	n	.	.	.	.	.	.	.	.	.	.	.	.
88	*E. uniglumis*	n	.	.	.	.	.	.	.	.	.	.	.	.	.
89	*Isoetes hystrix*	m	.	.	.	2	.	.	.	.	.	.	.	.	.
90	*Polygonum minus*	n	.	.	.	.	.	.	.	.	.	.	.	1	.
91	*Rorippa microphylla*	s	.	.	.	.	.	.	.	.	.	.	.	.	.
92	*Rumex hydrolapathum*	n + per	.	.	.	.	.	.	.	.	.	.	.	1	.
93	*R. maritimus*	n	.	.	.	.	.	.	.	.	.	.	.	6	.
94	*R. maritimus*	per	.	.	.	.	.	.	.	.	.	.	.	.	.
	Fen and reedswamp helophytes														
95	*Bidens tripartita*	cy	.	.	.	.	.	1	.	.	.	.	.	.	
96	*Carex* cf. *aquatilis*	n + u	.	.	.	.	.	.	.	.	.	.	.	.	.
97	*C. aquatilis/bigelowii*	n, n + u	.	.	.	.	.	65	.	.	.	.	.	.	.
98	*Carex* cf. *pseudocyperus*	n + u	.	.	.	.	.	3	.	.	.	.	.	.	.
99	*C. rostrata*	n + u	.	.	.	.	.	.	.	.	.	3	.	.	.
100	*Carex* cf. *rostrata*	n + u	.	.	.	.	1	.	.	.	.	.	.	.	.
101	*Cicuta virosa*	fr	.	.	.	.	.	.	.	.	.	3	.	2	.
102	cf. *C. virosa*	fr	.	.	.	.	.	.	.	.	1	.	.	.	.
103	cf. *Epilobium palustre*	s	.	.	.	.	.	.	.	.	.	.	.	.	.
104	*Filipendula ulmaria*	a	1	.	.	.	1	7	.	.	.	.	.	.	.
105	cf. *Glyceria* sp.	ca	.	.	.	.	.	.	.	.	.	.	.	.	.
106	*Hydrocotyle vulgaris*	fr	.	.	.	.	.	.	.	.	.	.	.	.	.
107	*Iris pseudacorus*	s	.	.	.	.	.	.	.	.	.	.	.	.	.
108	*Lycopus europaeus*	n	.	.	.	.	.	.	.	.	.	.	.	9	.
109	*Mentha* cf. *aquatica*	n	.	.	.	.	.	.	.	3	.	.	.	.	.
110	*Oenanthe aquatica*	fr	.	.	.	.	.	5	.	.	8	.	.	.	.
111	cf. *O. aquatica*	fr	.	.	.	.	.	.	.	.	3	.	.	.	.
112	*Osmunda* sp.	spo	.	1	.	.	.	.	.	.	.	.	.	.	.
113	*Osmunda* sp.	spo + ann	14	.	.	.	.	.	.	.	.	.	.	.	.
114	*Parnassia palustris*	s	.	.	.	.	.	.	.	.	.	.	.	.	.
115	*Phragmites communis*	r	.	.	.	.	.	.	.	.	.	.	.	.	.
116	cf. *P. communis*	r	.	.	.	.	.	.	1	.	.	.	.	.	.
117	*Potentilla palustris*	a	.	.	.	.	.	10	.	.	.	.	.	.	.

IIIb		IVa							IVc	An e An								
MAK +9–15	MAR 180–190	MAR 130–145	MAR 100–120	MAM 40–50	MAM 30–40	MAM 23	PA 200	PA 30	MAU 0–35	PH 1	MAH 10–30	MAJ	MAA	ME 5	ME 2	ME 4	PC	
.	.	.	.	.	.	.	.	.	1	.	.	.	.	.	.	.	.	79
.	.	.	.	.	.	.	.	.	.	.	.	.	.	13	.	7	1	80
.	.	.	.	.	.	.	.	.	.	.	.	.	.	.	.	.	1	81
.	.	.	.	.	.	.	.	1	.	.	.	2	.	.	.	.	.	82
.	.	.	.	.	.	.	.	.	.	.	.	.	.	.	.	.	.	83
2	.	.	2	.	.	.	.	.	56	.	.	.	.	.	1	.	.	84
.	.	.	.	.	.	.	.	.	.	.	.	1	.	.	.	.	.	85
.	.	1	.	.	.	.	.	.	.	.	.	.	.	.	.	.	.	86
.	.	.	.	.	.	.	.	.	3	.	1	.	.	.	.	.	.	87
.	.	.	.	.	.	.	.	.	.	.	1	.	.	.	.	.	.	88
.	.	.	.	.	.	.	.	.	.	.	.	.	.	.	.	.	.	89
.	.	.	.	.	.	.	.	.	.	.	.	.	.	.	.	.	.	90
.	.	.	.	.	.	.	.	.	.	.	.	.	.	.	1	.	.	91
.	.	.	.	.	.	.	.	.	.	.	.	.	.	.	.	.	.	92
.	.	.	.	.	.	.	.	.	.	.	.	.	.	.	.	.	.	93
.	.	.	.	.	.	.	.	.	5	.	.	.	.	.	.	.	.	94
.	.	.	.	.	.	.	.	.	.	.	.	.	.	.	.	.	.	95
.	.	.	.	.	.	.	.	.	.	.	.	.	.	.	15	.	.	96
.	.	1	.	2	.	.	.	.	.	.	24	.	.	.	.	.	.	97
.	.	.	.	.	.	.	.	.	.	.	.	.	.	.	.	.	.	98
.	.	.	.	.	.	.	.	.	.	.	.	.	.	.	.	.	.	99
.	.	.	.	.	.	.	.	.	.	.	.	.	.	.	.	.	.	100
.	.	.	.	.	.	.	.	.	1	.	.	.	.	.	.	.	.	101
.	.	.	.	.	.	.	.	.	.	.	.	.	.	.	.	.	.	102
.	1	.	.	.	.	.	.	.	.	.	.	.	.	.	.	.	.	103
.	.	1	1	.	.	.	.	.	2	.	.	.	.	.	.	.	.	104
.	.	.	.	.	.	.	.	.	.	.	.	.	.	.	.	3	.	105
.	.	.	.	.	.	.	.	.	.	.	.	2	.	.	.	.	.	106
.	.	.	.	.	.	.	.	.	.	.	.	.	.	.	1	.	.	107
.	.	1	1	.	.	.	.	.	11	.	.	.	.	.	.	.	.	108
.	.	.	.	.	.	.	.	.	.	.	.	.	.	.	.	.	.	109
.	.	.	.	.	.	.	.	.	3	.	.	.	.	.	.	.	.	110
.	.	.	.	.	.	.	.	.	.	.	.	.	.	.	.	.	.	111
.	.	.	.	.	.	.	.	.	.	.	.	.	.	.	.	.	.	112
.	.	.	.	.	.	.	.	.	.	.	.	.	.	.	.	.	.	113
.	.	.	.	.	.	.	.	.	.	.	.	.	.	.	.	1	.	114
.	.	.	.	.	10	.	.	.	.	.	.	.	.	.	.	.	.	115
.	.	.	.	.	.	.	.	.	.	.	.	.	.	.	.	.	.	116
.	.	.	.	.	.	.	.	.	.	.	1	.	.	.	.	.	.	117

Table 22 (*cont.*)

	Taxon	Type of re-mains*	No. of specimens at different depths												
			Pa							Cr					
			II	III	IVa				IVc	IIa		II			
			PG	PH 2	ML	MAQ	MPA 2	MPA 1	PE	MG 0–10	MU 25–45	MU 20–30	MU 0–15	MF 0–20	MAK 6–11
	Fen and reedswamp helophytes (*cont.*)														
118	*Ranunculus lingua*	a	.	.	.	.	1	2	.	.	.	.	.	.	.
119	*R. sceleratus*	a	4	.	.	.	.	.	.	17	68	6	1	42	.
120	cf. *Selinum carvifolia*	fr	.	.	.	.	.	.	.	.	.	.	.	.	.
121	*Swertia* cf. *perennis*	s	.	.	.	.	1	.	.	.	.	.	.	.	.
122	*Thalictrum flavum*	a	.	.	.	.	.	.	.	.	.	.	.	.	.
123	*Urtica dioica*	n	.	.	.	3	1	6	.	16	40	1	.	19	4
	Aquatics														
124	*Callitriche* sp.	fr	.	3	.	.	.	.	.	.	.	.	.	.	.
125	*C. hermaphroditica*	fr	.	.	.	.	.	.	.	.	.	.	.	.	.
126	*C. intermedia*	fr	4	.	.	.	.	.	.	.	.	.	.	.	.
127	*C. platycarpa*	fr	.	.	.	.	.	.	.	.	.	.	.	.	.
128	cf. *Caltha* sp.	a	.	.	.	.	.	.	.	.	.	.	.	.	.
129	*Ceratophyllum demersum*	fr	.	.	.	.	.	.	.	.	6	3	.	1	.
130	*C. submersum*	fr	.	.	.	.	1	.	.	.	8	.	.	1	.
131	*Chara* sp.	oos	1	.	.	1	3	.	.	.	.	.	.	.	8
132	*Elatine hydropiper*	s	1	.	.	.	.	.	.	.	.	.	.	.	.
133	*Hippuris vulgaris*	a	3	.	.	.	2	1	.	.	.	.	.	.	.
134	*Hydrocharis morsus-ranae*	s	.	.	.	.	.	.	.	2	20	1	3	4	.
135	*Isoetes lacustris*	m	.	.	.	.	.	.	.	.	.	.	.	.	.
136	*Lemna minor*	s	.	.	.	.	.	.	.	.	8	.	.	.	.
137	*Lemna* cf. *minor*	s	.	.	.	.	.	.	.	.	.	.	.	1	.
138	*Lemna* cf. *trisulca*	s	.	.	.	.	.	.	.	.	.	.	.	.	.
139	*Littorella uniflora*	n	.	.	.	.	.	.	.	.	.	.	.	.	.
140	*Menyanthes trifoliata*	s	.	.	.	.	4	4	.	1	.	.	.	.	.
141	cf. *Myriophyllum alterniflorum*	n	.	.	.	.	.	.	.	.	.	.	.	.	.
142	*M. spicatum*	n	.	.	.	.	.	.	.	.	.	1	.	.	.
143	*M. verticillatum*	n	.	.	.	.	.	.	.	.	.	.	.	.	1
144	*M. verticillatum* v. *pectinatum*	br	.	.	.	.	.	.	.	.	.	.	.	.	1
145	*M. verticillatum/spicatum*	n	.	.	.	.	.	.	.	.	.	.	.	.	.
146	*Myriophyllum* sp.	n	.	.	.	.	.	.	.	.	.	.	.	.	.
147	*Najas* sp.	fr	.	1	.	.	.	.	.	.	.	.	.	.	.
148	*N. marina*	fr	.	.	.	.	1	.	.	.	.	.	.	.	.
149	*N. minor*	fr	.	.	.	.	.	.	.	.	.	.	.	.	1
150	*N. tenuissima*	fr	1	.	.	.	1	.	.	.	.	.	.	.	.
151	*Nitella/Tolypella*	oos	.	.	.	.	5	.	.	.	.	.	.	.	.
152	*Nuphar lutea*	s	.	.	.	.	.	.	.	.	.	.	15	.	.
153	*N. lutea*	p	.	.	.	.	.	.	.	.	.	.	.	.	.
154	*Nymphaea alba*	s	.	.	.	.	.	1	.	.	.	.	.	.	.
155	*Potamogeton alpinus*	en	.	.	.	.	.	.	.	.	.	.	.	.	.
156	*Potamogeton* cf. *alpinus*	en	.	.	.	.	.	.	.	.	.	.	.	.	.
157	*P. berchtoldii*	en	.	.	.	.	.	.	.	.	.	.	.	.	.
158	*P. compressus*	en	.	.	.	.	1	.	.	.	.	.	.	.	.
159	*P. filiformis*	en	.	.	.	.	.	.	.	.	.	.	.	.	.

										An								
IIIb		IVa							IVc	e An								
MAK +9– 15	MAR 180– 190	MAR 130– 145	MAR 100– 120	MAM 40– 50	MAM 30– 40	MAM 23	PA 200	PA 30	MAU 0– 35	PH 1	MAH 10– 30	MAJ	MAA	ME 5	ME 2	ME 4	PC	
.	.	.	.	.	.	.	.	.	1	.	.	.	.	.	.	.	.	118
.	.	.	.	.	.	.	.	.	11	.	.	.	.	1	.	.	.	119
.	.	.	.	.	.	.	.	.	1	.	.	.	.	.	.	.	.	120
.	.	.	.	.	.	.	.	.	.	.	.	.	.	.	.	.	.	121
.	.	.	.	.	.	.	.	.	9	.	.	.	.	.	.	.	.	122
2	2	3	1	.	.	.	.	1	5	.	.	1	.	1	1	.	.	123
.	.	.	.	.	.	.	.	.	.	.	.	1	.	.	.	.	1	124
1	.	.	.	.	.	.	.	.	.	.	.	.	.	.	1	.	.	125
.	.	.	.	.	.	.	.	.	.	.	.	.	.	.	.	.	.	126
.	.	.	.	.	.	.	.	.	.	.	6	.	.	.	.	.	1	127
.	.	.	.	.	.	.	.	.	1	.	.	.	.	.	.	.	.	128
.	.	.	.	.	.	.	.	.	19	.	.	.	.	.	.	.	.	129
.	.	.	.	.	.	.	.	.	.	.	.	.	.	.	.	.	.	130
.	.	.	.	.	.	.	.	.	30	.	4	.	.	1	.	12	2	131
.	.	5	.	.	.	.	.	.	.	.	.	.	.	.	.	.	.	132
.	.	.	.	.	.	.	.	.	4	.	1	.	.	.	.	.	.	133
.	.	.	.	.	.	.	.	.	.	.	.	.	.	.	.	.	.	134
.	.	.	.	.	.	.	.	.	1	.	.	.	.	.	1	.	.	135
.	.	.	.	.	.	.	.	.	.	.	.	.	.	.	.	.	.	136
	.	.	.	.	.	.	.	.	.	.	.	.	.	.	.	.	.	137
.	.	.	1	.	.	.	.	.	.	.	.	.	.	.	.	.	.	138
.	.	.	.	.	.	.	.	.	.	.	.	1	.	.	.	.	.	139
.	.	4	.	.	3	.	.	.	.	.	.	.	.	.	.	.	.	140
.	.	.	.	.	.	.	.	.	1	.	.	.	.	.	.	.	.	141
.	.	.	.	.	.	.	.	.	6	1	2	.	.	.	.	.	.	142
.	.	.	.	.	.	.	.	.	.	.	.	.	.	.	.	.	.	143
.	.	.	.	.	.	.	.	.	.	.	.	.	.	.	.	.	.	144
.	.	.	.	.	.	.	.	.	1	.	1	.	.	.	.	.	.	145
1	.	.	.	.	.	.	.	.	.	.	.	.	.	.	.	.	.	146
.	.	.	.	.	.	.	.	.	.	.	.	.	.	.	.	.	.	147
.	.	.	.	.	.	.	.	.	.	.	.	.	.	.	.	.	.	148
.	.	.	.	.	.	.	.	.	.	.	.	.	.	.	.	.	.	149
.	.	.	.	.	.	.	.	.	.	.	.	.	.	.	.	.	.	150
.	.	.	.	.	.	.	.	.	.	.	.	2	.	8	2	.	.	151
.	.	.	.	.	.	.	.	.	42	.	.	.	.	.	.	.	.	152
.	.	.	.	.	.	.	.	.	4	.	.	.	.	.	.	.	.	153
.	.	.	.	.	.	.	.	.	.	.	.	.	.	.	.	.	.	154
.	.	.	.	.	.	.	.	.	.	.	.	.	.	.	1	.	.	155
.	.	.	.	.	.	.	.	.	1	.	.	.	.	.	.	.	.	156
.	.	.	.	.	.	.	.	.	.	.	.	1	.	.	.	.	.	157
.	.	.	.	.	.	.	.	.	.	.	.	.	.	.	.	.	.	158
.	.	.	.	.	.	.	.	.	1	.	.	.	.	3	5	.	1	159

Table 22 (*cont.*)

			No. of specimens at different depths												
			Pa							Cr					
			II	III	IVa				IVc	IIa		II			
	Taxon	Type of re-mains*	PG	PH 2	ML	MAQ	MPA 2	MPA 1	PE	MG 0–10	MU 25–45	MU 20–30	MU 0–15	MF 0–20	MAK 6–11
	Aquatics (*cont.*)														
160	*P. friesii*	en	1	.	.	.	.	.	.	.	.	.	.	.	.
161	*Potamogeton* cf. *friesii*	en	.	.	.	.	.	.	.	.	.	.	.	.	.
162	*P. friesii* (West Runton type)	en	.	.	.	.	.	.	.	.	.	.	.	.	.
163	*P. gramineus*	en	.	.	.	.	.	.	.	.	.	.	.	.	.
164	*P. natans*	en	.	.	.	.	.	.	.	.	.	.	.	1	.
165	*P. pectinatus*	en	.	.	.	.	.	.	.	.	.	.	.	.	.
166	*P. pectinatus*	li	.	1	.	.	.	.	.	.	.	.	.	.	.
167	*P. praelongus*	en	.	.	.	.	.	.	.	.	.	.	.	.	.
168	*Potamogeton* cf. *praelongus*	en	1	1	.	.	.	.	.	.	.	.	.	.	.
169	*P. trichoides*	en	.	.	.	.	.	.	.	.	.	.	.	1	.
170	*P. vaginatus*	en	.	.	.	.	.	.	.	.	.	.	.	.	.
171	*Potamogeton* cf. *vaginatus*	en	.	.	.	.	.	.	.	.	.	.	.	.	.
172	*P. vaginatus/filiformis*	en	.	.	.	.	.	.	.	.	.	.	.	.	.
173	*P. vaginatus/filiformis*	li	.	.	.	.	.	.	.	.	.	.	.	.	.
174	*Potamogeton* sp.	en	.	1	.	.	.	.	.	.	.	.	.	.	.
175	*Potamogeton* sp.	li	.	.	.	.	.	.	.	.	.	.	.	.	.
176	*Ranunculus* subg. *Batrachium* spp.	a	2	.	.	.	5	.	.	.	.	1	.	5	5
177	cf. *Sagittaria* sp.	fr	.	2	.	.	.	.	.	.	.	.	.	.	.
178	*Sagittaria sagittifolia*	fr	.	.	.	.	.	.	.	.	.	.	.	5	.
179	*Scirpus lacustris*	n	.	.	.	.	.	.	.	.	3	5	.	5	.
180	*Sparganium* sp.	n	.	.	.	.	.	4	.	.	.	.	.	.	.
181	cf. *Sparganium* sp.	n	.	.	.	.	1	.	.	.	.	.	.	.	.
182	*S. angustifolium*	n	.	.	.	.	1	.	.	.	1	.	.	2	.
183	*Sparganium* cf. *angustifolium*	n	.	.	.	.	.	.	.	.	.	.	.	.	.
184	*Stratiotes aloides*	ls	3	.	.	1	1	124	.	.	.	.	.	4	.
185	*Typha* spp.	s	2	.	.	.	.	.	.	.	.	.	.	.	.
186	*T. angustifolia*	s	.	.	.	4	.	.	.	.	.	.	.	.	.
187	*T. latifolia*	s	10	.	.	4	1	.	.	.	20	7	30	.	10
188	*Typha* cf. *latifolia*	s	.	.	.	.	.	.	.	12	.	.	.	35	.
189	*Zannichellia palustris*	a	.	.	.	.	1	.	.	.	.	.	.	.	.
	Water ferns														
190	*Azolla filiculoides*	m	4	v.a.	.	.	.	.	.	.	.	.	.	.	.
191	*Salvinia natans*	m	.	.	.	.	.	.	.	50	a.	a.	50	a.	15
192	*S. natans*	ma	.	.	.	.	.	.	.	.	.	.	.	.	5
	Unclassified														
193	cf. *Alyssum* sp.	c.v	.	.	.	.	.	.	.	.	.	.	.	.	.
194	*Carduus* sp.	cy	.	.	.	.	.	.	.	.	.	.	.	.	.
195	*Carex* spp.	n, u	8	8	.	1	20	27	1	4	3	5	1	3	11
196	Caryophyllaceae	s	.	.	.	.	.	.	.	.	.	.	.	.	.
197	cf. Caryophyllaceae	s	.	.	.	.	.	.	.	.	.	.	.	.	1
198	*Cirsium* sp.	cy	.	.	.	.	.	.	.	.	.	.	.	.	.
199	Compositae	cy	.	1	.	1	.	.	.	.	.	.	.	.	.

										An								
IIIb		IVa							IVc	e An								
MAK +9–15	MAR 180–190	MAR 130–145	MAR 100–120	MAM 40–50	MAM 30–40	MAM 23	PA 200	PA 30	MAU 0–35	PH 1	MAH 10–30	MAJ	MAA	ME 5	ME 2	ME 4	PC	
.	.	.	.	.	.	.	.	.	75	.	.	.	.	.	.	.	.	160
.	.	.	.	.	.	.	.	.	.	.	1	.	.	.	.	.	.	161
.	.	.	.	.	.	.	.	.	5	.	.	.	.	.	.	.	.	162
.	.	.	.	.	.	.	.	.	1	.	.	.	.	.	1	.	.	163
.	.	.	.	.	.	.	.	.	.	.	.	.	.	1	.	.	.	164
.	.	.	.	.	.	.	.	.	26	.	.	.	.	.	.	2	.	165
.	.	.	.	.	.	.	.	.	.	.	.	.	.	.	.	1	.	166
.	.	.	.	.	.	.	.	.	2	.	.	.	.	.	.	.	.	167
.	.	.	.	.	.	.	.	.	.	.	.	.	.	.	.	.	.	168
.	.	.	.	.	.	.	.	.	2	.	.	.	.	.	.	.	.	169
.	.	.	.	.	.	.	.	.	.	.	13	.	.	4	.	1	.	170
.	.	.	.	.	.	.	.	.	.	.	1	.	.	2	.	.	.	171
.	.	.	.	.	.	.	.	.	.	.	.	.	.	2	6	.	.	172
.	.	.	.	.	.	.	.	.	.	.	.	.	.	.	4	.	.	173
.	.	.	.	.	.	.	.	.	12	.	.	.	.	3	1	.	.	174
.	.	.	.	.	.	.	.	.	.	.	.	.	.	.	1	.	.	175
1	.	.	.	.	.	.	1	.	131	.	7	2	.	1	1	.	1	176
.	.	.	.	.	.	.	.	.	.	.	.	.	.	.	.	.	.	177
.	.	.	.	.	.	.	.	.	1	.	.	.	.	.	2	.	.	178
.	.	2	.	3	.	.	.	.	28	2	.	.	.	.	.	.	.	179
.	.	.	.	.	.	.	.	.	.	.	.	.	.	.	.	.	.	180
.	.	.	.	.	.	.	.	.	.	.	.	1	.	.	.	.	.	181
.	.	.	.	.	.	.	.	.	.	.	.	.	.	.	.	.	.	182
.	.	.	.	1	.	.	.	.	1	.	.	.	.	.	.	.	.	183
.	.	4	1	.	.	1	.	.	1	.	.	.	.	.	1	.	.	184
.	.	.	.	.	.	.	.	.	.	.	.	.	.	.	.	.	.	185
.	.	.	.	.	.	.	.	.	.	.	.	.	.	.	.	.	.	186
1	4	11	1	.	.	1	.	1	18	1	1	.	.	.	.	.	.	187
.	.	.	.	.	.	.	.	.	.	.	.	.	.	1	.	.	.	188
.	.	.	.	.	.	.	.	.	64	.	.	.	.	.	1	.	1	189
.	1	.	.	.	.	.	.	.	.	.	.	.	.	.	.	.	.	190
.	.	.	.	.	.	.	.	.	.	.	.	.	.	.	.	.	.	191
.	.	.	.	.	.	.	.	.	.	.	.	.	.	.	.	.	.	192
.	.	.	.	.	.	.	.	.	.	.	.	.	.	.	1	.	.	193
.	.	.	.	.	.	.	.	.	1	.	.	.	.	.	.	.	.	194
.	.	11	3	6	5	1	.	.	65	.	.	10	.	7	11	4	2	195
.	.	1	.	.	.	.	.	.	.	.	.	.	.	.	.	.	.	196
.	2	.	.	.	.	.	.	.	.	.	.	.	.	.	.	.	.	197
.	.	.	.	.	.	.	.	.	7	.	.	.	.	.	.	.	.	198
.	.	.	.	.	.	.	.	.	.	.	.	.	.	.	.	.	.	199

Table 22 (*cont.*)

		No. of specimens at different depths												
		Pa							Cr					
		II	III	IVa				IVc	IIa		II			
Taxon		Type of re-mains*	PG	PH 2	ML	MAQ	MPA 2	MPA 1	PE	MG 0–10	MU 25–45	MU 20–30	MU 0–15	MF 0–20	MAK 6–11
	Unclassified (*cont.*)														
200	cf. Compositae	cy	.	.	.	.	.	.	.	.	.	.	.	4	.
201	Cruciferae	s	.	1	.	.	8	.	.	.	.	.	.	.	.
202	Cruciferae	c.v	.	.	.	.	.	.	.	.	.	.	.	.	.
203	Cyperaceae	n	.	.	.	.	.	.	3	.	3	.	.	2	.
204	*Dryopteris*-type	spo	21	.	.	30	.	11	.	.	.	1	4	9	.
205	Gramineae	ca	3	.	.	5	.	.	.	.	.	.	.	.	.
206	*Juncus* sp.	s	.	.	.	.	.	.	.	.	.	.	.	.	.
207	cf. Labiatae	n	.	1	.	.	.	.	.	.	.	.	.	.	.
208	*Mentha* sp.	n	.	.	.	.	1	.	.	.	.	.	.	.	.
209	cf. *Mentha* sp.	n	.	.	.	.	.	.	.	.	.	.	.	.	.
210	*Mentha arvensis/aquatica*	n	3	.	.	.	.	.	.	.	2	1	.	15	.
211	Polygonaceae	s	.	1	.	.	.	.	.	.	.	.	.	.	.
212	*Potentilla* sp.	a	2	100	.	.	.	3	.	.	.	.	.	.	.
213	*P. erecta*	a	.	.	.	.	.	.	.	.	.	.	.	.	.
214	*Ranunculus* subg. *Ranunculus*	a	2	1	.	.	.	4	.	.	.	.	.	.	.
215	*Rubus/Cirsium* sp.	th	.	.	.	.	.	.	.	.	.	.	.	.	.
216	*Rumex* sp.	n	.	.	.	.	.	.	.	.	.	.	.	.	.
217	cf. *Rumex* sp.	per	.	.	.	2	.	.	.	.	.	.	.	.	.
218	*Rumex* subg. *Rumex*	tub	.	.	.	1	.	.	.	.	.	.	.	.	.
219	*Scirpus* sp.	n	1	.	.	.	.	.	.	.	.	.	.	.	.
220	cf. *Scirpus* sp.	n	.	.	.	.	1	.	.	.	.	.	2	.	.
221	*Solanum* sp.	s	.	.	.	1	.	.	.	.	.	.	.	.	.
222	*Stellaria* sp.	s	.	.	.	.	1	.	.	.	.	.	.	.	.
223	Umbelliferae	fr	1	.	.	.	2	.	.	.	.	.	.	.	.
224	*Viola* sp.	s	.	.	.	.	.	2	.	.	.	.	.	.	.
225	Total (excluding water ferns)		162	222	1	98	115	414	5	57	197	40	61	195	91
226	Derived megaspores		36	62	.	2	61	7	.	.	.	.	.	2	.

* Abbreviations as in Table 2.

										An								
IIIb		IVa							IVc	e An								
MAK +9–15	MAR 180–190	MAR 130–145	MAR 100–120	MAM 40–50	MAM 30–40	MAM 23	PA 200	PA 30	MAU 35	PH 1	MAH 10–30	MAJ	MAA	ME 5	ME 2	ME 4	PC	
.	.	.	.	.	.	.	.	.	.	.	.	.	.	.	.	.	.	200
.	.	.	.	.	.	.	.	.	.	.	.	8	.	.	25	10	.	201
.	.	.	.	.	.	.	.	.	.	.	.	.	.	.	.	3	.	202
.	.	3	.	.	.	1	.	.	3	.	.	.	.	.	.	.	3	203
.	.	2	.	7	40	7	.	30	1	5	.	.	.	.	.	.	.	204
.	2	4	2	.	.	.	.	.	1	.	.	.	.	.	3	2	3	205
.	1	.	.	.	.	1	.	1	.	.	.	.	.	.	.	.	.	206
.	.	.	.	.	.	.	.	.	.	.	.	.	.	.	.	.	.	207
1	.	.	.	.	.	.	.	.	.	.	.	.	.	.	.	.	.	208
.	2	.	.	.	.	.	.	.	.	.	.	.	.	.	.	.	.	209
.	.	.	.	.	.	.	.	.	.	.	.	.	.	.	.	.	.	210
.	.	.	.	.	.	.	.	.	.	.	.	.	.	.	.	.	.	211
.	.	.	.	.	.	.	.	.	.	.	.	.	.	2	1	.	.	212
.	.	.	.	.	.	.	.	.	3	.	.	.	.	.	.	.	.	213
.	.	.	.	.	1	.	.	.	1	.	.	1	.	.	1	.	.	214
.	.	.	.	.	.	.	.	.	1	.	.	.	.	.	.	.	.	215
.	.	1	.	.	.	.	.	.	.	1	1	.	.	.	.	.	.	216
.	.	.	.	.	.	.	.	.	.	.	.	1	.	.	.	.	1	217
.	.	.	1	.	.	.	.	.	.	.	.	.	.	.	.	.	.	218
.	.	.	.	.	.	.	.	.	.	.	.	.	.	.	.	.	.	219
.	.	.	.	.	.	.	.	.	.	.	.	.	.	.	.	.	.	220
.	.	.	.	.	.	.	.	.	.	.	.	.	.	.	.	.	.	221
.	.	.	.	.	.	.	.	.	.	.	.	.	.	.	.	.	.	222
.	.	.	1	.	.	.	.	.	.	.	.	.	.	1	.	.	.	223
.	.	.	.	.	.	.	.	.	.	.	.	4	.	3	17	5	.	224
17	20	82	33	21	60	12	1	42	687	14	81	39	1	57	128	67	28	225
6	28	.	4	.	.	.	.	1	.	4	4	60	.	50	39	50	16	226

Table 23. *Paston–Mundesley: summary of environmental history*

Bed	Sediment	Hiatus	Sedimentary regime	Sea-level	Periglacial: Ice-wedge casts	Periglacial: Involutions	Stage	Pollen assemblage biozone	Regional vegetation	Climate
p	till		glacial	.			An			
o	laminated silty clay		freshwater	.		+				
					+					
n	cross-bedded sand		fluviatile							
						+				
m	laminated silty clay		freshwater pool	.				Gramineae–Cyperaceae	grassland	cold
					+					
l	laminated silt and stony detritus mud		freshwater	.			e An	Gramineae–Cyperaceae–*Betula–Pinus*	grassland	cold
j, j i	cross-bedded sand		fluviatile	.	+	+				
k	clay-mud		slowly-running freshwater	.			Cr IVc	*Betula*	woodland	cool
j	cross-bedded sand		fluviatile	.			Cr IVb	*Pinus–Alnus–Betula*	woodland	cool
				regression						
i (o)	detritus mud		freshwater/brackish	.				Gramineae–*Pinus–Betula*	fen, woodland	cool
							Cr IVa	*Pinus–Alnus–Picea–Betula*	woodland	cool
i(lc)	laminated clay		tidal							
							Cr IIIb	*Abies–Carpinus*	woodland	temperate
							Cr IIIa	*Carpinus*	woodland	temperate
i(b)	beach gravel		beach	.			Cr IIb	*Quercus–Ulmus–Tilia*	woodland	temperate
i(s)	sand with silt laminae		tidal							
				marine transgression						
								Pinus–Betula–Gramineae	fen, woodland	temperate
h	detritus mud		freshwater	.			Cr II	Gramineae–*Betula–Pinus*	fen, woodland	temperate
							Cr IIa	*Quercus–Ulmus–Pinus*	woodland	temperate
		+			+					
g	detritus mud		freshwater							
		+								
f	ferruginous sand and gravel, clay conglomerate, cross-bedded sand		land surface marine	regression						
								Myrica–Pinus–Betula	woodland	cool
							Pa IVc	*Pinus–Betula*	woodland	cool
e	sand with laminae of silty clay		marine	.			PaIVb	*Pinus–Picea*	woodland	cool
								Betula–Alnus	woodland	cool
							Pa IVa	*Pinus–Betula–Alnus*	woodland	cool
								Pinus–Alnus–Picea–Betula	woodland	cool
							Pa III	*Pinus–Betula–Alnus*	woodland	temperate
d	sandy gravel		marine	.				*Alnus–Betula–Quercus*–Ericales	woodland	temperate
c	sand with laminae of silty clay		marine	.			Pa II	*Pinus–Quercus–Carpinus–Ulmus*	woodland	temperate
		+								
b	laminated silty clay		marine	.			Pre-Pa a	*Pinus*–Gramineae–Ericales		
									grassland (heath?), some woodland	cold
a	stone bed		marine	.			Pre-Pa a	*Pinus*–Gramineae–Ericales		

Table 24. *Bacton: pollen assemblage biozones*

Site	Depth (cm)	Location (m) (Fig. 39)	Bed	Designation of p.a.b.	Biozone	Substage
BI	5	358	f	Gramineae–*Betula–Pinus–Alnus*	d	e An
BB	80	445	e	*Pinus–Alnus–Picea–Betula*	c	Cr IVa
BI	135–220	358	e, f	*Pinus–Alnus–Picea–Betula*	c	Cr IVa
BE	0–70	0	e	*Pinus–Alnus–Picea–Betula*	c	Cr IVa
BH	325–400	494	e	*Pinus–Alnus–Picea–Betula*	c	Cr IVa
BI	500	358	a	*Pinus–Quercus–Carpinus–Ulmus*	b	Pa II
BH	545–575	494	a	*Pinus–Quercus–Carpinus–Ulmus*	b	Pa II
BH	579	494	a	*Pinus–Betula*–Gramineae	a	Pa I

Table 25. *Bacton, Ostend and Happisburgh: macroscopic plant remains*

		No. of specimens at different depths												
		Pa												
			II			IVa	II			II		Cr IVa		e An
	Type of re-	Ia	HC	HC	HC	HC				BH				
Taxon	mains*	HC 381	317–321	306–309	265–268	223–233	HF 15	HE 20	HE 0–10	545–575	IV OE	OC 20–25	OA 70–80	BH 70–100
Trees														
Acer campestre	e	.	.	.	.	.	.	.	.	.	.	1	2	.
cf. *Alnus* sp.	fr	.	.	.	.	.	.	.	3	.	.	127	12	.
Alnus sp.	c	.	.	2	.	.	.	.	1	.	1	165	21	.
cf. *Alnus* sp.	c	.	1	.	.	.	.	.	.	.	.	.	.	.
Alnus sp.	c.sc	.	.	.	.	.	.	.	.	.	.	.	12	.
Alnus glutinosa	fr	.	.	.	.	.	.	.	.	.	11	539	245	.
Alnus cf. *glutinosa*	fr	.	.	4	.	.	.	.	24	.	.	.	.	.
Alnus cf. *incana*	fr	.	.	.	.	.	.	.	.	.	.	.	15	.
Betula sp.	fr	.	.	.	.	.	.	.	.	.	23	1	.	.
Betula sp.	c.sc	.	.	.	.	.	.	.	.	.	1	.	5	.
Betula pendula	fr	.	.	.	.	.	1	.	.	.	.	.	.	.
B. pendula/pubescens	fr	.	.	.	.	.	.	.	.	.	16	.	6	.
Betula cf. *pubescens*	c.sc	.	.	.	.	.	.	.	.	.	.	.	1	.
Betula cf. *nana* × *Betula* tree sp.	fr	.	.	.	.	.	.	.	.	.	50	.	.	.
Coniferae	l	.	.	.	.	.	.	.	.	.	.	1	.	.
cf. Coniferae	l	.	.	.	.	.	.	.	.	1	.	.	.	.
cf. *Corylus avellana*	n	.	.	.	.	.	.	.	.	.	.	5	.	.
Picea sp.	s	.	.	.	.	.	.	.	.	.	.	1	.	.
cf. *Picea* sp.	s	.	.	.	.	.	.	.	.	.	.	1	.	.
Picea sp.	l	.	.	.	.	.	.	.	.	.	.	4	.	.
Picea abies	l	.	.	.	.	.	.	.	.	.	87	.	.	.
P. abies	c	.	.	.	.	.	.	.	.	.	1	.	.	.
P. abies	s	.	.	.	.	.	.	.	.	.	1	2	.	.
Taxus baccata	s	.	.	.	.	.	.	.	.	.	.	1	.	.
T. baccata	l	.	.	.	.	.	.	.	.	.	.	1	.	.
Shrubs														
Betula cf. *nana*	fr	.	.	.	.	.	.	.	.	.	10	.	.	.
Salix sp.	b	.	.	.	.	.	.	.	.	2	.	.	.	.

Table 25 (*cont.*)

		Pa												
			II			IVa	II			II		Cr IVa		e An
	Type of re-	Ia	HC	HC	HC	HC				BH				
Taxon	mains*	HC 381	317–321	306–309	265–268	223–233	HF 15	HE 20	HE 0–10	545–575	IV OE	OC 20–25	OA 70–80	BH 70–100
Shrubs (*cont.*)														
Thelycrania sanguinea	fr.st	.	.	.	.	.	.	.	.	.	.	16	.	.
Viburnum opulus	fr.st	.	.	.	.	.	.	.	.	.	.	1	.	.
Maritime														
cf. *Aster tripolium*	cy	.	.	.	.	.	.	.	.	.	1	.	.	.
Salicornia sp.	s	.	2	.	.	8	1	.	.	8	.	.	.	.
cf. *Salicornia* sp.	s	.	1	.	.	.	.	.	.	.	.	.	.	.
Heath														
cf. Ericaceae	s	.	.	.	.	.	.	.	.	1	.	.	.	.
Erica sp.	s	.	1	.	.	5	.	.	.	.	.	.	.	.
Eriophorum vaginatum	s.s	.	.	.	.	2	.	.	.	.	.	.	.	.
Dry-ground herbs														
Campanula glomerata	s	.	.	.	.	1	.	.	.	.	.	.	.	.
Potentilla crantzii/ tabernaemontani	a	.	.	.	.	.	.	.	.	.	.	.	.	4
Ranunculus cf. *bulbosus*	a	.	.	.	.	.	.	.	.	.	.	.	2	.
Rumex acetosa	n	.	.	.	.	.	.	.	.	.	.	.	2	.
Rumex cf. *acetosa*	n	.	.	.	.	.	.	.	.	.	.	11	.	.
Rumex acetosella agg.	n	.	.	.	.	.	.	.	.	.	.	.	1	.
Thalictrum minus	a	.	.	.	1	.	.	.	.	.	.	.	.	.
Trinia glauca	fr	.	.	.	.	.	.	.	1	.	.	.	.	.
Urtica urens	n	.	.	.	.	.	.	.	.	1	.	.	.	.
Damp-ground herbs														
Carex cf. *remota*	n+u	.	.	.	.	.	.	.	.	.	.	.	1	.
Ranunculus repens	a	.	.	.	.	.	.	.	.	.	.	5	.	.
cf. *Selaginella selaginoides*	m	.	.	1	.	.	.	.	.	4	.	.	.	.
Solanum dulcamara	s	.	.	.	.	.	.	.	.	.	.	.	1	.
Muddy substrates, shallow water														
Alisma sp.	fr	.	.	.	.	.	.	.	.	.	.	6	5	.
Alisma plantago-aquatica	fr	.	2	24	.	.	.	.	.	.	.	.	.	.
Alisma cf. *plantago-aquatica*	fr	.	.	.	.	.	2	.	.	.	.	.	.	.
Elatine hexandra	s	.	.	.	.	1	.	.	.	.	.	.	.	.
Eleocharis cf. *acicularis*	n	.	.	84	.	.	.	.	.	.	.	.	.	.
Eleocharis palustris ssp. *palustris*	n	.	.	1	.	.	.	.	1	.	1	.	.	4
Eleocharis sp.	n	.	.	.	.	1	.	.	.	.	.	.	1	.
Polygonum hydropiper	n	.	.	11	.	.	.	.	.	.	.	.	.	.
Rumex palustris	n+per.	.	.	1	.	.	.	.	.	.	.	.	.	.

Table 25 (*cont.*)

		Pa												
			II			IVa	II			II		Cr IVa		e An
	Type of re-	Ia	HC	HC	HC	HC				BH				
Taxon	mains*	HC 381	317–321	306–309	265–268	223–233	HF 15	HE 20	HE 0–10	545–575	IV OE	OC 20–25	OA 70–80	BH 70–100
Fen and reed swamp helophytes														
Bidens sp.	cy	.	.	.	.	.	.	.	.	.	.	.	1	.
Bidens sp.	te	.	.	.	.	.	.	.	.	.	.	1	.	.
Bidens cf. *tripartita*	cy	.	.	1	.	.	.	.	.	.	.	.	.	.
Carex aquatilis/bigelowii	n+u	.	.	.	.	.	.	.	1	.	.	.	.	9
C. laevigata	n+u	.	.	.	.	.	.	.	1	.	.	.	.	.
C. riparia	n	.	.	.	.	.	.	.	1	.	.	.	.	.
Cicuta virosa	fr	.	.	.	.	.	.	.	1	.	.	1	.	.
Cyperus longus	n	.	.	15	.	.	.	.	11	.	.	.	.	.
Filipendula ulmaria	a	.	.	.	.	.	.	.	.	.	.	.	1	.
Hydrocotyle vulgaris	fr	.	.	.	.	.	1	.	.	.	.	.	.	.
Lycopus europaeus	n	.	.	2	.	.	.	.	.	.	.	16	8	.
Lysimachia vulgaris	s	.	.	.	.	.	.	.	.	.	.	3	.	.
Lythrum salicaria	s	.	.	.	.	.	.	.	.	.	.	.	1	.
Oenanthe sp.	fr	.	.	20	.	.	.	.	.	.	.	.	.	.
Oenanthe aquatica	fr	.	.	17	.	.	.	.	.	.	.	2	.	.
Oenanthe cf. *aquatica*	fr	.	.	10	.	.	.	.	.	.	.	.	.	.
Osmunda sp.	spo	.	.	.	.	.	.	.	100	1	.	.	.	.
Osmunda sp.	ann	.	2	1	1	.	6	.	100	.	.	.	.	.
Ranunculus sceleratus	a	.	1	4	.	.	.	.	.	.	.	.	.	.
Urtica dioica	n	.	2	5	.	1	3	.	.	.	.	61	50	.
Valeriana dioica	n	.	.	.	.	.	.	.	.	.	.	2	.	.
Aquatics														
Callitriche sp.	fr	.	.	.	.	.	.	.	.	.	.	10	1	.
Callitriche platycarpa	fr	.	1	.	.	.	.	.	.	.	.	.	.	.
Callitriche stagnalis	fr	.	.	.	.	.	1	.	.	.	.	.	.	.
Ceratophyllum sp.	fr	.	.	1	.	.	.	.	.	.	.	.	.	.
Chara sp.	oos	.	.	.	.	.	.	.	.	.	.	.	3	25
Elatine hydropiper	s	.	.	.	.	.	1	.	.	.	.	16	43	.
Hippuris vulgaris	a	.	.	1	.	.	.	.	.	1	.	.	.	.
Hydrocharis morsus-ranae	s	.	.	3	.	.	.	.	.	.	.	.	.	.
Lemna sp.	s	.	.	1	.	.	.	.	.	.	.	.	.	.
Lemna cf. *trisulca*	s	.	1	.	.	.	.	.	.	.	.	.	.	.
Menyanthes trifoliata	s	.	.	.	.	.	.	.	.	.	3	.	.	.
cf. *Najas* sp.	fr	.	.	.	.	.	.	.	.	1	.	.	.	.
Najas flexilis	fr	.	.	.	.	.	1	.	.	.	.	1	3	.
N. minor	fr	.	.	.	.	.	1	.	.	.	.	.	2	.
Nitella/Tolypella	oos	.	.	22	.	3	.	.	.	.	.	.	.	18
Nuphar sp.	s	.	.	.	.	.	.	.	.	.	.	1	.	.
Nymphaea alba	s	.	.	.	.	.	.	.	.	.	.	1	2	.
cf. *Nymphaea alba*	s	.	.	.	.	.	.	.	.	.	.	4	.	.
Potamogeton cf. *lucens*	en	.	.	.	.	.	.	.	.	.	.	.	.	5
Potamogeton cf. *lucens*	li	.	.	.	.	.	.	.	.	.	.	.	.	4

Table 25 (*cont.*)

		Pa												
			II			IVa	II			II		Cr IVa		e An
	Type of re-	Ia	HC	HC	HC	HC				BH				
Taxon	mains*	HC 381	317–321	306–309	265–268	223–233	HF 15	HE 20	HE 0–10	545–575	IV OE	OC 20–25	OA 70–80	BH 70–100
Aquatics (*cont.*)														
Potamogeton perfoliatus	en	.	.	.	.	.	.	.	36	.	.	.	.	1
Potamogeton cf. *perfoliatus*	li	.	.	.	.	.	.	.	.	.	.	.	.	5
Potamogeton trichoides	en	.	.	10	.	.	.	.	.	.	.	.	.	.
Potamogeton sp.	en	.	.	1	.	.	1	.	1	.	.	.	.	1
Potamogeton sp.	li	.	.	.	.	.	.	.	.	.	.	.	.	3
Ranunculus subg. *Batrachium* spp.	a	.	.	.	.	.	.	.	.	1	.	1	7	1
Sagittaria sagittifolia	fr	.	.	5	1	.	.	.	.	.	.	.	.	.
Scirpus tabernaemontani	n	.	.	.	.	1	.	.	.	.	.	.	.	.
Sparganium sp.	n	.	.	.	.	.	.	.	.	.	.	1	2	.
Sparganium erectum	n	.	.	.	.	.	.	.	.	.	.	1	.	.
Stratiotes aloides	l	.	.	.	.	1	.	.	.	.	.	.	.	.
S. aloides	ls	.	4	20	3	4	.	.	.	.	.	.	.	.
Trapa natans	bar	.	.	.	.	.	.	.	.	.	.	.	1	.
Typha spp.	s	.	2	.	3	.	.	.	.	.	.	.	3	.
Typha cf. *angustifolia*	s	.	2	.	.	.	1	.	.	.	.	.	.	.
Typha latifolia	s	.	.	42	.	3	35	.	31	7	.	.	.	.
Zannichellia palustris	a	.	.	.	.	.	.	.	.	.	.	.	16	.
Water ferns														
Azolla filiculoides	m	3	66	v.a.	70	.	.	.	1	3	.	3	.	.
A. filiculoides	m+ma	.	.	.	.	.	.	v.a.	.	.	.	.	.	.
Unclassified														
Carex spp.	n	.	1	2	1	.	.	.	11	.	.	32	.	.
Cyperaceae	n	.	.	.	5	.	.	.	.	.	.	.	.	.
Cyperus sp.	n	.	2	.	.	.	.	.	.	.	.	.	.	.
cf. *Cyperus* sp.	n	.	.	1	.	.	2	.	.	.	.	.	.	.
Dryopteris-type	spo	.	7	.	1	14	40	.	.	9	8	.	10	.
cf. *Epilobium*	s	.	.	.	.	.	.	.	1	.	.	.	.	.
Eriophorum sp.	n	.	.	1	.	.	.	.	.	.	.	.	.	.
Gramineae	ca	.	.	22	.	.	.	.	.	.	.	1	.	.
cf. Gramineae	ca	.	.	.	.	.	2	.	.	.	.	.	.	.
Juncus sp.	s	.	.	.	.	.	.	.	.	1	.	.	1	.
Labiatae	n	.	.	.	.	.	.	.	.	.	.	.	1	.
Mentha arvensis/ aquatica	n	.	.	5	.	.	.	.	.	.	.	.	3	.
Potentilla sp.	a	.	.	.	.	.	.	.	.	1	.	.	.	.
Ranunculus subg. *Ranunculus*	a	.	.	.	.	.	.	.	.	.	.	10	10	.
Rumex sp.	n	.	.	1	.	.	.	.	.	.	.	.	.	.
Umbelliferae	fr	.	1	5	.	.	.	.	.	.	.	.	.	.
Total (excluding water ferns)		3	33	346	16	45	99	.	325	39	214	1053	501	80
Derived megaspores		.	8	.	1	.	.	.	.	11	.	.	.	8

* Abbreviations as in Table 2.

Table 26. *Bacton and Ostend: summary of environmental history*

Bed	Sediment	Hiatus	Sedimentary regime	Sea-level	Periglacial: Ice-wedge casts	Periglacial: Involutions	Stage	Pollen assemblage biozone	Regional vegetation	Climate
Bacton										
g	till		glacial	.			An			
					+	+				
f	laminated clay, sand, silty mud		freshwater in part	regression			e An	Gramineae–*Betula–Pinus–Alnus*	grassland, some woodland	cold
e	laminated sand, with silty clay and organic laminae		tidal	.			Cr IVa	*Pinus–Alnus–Picea–Betula*	woodland	cool
d	sand, silty laminations		tidal							
				marine transgression						
		+								
c	ferruginous gravel, bedded sand			regression						
		+ ?								
b	ferruginous gravel									
a	silty sand, laminated silty clay		tidal	.			Pa II	*Pinus–Quercus–Ulmus–Carpinus*	woodland	temperate
							Pa I	*Pinus–Betula*–Gramineae	woodland	temperate
Ostend										
e	till		glacial	.			An			
d	detritus mud, clay-mud		freshwater	.			Cr IVa	*Pinus–Alnus–Picea–Betula*	woodland	cool
c	clay with rootlets and branches		land surface	.			Cr IV a	*Alnus–Pinus–Picea–Quercus–Ulmus*	woodland, carr	cool
b	clay with race		alluvium	.			Cr ?			
		+								
a	sand		.	.			Pre-Pa b/d	Gramineae–*Betula–Pinus*	grassland	cold

Table 27. *Ostend: pollen assemblage biozones*

Site	Depth (cm)	Location (m) (Fig. 39)	Bed	Designation of p.a.b.	Biozone	Substage
OA	0–88	60	d	*Pinus–Alnus–Picea–Betula*	c	Cr IVa
OB		60	d	*Pinus–Alnus–Picea–Betula*	c	Cr IVa
OA	94–104	60	c	*Alnus–Pinus–Picea–Quercus–Ulmus*	b	Cr IVa
OD	0	80	c	*Alnus–Pinus–Picea–Quercus–Ulmus*	b	Cr IVa
OD	300	80	a	Gramineae–*Betula–Pinus*	a	Pre-Pa d
OE		mud erratic		*Betula–Pinus–Alnus–Picea*	d	Pa IVa

Table 28. *Happisburgh: pollen assemblage biozones*

Site	Depth (cm)	Location (m) (Fig. 39)	Bed	Designation of p.a.b.	Biozone	Substage
HA	40–320	272 N	j	*Pinus–Alnus–Picea–Betula*	m	Pa IVa
HC	188–233	0	j	*Pinus–Alnus–Picea–Betula*	m	Pa IVa
HC	250–255	0	j	*Pinus–Carpinus–Quercus–Ulmus*	l	Pa III
HD		326	j	*Pinus–Quercus–Carpinus–Ulmus*	i	Pa II
HF	6	278	j	*Pinus–Quercus–Carpinus–Ulmus*	i	Pa II
HF	10	278	j (o)	*Alnus–Quercus–Osmunda*	(l) k	Pa II
HF	15	278	j	*Alnus–Quercus–Carpinus–Ulmus*	i	Pa II
HC	260–366	0	j	*Alnus–Quercus–Carpinus–Ulmus*	i	Pa II
HE	0	326	j (o)	*Alnus–Quercus–Pinus–Carpinus–Osmunda*	(l) j	Pa II
HE	20	326	j	*Alnus–Quercus–Carpinus–Ulmus*	i	Pa II
HC	381	0	j	*Betula–Pinus*–Gramineae	h	Pa Ia
HC	427	0	h	*Betula*–Gramineae	g	Pre-Pa d
HC	457	0	h	Gramineae–*Betula–Pinus*	f	Pre-Pa b/d
HC	601	0	g	*Pinus*–Gramineae	e	Pre-Pa
HC	712–1859	0	f, e	*Pinus*–Gramineae–Ericales	d	Pre-Pa a
HC	1968–2013	0	c	Gramineae–Ericales	c	Lu 2, 4?
HC	2225–2651	0	a	*Pinus*–Gramineae–Ericales	b }	Lu 1/2?
HC	2743	0	a	*Pinus–Picea*	a }	

Table 29. *Happisburgh: summary of environmental history*

Bed	Sediment	Hiatus	Sedimentary regime	Sea-level	Stage	Pollen assemblage biozone	Regional, local vegetation	Climate
k	till	.	glacial	.	An	.	.	.
				regression				
j	laminated silty clay and sand	.	tidal	.	Pa IVa	*Pinus–Alnus–Picea–Betula*	woodland	cool
					Pa III	*Pinus–Carpinus–Quercus–Ulmus*	woodland	temperate
j (o)	detritus mud with wood	.	freshwater/brackish ?	.	.			.
						Alnus–Quercus–Osmunda	fen, carr	temperate
					Pa II	*Alnus–Quercus–Pinus–Carpinus–Osmunda*	fen, carr	
j	laminated silty clay and sand	.	tidal	.		*Alnus–Quercus–Carpinus–Alnus*	woodland	temperate
					Pa Ia	*Betula–Pinus*–Gramineae	woodland	temperate
				marine transgression?				
	gravel	.	.	.	.	.	.	.
h	sand with mud erratics	.	fluviatile ?	.	Pre-Pa d	*Betula*–Gramineae	grassland, some woodland	
						Gramineae–*Betula–Pinus*		cold
	sand, silty clay laminae, cemented silty clay fragments	.	.	regression?	Pre-Pa	*Pinus*–Gramineae	grassland, some woodland	cold
		+ ?						
	laminated silty clay	.	marine	.	↑			
					Pre-Pa a	*Pinus*–Gramineae–Ericales	grassland (heath ?) some woodland	cold
e	sand with silty clay laminae	.	marine	.	↓	.		.
d	gravelly sand	.	marine	marine transgression	.	.	.	.
c	silty clay, sand and gravel	.	tidal/freshwater ?	.	Lu 2 or 4 ?	Gramineae–Ericales	grassland	cold
		+ ?		regression				
b	sandy gravel	.	marine	.	.	.	.	.
a	sand with silty clay laminae	.	marine	.	Lu 1/2 ?	*Pinus*–Gramineae–Ericales	grassland (heath ?), some woodland	cold
						Pinus–Picea	woodland	cool/ temperate?

Table 30. *Corton: pollen assemblage biozones*

Site	Depth (cm)	Location (Fig. 43) CW, Corton Woods CC, Corton Cliffs	Bed	Designation of p.a.b.	Biozone	Substage
CBb		410 m, CW	h	*Abies–Carpinus*	k	Cr IIIb
CC	0–86	0 m, CW	h, g	*Abies–Carpinus*	k	Cr IIIb
CK	15–37	80 m S, CC	h, f	*Carpinus*	j	Cr IIIa
CA	0	0 m, CC	f	*Carpinus*	j	Cr IIIa
CK	42–45	80 m S, CC	f	*Quercus–Ulmus–Tilia*	i	Cr IIb
CB		205 m S, CC	f	*Quercus–Ulmus–Tilia*	i	Cr IIb
CA	6–20	0 m, CC	f, e	*Quercus–Ulmus–Tilia*	i	Cr IIb
CK	45–63	80 m S, CC	e	*Pinus–Alnus–Picea–Tilia*	h	Cr IIb
CC	100	0 m, CW	e	*Alnus–Pinus–Tilia*	g	Cr IIb
BH 2/6	45–137	110 m S, CC	e	*Pinus–Picea*	f	Cr IIb
BH 2/6	686–884	110 m S, CC	b, a	*Quercus–Ulmus–Tilia*	e	Cr IIb
BH 3/6	457	840 m S, CC	b	*Pinus–Alnus–Quercus–Ulmus*	d	Cr IIb
BH 3/10	381	970 m S, CC	b	*Alnus–Quercus–Ulmus–Pinus*	c	Cr IIa
BH 3/5	762	710 m S, CC	b	*Alnus–Quercus–Ulmus–Pinus*	c	Cr IIa
CD		130 m, CW	d	*Alnus–Quercus–Ulmus–Pinus*	c	Cr IIa
CF	10	180 m, CW	d	*Alnus–Quercus–Ulmus–Pinus*	c	Cr IIa
CG		225 m, CW	a	*Betula–Alnus*	b	?
BH 3/11	762	1100 m S, CC	a	*Pinus–Quercus–Picea–Ulmus–Carpinus*	a	Pa II/III

Table 31. *Corton and Pakefield: macroscopic plant remains*

		No. of specimens at different depths														
		Pakefield	Corton Cr									Pakefield Cr IIIa				
Taxon	Type of remains*	Pa II PL 823	IIb CK 63	IIb CK 42	IIb CA 16–22	IIb/IIIa CA 0–13	IIb CB 2	IIb CI 0–10	IIb CI 12	IIIa CI 13–17	IIIa CI 17–20	PC 210–230	PC 80–90	PH 155–165	PD	PD a
Trees																
Acer campestre	fr	.	.	.	.	.	.	.	.	.	.	.	.	8	10	.
Acer sp.	fr	.	.	.	.	.	.	.	.	.	.	.	.	.	1	.
Alnus sp.	fr	.	.	.	.	.	.	.	.	.	.	2	.	.	.	.
Alnus sp.	c	1	.	.	.	30	.	.	.	.	.	.	.	3	.	.
Alnus sp.	c.sc	.	.	.	1	.	.	.	.	.	.	.	.	12	.	.
Alnus glutinosa	fr	.	.	.	.	269	8	.	.	.	.	.	.	203	26	.
Alnus cf. *glutinosa*	fr	.	.	.	4	.	.	.	.	.	.	.	.	.	.	.
Alnus incana	fr	.	.	.	.	.	.	.	.	.	.	1	.	.	.	.
Betulaceae	fr	.	.	.	9	17	.	.	.	.	.	.	.	.	.	.
cf. *Betula* sp.	fr	.	.	.	2	.	.	1	.	.	.	.	.	.	.	.
Betula pendula	fr	1	.	.	.	.	.	.	.	.	.	.	.	.	.	.
B. pendula/pubescens	fr	.	.	.	.	2	.	.	.	.	.	.	.	7	.	.
B. pubescens	fr	.	.	.	.	.	.	.	.	.	.	.	.	.	7	.
Carpinus betulus	n	.	.	.	.	.	.	.	.	.	.	.	.	.	11	2
cf. *Picea* sp.	l	.	.	.	.	.	.	.	.	.	.	.	.	2	.	.
Pinus sp.	s	.	.	.	.	.	.	.	.	.	.	.	.	.	2	.
Coniferae	l	.	.	.	.	.	.	.	.	.	.	.	.	1	.	.
cf. *Quercus* sp.	l	.	.	.	.	.	.	.	.	.	.	.	.	1	.	.
Populus sp.	b.sc	.	.	.	.	.	.	.	.	.	.	.	.	.	2	.
Shrubs																
Frangula alnus	s	.	.	.	.	5	.	.	.	.	.	.	.	.	.	.
Rhamnus catharticus	s	.	.	.	.	.	.	.	.	.	.	.	.	1	.	.
Rosa sp.	th	.	.	.	.	.	.	.	.	.	.	.	.	2	.	.
Salix sp.	b+b.sc	.	.	.	.	31	.	.	.	.	.	.	.	.	.	.
cf. *Staphylea* sp.	s	.	.	.	.	.	.	.	.	.	.	.	.	.	1	.
Viburnum opulus	fr.st	.	.	.	.	.	.	.	.	.	.	.	.	.	.	2
cf. *Viburnum* sp.	fr.st	.	.	.	.	.	.	.	.	.	.	.	.	.	1	.
Maritime																
Corema album	s	.	.	.	.	.	.	.	.	.	.	.	.	.	2	.
Salicornia sp.	s	3	.	.	.	.	.	.	.	.	.	1	1	.	.	.
Heath																
Erica ciliaris/lusitanica	s	.	.	.	.	.	.	.	.	.	.	.	1	.	.	.
Erica sp.	s	.	.	.	.	.	.	.	.	.	.	1	60	18	16	.
cf. *Erica* sp.	s	.	.	.	.	.	.	.	.	.	.	1	.	.	.	.
Dry-ground herbs																
Aphanes microcarpa	a	.	.	.	.	.	.	.	.	.	.	.	.	2	.	.
cf. *Atriplex* sp.	s	.	.	.	.	.	.	.	.	.	.	.	.	1	.	.
Calamintha sp.	n	.	.	.	.	.	.	.	.	.	.	1	.	.	.	.
Chenopodiaceae	s	.	.	.	.	.	.	.	.	.	.	.	.	1	.	.
Chenopodium album	s	.	.	.	.	.	.	.	.	.	.	.	.	1	.	.
Herniaria sp.	s	.	.	.	.	.	.	.	.	.	.	.	.	.	1	.

Table 31 (*cont.*)

Taxon	Type of re-mains*	No. of specimens at different depths														
		Pake-field	Corton Cr									Pakefield Cr IIIa				
		Pa II PL 823	IIb CK 63	IIb CK 42	IIb CA 16–22	IIb/IIIa CA 0–13	IIb CB 2	IIb CI 0–10	IIb CI 12	IIIa CI 13–17	IIIa CI 17–20	PC 210–230	PC 80–90	PH 155–165	PD	PD a
Dry-ground herbs (*cont.*)																
Hypericum montanum	s	.	.	.	.	.	.	.	.	.	.	.	.	.	1	.
Picris hieracioides	cy	.	.	.	.	.	.	.	.	.	.	.	.	2	1	.
Plantago major	s	.	.	.	.	.	.	.	.	.	.	1	.	.	.	.
Polygonum aviculare	n	.	.	.	.	.	.	.	.	.	.	.	.	1	.	.
Prunella vulgaris	n	.	.	.	.	.	.	.	.	.	.	.	.	1	.	.
Ranunculus bulbosus	a	.	.	.	.	.	.	.	.	.	.	.	.	.	4	.
Senecio cf. *sylvaticus*	cy	.	.	.	.	.	.	.	.	.	.	.	.	2	.	.
Thalictrum minus	a	.	.	.	.	.	.	.	.	.	.	.	.	1	1	.
cf. *Tussilago farfara*	cy	.	.	.	.	.	.	.	.	.	.	.	.	.	2	.
Urtica urens	s	.	.	.	.	.	.	.	.	.	.	.	.	2	3	.
Damp-ground herbs																
Lythrum cf. *borysthenicum*	s	1	.	.	.	.	.	.	.	.	.	.	.	.	.	.
Juncus cf. *inflexus*	s	.	.	.	.	.	.	.	.	.	.	1	.	.	.	.
Ranunculus repens	a	.	.	.	.	.	.	.	.	.	.	.	.	.	1	.
Solanum dulcamara	s	.	.	.	12	7	.	.	.	.	.	.	.	2	1	.
Muddy substrates, shallow water																
Alisma sp.	fr	6	.	.	.	.	.	.	.	.	.	1	.	11	.	.
Alisma plantago-aquatica	.	.	.	.	.	.	.	.	.	.	.	.	.	.	2	.
Eleocharis sp.	n	.	.	.	.	.	.	.	.	.	.	.	.	1	.	.
Rumex maritimus	n	.	.	.	.	.	.	.	.	.	.	.	.	1	.	.
Rumex maritimus	per	.	.	.	.	.	.	.	.	.	.	.	.	1	.	.
Fen and reedswamp helophytes																
Bidens sp.	cy	.	.	.	.	.	.	.	.	.	.	1	1	.	1	.
Bidens sp.	te	.	.	.	.	.	.	.	.	1	1	.	4	.	.	.
Bidens tripartita	cy	.	.	.	.	1	.	.	.	.	.	.	.	5	.	.
Carex riparia	n	.	.	.	11	.	.	.	.	.	.	.	.	.	.	.
Cicuta virosa	fr	.	.	.	.	3	.	.	.	.	.	.	.	3	1	.
Cyperus longus	n	1	.	.	2	.	.	.	.	.	.	4	.	17	.	.
Eupatorium cannabinum	cy	.	.	.	.	.	.	.	.	.	.	.	.	.	1	.
Hydrocotyle vulgaris	fr	.	.	.	.	.	.	.	.	.	.	.	1	.	.	.
Lycopus europaeus	n	.	.	.	6	12	.	.	.	.	.	.	.	2	2	.
cf. *Lysimachia vulgaris*	s	.	.	.	.	.	.	.	.	.	.	.	.	3	.	.
Lythrum salicaria	s	.	.	.	.	.	.	.	.	.	.	.	.	.	3	.
Oenanthe lachenalii	fr	.	.	.	.	.	.	.	.	.	.	.	.	3	.	.
Osmunda sp.	spo	.	.	.	.	1	3	.	.	.	.	.	4	4	7	.
Osmunda sp.	ann	.	.	.	.	.	.	.	.	1	.	.	.	.	.	.
cf. *Phragmites communis*	r	.	.	.	.	1	.	.	.	.	.	.	.	.	.	.
Ranunculus lingua	a	.	.	.	.	2	.	.	.	.	.	.	.	.	.	.
Ranunculus sceleratus	a	2	.	.	2	1	.	.	.	1	.	1	3	5	4	.
Thalictrum flavum	a	.	.	.	.	.	.	.	.	.	.	.	.	.	2	.
Urtica dioica	n	1	.	.	137	37	.	.	.	.	.	7	.	22	33	.

Table 31 (*cont.*)

Taxon	Type of remains*	Pakefield Pa II PL 823	Corton Cr IIb CK 63	Corton Cr IIb CK 42	Corton Cr IIb CA 16–22	Corton Cr IIb/IIIa CA 0–13	Corton Cr IIb CB 2	Corton Cr IIb CI 0–10	Corton Cr IIb CI 12	Corton Cr IIIa CI 13–17	Corton Cr IIIa CI 17–20	Pakefield Cr IIIa PC 210–230	Pakefield Cr IIIa PC 80–90	Pakefield Cr IIIa PH 155–165	Pakefield Cr IIIa PD	Pakefield Cr IIIa PD a
		No. of specimens at different depths														
Aquatics																
Callitriche sp.	fr	.	.	.	.	.	.	.	.	.	.	.	2	107	30	.
Callitriche platycarpa	fr	.	.	.	.	.	.	.	.	.	.	5	.	.	.	.
Callitriche truncata	fr	1	.	.	.	.	.	.	.	.	.	.	.	.	.	.
Ceratophyllum demersum	fr	.	.	.	.	.	.	.	.	.	.	.	.	.	1	.
Chara sp.	oos	1	.	.	.	.	.	.	.	.	.	.	.	10	3	.
Characeae	oos	.	.	.	.	.	.	.	.	.	.	.	6	.	.	.
Elatine hydropiper	s	.	.	.	.	.	.	.	.	.	.	1	12	3	2	.
Hippuris vulgaris	p	.	.	.	.	1	.	.	.	.	.	.	.	.	.	.
Hydrocharis morsus-ranae	s	.	.	.	.	8	.	2	.	.	.	.	.	.	2	.
Myriophyllum spicatum	n	.	.	.	.	.	.	.	.	.	.	.	.	6	5	.
Najas marina	fr	.	.	.	.	.	1	.	.	1	.	2	.	3	12	.
Najas minor	fr	.	.	.	.	.	.	.	.	.	.	10	7	107	47	.
Nuphar sp.	s	.	.	.	.	.	.	.	.	.	.	.	.	2	1	.
Nuphar sp.	p	.	.	.	.	.	.	.	.	.	.	.	.	2	1	.
Nymphaea alba	s	.	.	.	.	.	.	.	.	.	.	2	.	.	2	.
Nymphaea sp.	s	11	.	.	.	.	.	.	.	.	.	.	7	.	.	.
Potamogeton crispus	en	.	.	.	.	.	.	.	.	.	.	.	.	2	.	.
P. crispus	li	.	.	.	.	.	.	.	.	.	.	.	.	2	.	.
Potamogeton pectinatus	en	.	.	.	.	.	.	.	.	.	.	.	.	.	2	.
P. pectinatus	li	.	.	.	.	.	.	.	.	.	.	.	.	.	2	.
Potamogeton cf. *pectinatus*	en	.	.	.	.	.	.	.	.	.	.	1	.	.	1	.
Potamogeton cf. *perfoliatus*	en	.	.	.	.	.	.	.	.	.	.	.	.	.	5	.
Potamogeton sp.	en	.	.	.	.	1	.	.	.	.	.	.	.	1	3	.
Potamogeton sp.	li	.	.	.	.	.	.	.	.	.	.	.	.	.	3	.
Ranunculus subg. *Batrachium* spp.	a	2	.	.	.	.	.	.	.	.	.	6	.	87	68	.
Sagittaria sp.	fr	.	.	.	.	.	.	.	.	.	.	.	.	.	1	.
cf. *Sagittaria* sp.	fr	.	.	.	.	.	.	.	.	.	.	2	.	.	.	.
Scirpus lacustris	n	.	.	.	.	.	.	.	.	.	.	.	.	.	4	.
Scirpus lacustris + *Scirpus tabernaemontani*	n	.	.	.	.	.	.	.	.	.	.	.	.	9	.	.
Sparganium sp.	n	.	.	.	.	.	.	.	.	.	.	.	.	.	1	.
Sparganium cf. *erectum* ssp. *neglectum*	n	.	.	.	.	.	.	.	.	.	.	.	.	5	.	.
Stratiotes aloides	ls	.	.	.	7	417	10	.	.	15	.	10	26	17	13	.
Trapa natans	bar	.	.	.	.	.	.	.	.	.	.	.	.	4	2	.
cf. *Trapa natans*	fr	.	.	.	.	.	.	.	.	.	.	.	.	.	.	1
Typha latifolia	s	.	.	.	.	.	2	.	.	.	.	194	.	.	.	.
Typha angustifolia + *Typha latifolia*	s	39	.	.	.	.	.	.	.	.	.	.	.	168	271	.
Typha spp.	s	.	1	6	7	79	.	4	1	25	3	1	54	.	.	.
Zannichellia palustris	a	.	.	.	.	.	.	.	.	.	.	2	2	13	11	.
Water ferns																
Azolla filiculoides	m	.	.	1	v.a.	.	.	.	.	.	7	.	.	.	.	.
Azolla filiculoides	ma	.	.	7	.	.	.	.	.	.	12	.	.	.	.	.

Table 31 (*cont.*)

		No. of specimens at different depths														
		Pakefield	Corton Cr									Pakefield Cr IIIa				
	Type of remains*	Pa II	IIb	IIb	IIb	IIb/IIIa	IIb	IIb	IIb	IIIa	IIIa					
Taxon		PL 823	CK 63	CK 42	CA 16–22	CA 0–13	CB 2	CI 0–10	CI 12	CI 13–17	CI 17–20	PC 210–230	PC 80–90	PH 155–165	PD	PD a
Water ferns (*cont.*)																
Azolla filiculoides	m+ma	5	v.a.	9	.	31	49	v.a.	v.a.	a.	.	v.a.	v.a.	v.a.	v.a.	.
Salvinia natans	m	1	1	10	.	.	2	a.	.	10	.	.	1	.	.	.
Salvinia natans	ma	.	.	60	.	.	1	.	.	.	.	.	.	.	.	.
Salvinia natans	m+ma	.	.	.	81	198	.	.	20	.	.	57	.	44	55	.
Unclassified																
Carex spp.	n	.	.	2	.	7	.	.	.	.	.	.	.	22	21	.
Carex spp.	u	.	.	.	.	1	.	.	.	.	.	.	.	.	.	.
Cirsium/Carduus sp.	cy	.	.	.	.	.	.	.	.	.	.	.	.	2	.	.
Compositae	cy	.	.	.	.	.	.	.	.	.	.	.	.	3	2	.
Cruciferae	s	.	.	.	.	.	.	.	.	.	.	.	.	3	.	.
Cruciferae tribe Lepidieae	c.v	.	.	.	.	.	.	.	.	.	.	.	.	3	.	.
Cyperaceae	n	.	.	.	.	.	.	.	.	.	.	1	1	.	.	.
Cyperus sp.	n	.	.	.	.	9	.	.	.	.	.	.	.	.	23	.
cf. *Cyperus* sp.	n	.	.	.	.	.	.	.	.	2	.	.	.	.	.	.
Dryopteris-type	spo	28	.	.	.	149	70	.	.	100	.	6	9	.	20	.
Gramineae	ca	3	.	.	.	9	.	.	.	.	.	2	16	51	102	.
Juncus sp.	s	.	.	.	.	.	.	.	.	.	.	.	.	2	.	.
Labiatae	n	.	.	.	1	.	.	.	.	.	.	.	.	1	4	.
cf. Labiatae	n	.	.	.	1	.	.	.	.	.	.	.	.	.	.	.
Mentha sp.	n	.	.	.	.	.	.	.	.	.	.	6	.	.	.	.
Mentha arvensis/aquatica	n	.	.	.	.	.	.	.	.	.	.	.	1	9	6	.
Orchidaceae	s	.	.	.	.	.	.	.	.	.	.	1	.	.	.	.
Polygonum sect. *Persicaria*	n	.	.	.	.	.	.	.	.	.	.	.	.	4	.	.
Potentilla sp.	a	.	.	.	.	.	.	.	.	.	.	1	.	.	.	.
Ranunculus subg. *Ranunculus*	a	.	.	.	.	.	.	.	.	.	.	.	.	2	2	.
cf. *Ranunculus* subg. *Ranunculus*	a	.	.	.	.	9	.	.	.	.	.	.	.	.	.	.
Rumex sp.	n	.	.	.	.	.	1	.	.	.	.	.	.	2	.	.
Rumex sp.	per	.	.	.	.	.	.	.	.	.	.	.	.	2	.	.
Umbelliferae	fr	1	.	.	.	.	.	.	.	.	.	.	.	.	.	.
Urtica sp.	n	.	.	.	.	.	.	.	.	.	.	.	4	.	.	.
cf. *Veronica* sp.	s	.	.	.	.	.	.	.	.	.	.	1	.	.	.	.
Viola sp.	s	.	.	.	.	.	.	.	.	.	.	1	.	.	1	.
Total (excluding water ferns)		102	1	8	202	1109	95	7	1	146	4	278	222	1006	824	5
Derived megaspores		4	.	.	.	.	.	.	.	.	.	.	.	.	.	.

* Abbreviations as in Table 2.

Table 32. *Corton: summary of environmental history*

Bed	Sediment	Hiatus	Sedimentary regime	Sea-level	Periglacial Ice-wedge cast	Stage	Pollen assemblage biozone	Regional, local vegetation	Climate
j	till		glacial	.		An	.	.	
					+				cold
i	sand, drift mud laminae		fluviatile	regression					
h	laminated silty clay		tidal	.		↑ Cr IIIb	*Abies–Carpinus*	woodland	temperate
g	gravel and sand		beach						
				marine transgression					
f	detritus mud and wood		freshwater	.		Cr IIIa	*Carpinus*	woodland, fen, carr	temperate
						Cr IIb	*Quercus–Ulmus–Tilia*	woodland, fen, carr	temperate
e	Rootlet Bed (clay with race)		alluvium	.		Cr IIb	*Pinus–Alnus–Picea–Tilia*		
							Alnus–Pinus–Tilia		
							Pinus–Picea		
d	detritus mud		freshwater			↑			
c	sand and gravel		fluviatile or marine?						
b	laminated silty clay		marine in part	.		↓	*Quercus–Ulmus–Tilia*	woodland	temperate
						Cr IIa	*Alnus–Quercus–Ulmus–Pinus*	woodland, carr	temperate
		+							
a	sand with silty clay laminae		marine	.		Pa II	*Pinus–Quercus–Picea–Ulmus–Carpinus*	woodland	temperate

Table 33. *Pakefield, Kessingland, Covehithe and Skelding Hill: pollen assemblage biozones*

Site	Depth (cm)	Location	Bed	Designation of p.a.b.	Biozone	Substage
Pakefield, Kessingland (Fig. 46)						
PC	80–240	208 m S	g	*Carpinus*	d	Cr IIIa
PE	35, 200	435 m S	g	*Carpinus*	d	Cr IIIa
K	2–24		d	*Pinus*–Gramineae	c	Cr II
PF	B, 140	507 m S	d	*Pinus*–Gramineae	c	Cr II
PL	823	510 m S	b	*Pinus–Quercus–Ulmus–Carpinus*	b	Pa II
PL	854–1006	510 m S	a	*Pinus*–Ericales	a	Pre-Pa a
PK	884	211 m S	a	*Pinus*–Ericales	a	Pre-Pa a
Covehithe (Fig. 48)						
LC 3	1860	0 m	.	Ericales	.	Ba (Lu4b)
LC 5	1450	425 m	.	Ericales–Gramineae	.	Ba (Lu4b)
Skelding Hill, Sheringham (Figs. 2, 49)						
WL	.	.	f	Gramineae–*Pinus*	b	Pa Ib
WM	8–50	.	g	*Pinus*	a	Pa Ib
WP	10–80	.	e	*Pinus*	a	Pa Ib
WQ	.	.	b	*Pinus*	a	Pa Ib

Table 34. *Pakefield, Kessingland and Covehithe: summary of environmental history*

Bed	Sediment	Hiatus	Sedimentary regime	Sea-level	Stage	Pollen assemblage biozone	Regional vegetation	Climate
i	cross-bedded sand (Corton Beds)	.	fluviatile	.	An	.	.	.
		+						
h	laminated clays	.	tidal	.	.	.	.	.
g	silty mud	.	freshwater/brackish	marine transgression	Cr IIIa	*Carpinus*	woodland	temperate
f	ferrugineous sand and gravel	.	freshwater	.	.	.	.	.
		+						
e	'peat' (of Blake 1884, 1890)	.	freshwater	.	(Cr II)	.	.	.
d	Rootlet Bed (clay with race)	.	alluvium	.	Cr II	*Pinus*–Gramineae	.	temperate
			.	\				
c	sand with small gravel	.		? regression	.	.	.	.
				/				
b	laminated silty clay	.	marine	.	Pa II	*Pinus*–*Quercus*–*Ulmus*–*Carpinus*	woodland	temperate
		+						
a	sandy silt with silty clay laminae	.	marine	.	Pre-Pa a	*Pinus*–Ericales	grassland (heath ?) some woodland	cold
	laminated clay at Covehithe	.	marine	.	Ba (Lu 4b)	Ericales–Gramineae Ericales	grassland, heath	cold

Table 35. *Level and age of Cromerian transgression contacts and brackish muds*

Area, site	Age of freshwater sediment	Age of marine sediment	OD level of transgression contact (m) and nature of contact
West Runton BCa	.	Cr IIIa	5·1 sharp
BZb	Cr IIb	Cr IIIb	4·3 sharp
BQ	Cr IIb	Cr IIIb	2·0 sharp
Sidestrand SST	Cr Ib	Cr Ib	level disturbed by glacitectonics
Paston–Mundesley MU	Cr II	Cr IIIa	*c.* 2·0 sharp
MAK	Cr II	Cr IIIb	2·6 sharp
MAM	brackish muds at Cr IIIb/IVa transition		2·8 sharp
MAR			2·9 sharp
Corton CI, CK	Cr IIIa	Cr IIIa	0·5 transition
Pakefield PC	Cr IIIa	.	2·2 transition

Table 36. *Levels of Cromerian freshwater and marine sediments*

Area, site	Sediment and age	OD limit (m) Base	OD limit (m) Top
Beeston	laminated clay (Cr IVa)	at *c.* 8·0	
West Runton	marine gravel and laminated clay (Cr IIIa, b; Cr IVa)	2·0	7·5
	West Runton Freshwater Bed (Cr IIb)	.	4·8
Trimingham BHT 14A	marine (?) mud (Cr IIa)	at −3·5	
Paston–Mundesley	marine gravel and laminated clay (Cr IIb; Cr IIIa, b; Cr IVa)	1·6	4·3
	West Runton Freshwater Bed (Cr II)	.	2·6
Bacton BH	marine sand and clay (Cr IVa)	*c.* 2·0	*c.* 4·0
Ostend OA	freshwater mud (Cr IVa)	*c.* 2·0	*c.* 3·0
Corton	marine gravel and laminated clay (Cr IIIa, b)	0·5	*c.* 2·5*
	detritus mud (Cr IIb/IIIa)	0·3	0·5
	Rootlet Bed (Cr II)	−1·3	2·0
	marine clay (Cr IIb)	*c.* −6·0	*c.* −4·0
Pakefield	marine laminated clay	2·3	*c.* 4·0*
	West Runton Freshwater Bed (Cr IIIa)	0·6	2·2
	Rootlet Bed (Cr II)	0·4	*c.* 3·2

* Estimated from sections drawn by Blake (1884).

Table 37. *Level and age of Cromerian regression contacts*

Area, site	Age of Marine sediment	Age of Freshwater sediment	OD level of regression contact (m)
West Runton WRAYa	Cr IVa	Cr IVc	7·5
Overstrand OB	Cr IIIb	Cr IVb	*c.* 5·3
Paston–Mundesley			
MAU	Cr IIIb	Cr IVc	3·7
MB	Cr IIIb	Cr IVb	2·5
MAD		Cr IVa	3·0

Table 38. *Levels of Beestonian marine sediments*

Area, site	Sediment	OD limit (m) Base	OD limit (m) Top
Beeston, BV to BC	gravel, sand	4·4	7·8
West Runton			
25 m west of datum	gravel, sand	3·0	4·9
105 m west of datum	gravel, sand	5·0	6·7

Table 39. *Level and age of Pastonian regression and transgression contacts and sediments*

	Area, site	Age of Marine sediment	Age of Freshwater sediment	OD level of contact (m) and nature of contact
Regression	Beeston, BR–S	Pa III	Pa IVa	5·1 sharp
	Paston–Mundesley			
	PE	Pa IVa	Pa IVc	2·0 sharp
	MPA	.	Pa IVa	−0·8
Transgression	Beeston			
	BQ	Pa II	Pa Ib	4·8 transition
	BW	Pa II	Pa Ib	5·7 transition

Table 40. *Levels of Pastonian transgression and associated sediments*

Area, site	Sediment and age	OD limit (m) Base	OD limit (m) Top
Skelding Hill	stone bed, clay (Pa Ib at base)	6·5	9·5
Beeston	gravels, clay (Pa II, III, IVa)	4·4	6·2
West Runton			
WRCNA	sand (Pa II, III)	.	3·9
50 m west of datum	sand (Pa II, III)	.	4·4
105 m west of datum	sand (Pa II, III)	.	5·0
WRF	brackish mud (Pa II)	.	*c.* −1·5
Overstrand			
Boreholes	sand, clay (Pa II, IVb)	−2·0	1·2
OB	clay (Pa IVb)	.	*c.* 2·5
Trimingham, BHT 16	sand, clay (Pa Ib)	−9·0	−1·5
Paston–Mundesley	sand, clay (Pa II, III, IVa)	−3·2	2·4
Happisburgh			
HC, HA	clays (Pa I–IV)	−1·5	2·4
HF	mud intercalation in tidal sediments (Pa II)	.	−0·5
Corton, borehole 3/11	sand, clay (Pa II)	.	−6·7
Pakefield, borehole PL	clay (Pa II)	−2·9	−2·2

Table 41. *Levels of Pre-Pastonian freshwater and marine sediments, and of Baventian (Lu 4b) sediments*

	Area, site	Sediment and age	OD limit (m) Base	Top
Pre-Pastonian Freshwater	Beeston, BR–S	sand, clay, mud (Pre-Pa b–d)	3·2	4·7
	East Runton, ERA	mud (Pre-Pa d)	−2·2	−1·0
Marine	Beeston, BE	stone bed (Pre-Pa a)	3·4	.
	West Runton, WRBN	stone bed (Pre-Pa a)	1·9	.
	Overstrand, borehole 1B	stone bed (Pre-Pa a)	−4·2	.
	Paston–Mundesley, MAT	stone bed, clay (Pre-Pa a)	−11·5	−4·8
	Happisburgh, HC	sand, clay (Pre-Pa a)	−16·5	−3·0
	Pakefield	sand, clay (Pre-Pa a)	−5·7	−3·0
Baventian	Covehithe	clay (Baventian, Lu 4b)	−10·0	5·0
	Easton Bavents	clay (Baventian, Lu 4b)	4·0	6·0

Table 42. *Principal pollen assemblage biozones for each stretch of coast.* See end of figures in the separate booklet.

Table 43. *Stratigraphical classification and nomenclature*

Formation	Member	Bed	Type site	Bed	Fig. no.	Stage
Cromer Forest-bed Formation	*Bacton Member (freshwater)* Clays and muds, to *c.* 1·5 m thick on Norfolk coast	*Arctic Freshwater Bed* of Reid (1882) at Mundesley and Bacton	Mundesley ME	m	33	e An
		(Other freshwater beds of Cr IV age at West Runton, Overstrand, Mundesley, Ostend)				Cr IV
	Mundesley Member (marine) Laminated clays and sands, beach gravels and sands, to *c.* 2·5 m thick, on Norfolk and Suffolk coasts	Yoldia (Leda) myalis *Bed* of Reid (1882) at West Runton. Marine gravel, sand and laminated clay	West Runton WRBZ to WRAW	post-WRFB a, b, c in part	7	Cr III, IV
		Mundesley Clay. Laminated clay at base of Bacton–Paston–Mundesley Cliffs	Mundesley MAR	i (lc)	33	Cr III, IV
	West Runton Member (freshwater) Muds and alluvial clays to *c.* 2 m thick, on Norfolk and Suffolk coasts	*West Runton Freshwater Bed (Upper Freshwater Bed* of Reid (1882) at West Runton). Shelly muds, marls and sands. Also at Mundesley, Corton	West Runton WRAQ, WRU	WRFB a to g	6	lBe, Cr I, II
		Corton Rootlet Bed. Alluvial clays of Corton and Pakefield–Kessingland coasts	Corton Cliffs, CK, CKG	e	43	Cr II
	Runton Member Beach gravels and sands, and freshwater clays of West Runton and Beeston, to *c.* 2 m thick	*Woman Hithe Gravel*. Beach gravels with shelly sand and laminated clays at West Runton. Gravels at Beeston	West Runton WRAW to WRCS	pre-WRFB i	7, 8	Be
		Woman Hithe Clay. Freshwater laminated clay at West Runton	West Runton WRCO	pre-WRFB h	7	Be
		Beeston Beds. Freshwater muds, clays and sands at Beeston	Beeston BA–B, BF	n, p, q, r	17, 18	Be
		(Freshwater muds of Pa IV age at Beeston, Paston–Mundesley)				Pa IV
	Paston Member (marine) Laminated silts and clays, clay conglomerates, shelly sands and gravels, to *c.* 8 m thick, on Norfolk and Suffolk coasts	*West Runton Sand*. Cross-bedded sand with marine shells at West Runton	West Runton WRCNA	pre-WRFB f	4	Pa
		West Runton Clay. Laminated clay at West Runton, Happisburgh	West Runton WRAX	pre-WRFB d i	7	Pa II, III, IV
		Paston Beds. Marine sands, clay conglomerates and gravel (includes most of the Forest-bed (Estuarine) of Reid (1882))	Paston MN	e	31	Pa
	Sheringham Member (freshwater) Clays, sands and muds, to 2·5 m thick on Norfolk coast	*Sheringham Beds*. Clay, sand and mud at Beeston between stone bed and marine Pastonian, mud at East Runton (includes some of Lower Freshwater Bed of Reid (1882))	Beeston BR/S	c to i	18	Pre-Pa, b, c, d, Pa I
Norwich Crag Formation	*Sidestrand Member* Marine laminated clays, sands and gravels, to *c.* 13 m thick, on Norfolk and Suffolk coasts	*Sidestrand Sands*. Cross-bedded sands (to *c.* 5 m thick) at Sidestrand	Sidestrand SSB	b	22	Pre-Pa a
		Cromer Stone Bed. Shelly stone bed and associated crag and silt of Beeston to Cromer area	Beeston BE, BKC	b	16, 18	Pre-Pa a
		Easton Bavents Clay. Laminated clay at Easton Bavents, Covehithe, Aldeby, Ludham	Easton Bavents	.	.	Ba (Lu 4b)

Table 44. *Distribution of weathered and rootlet horizons in the Cromer Forest-bed Formation*

Site	Bed	Weathered	Rootlets	Age of weathering or rootlets
Beeston				
BAX	k	+	.	pre-bed o (Beestonian)? post-Pastonian II/III
85 m	r	+	.	Cromerian?
BV, BP, BL, BAM	r	.	+	Cromerian?
BO	g	.	+	Pre-Pastonian d
West Runton				
WRBT (Plate 2)	h	.	+	Cromerian
WRBZb	i	.	+	Cromerian
255 m W, WRAZ	f or i	+ (palaeosol)	.	Cromerian
WRBCb (Plate 3)	?	.	+	Cromerian
Overstrand				
OB	g	.	+	late Pastonian?
Ostend				
OA, OD (Plate 32)	b, c	+	+	Cromerian
Corton				
CJ (Plate 35)	e	+	+	Cromerian
Pakefield				
PJ	d	+	+	Cromerian

Table 45. *Comparison of substages and pollen assemblage biozones of the Pastonian and Cromerian temperate stages*

Sub-stage	Pastonian biozones	Cromerian biozones
IV	c *Betula–Pinus*	c *Betula*
	b *Pinus–Picea*	b *Pinus–Betula*
	a *Pinus–Alnus–Picea–Betula*	a *Pinus–Alnus–Picea–Betula*
III	*Pinus–Betula–Quercus* a.o.	b *Abies–Carpinus*
		a *Carpinus*
II	*Pinus–Quercus–Carpinus–Ulmus*	b *Quercus–Ulmus–Tilia*
		a *Pinus–Quercus–Ulmus*
I	b *Pinus*	b *Pinus–Ulmus*
	a *Betula–Pinus*–Gramineae	a *Betula–Pinus*

Table 46. *Exotic taxa of the Pre-Pa to An stages*

Taxon	Type of remains	Records
**Azolla filiculoides* Lam.	ms	a
Salvinia natans (L.) All.	ms	a
Osmunda claytoniana-type	p	a
Abies alba Mill.	p	a
Ephedra distachya-type	p	r
Picea abies (L.) Karst	m, p	a
*cf. *P. omorika* (Panc.) Purkyne	m	r
Tsuga	p	r
T. canadensis-type	p	r
T. diversifolia-type	p	r
Cupressaceae b (*Chamaecyparis* ?)	p	a
**Alnus incana* (L.) Moench	m	r
Bruckenthalia spiculifolia (Salisb.) Reichenb.	p	r
Carya	p	r
**Chamaedaphne calyculata* (L.) Moench	m	r
**Corema album* (L.) D. Don	m	r
**Corispermum*	m	r
Eleocharis carniolica C. Koch	m	a
Eleocharis mamillata Lindb. f.	m	r
Erica lusitanica Rudolphi	m	r
E. umbellata L.	m	r
Eucommia	p	r
Lythrum cf. *borysthenicum* (Schrank) Litv. in Majevski	m	r
Naias minor All.	m	a
**N. tenuissima* (A. Br.) Magnus	m	r
Pterocarya	p	r
Ranunculus hyperboreus Rottb.	m	r
Salix cf. *polaris*	m	r
*cf. *Staphylea* sp.	m	r
**Swertia perennis* L.	m	r
Trapa natans L.	m	r
Monocolpate X (*Sabal*?)	p	a

*, Discussed by Wilson, 1973; m, macroscopic remains; ms, megaspores and massulae; p, pollen or spores; r, single or few records; a, many records.

Table 47. *Records of* Ilex, Hedera *and* Taxus

Substage	*Ilex*	*Hedera*	*Taxus*
e An	.	.	1x
Cr IVa	1	3	3x
IIIa	.	2	2
II	2x	4	4
I	.	1	.
Pa IV	.	1?	1?
III	.	1	.
II	5	7	4
I	1	.	.
Lu 4b	1	.	.
Lu 1/2	1	.	.

Numbers indicate the number of pollen analyses containing a record of the taxon in the substage. x indicates presence of macroscopic remains in the substage.

Table 48. *Previous classifications of the pre-glacial sequence* (*Woodward, 1891*)

	J. Gunn		J. Prestwich		C. Reid
Glacial	Lower Boulder Clay		Lower Boulder Clay	Glacial	Cromer Till Arctic Fresh-water Bed
	Laminated Beds	Westleton and Mundesley Beds	Laminated clays, sands, and shingle with plant-remains and fresh-water shells Sand and quartzose shingle with marine shells Carbonaceous clays and sands, with flint-gravel and pebbles of clay, drift wood, land- and lacustrine shells and seeds		*Leda-myalis* Bed
Forest Bed Series	Unio and Rootlet Beds Re-constructed Forest-Bed True Forest Bed Estuarine soil: Blue clay with stools of trees, and indurated ferrugineous gravel (Elephant Bed)			Forest-Bed Series	Upper Fresh-water Bed 'Forest-bed' (Estuarine) Lower Fresh-water Bed
		Forest-Bed Group	Greenish clay, sandy and laminated in places, containing abundant Mammalian remains and driftwood, with stumps of trees standing on its surface Ferrugineous clay, peat with land and fresh-water remains, and gravel		Weybourne Crag and Chillesford Clay?
	Norwich Crag		Norwich Crag		Chalk
	Chalk		Chalk		

Table 49. *Correlation of stages of the coastal pre-glacial in Britain*

Norfolk and Suffolk coasts	Ludham	Covehithe	Easton Bavents	Suffolk, Essex	Other sites
Cromerian				? Valley Farm Soil	Sugworth, Oxfordshire Westbury-sub-Mendip, Somerset
Beestonian				Kesgrave sands and gravels	Plateau Drift of Oxford area in part
Pastonian	Lu 5				
Pre-Pa a	Lu 4c				
		? Westleton Beds	? Westleton Beds	Chillesford pollen assemblage Crag at Chillesford, Sizewell, etc.	
Baventian	Lu 4b	Lu 4b	Lu 4b Lu 4a		

THE PLATES

Plate 1
Black shelly mud with breccia (bed d) of the West Runton Freshwater Bed, at West Runton. The shelly sediment contains angular fragments of grey marl from the underlying bed a.

Plate 2
Section WRBT at West Runton, 515 m east of Woman Hithe, 1965 (Fig. 4). The black organic horizon of the Cromerian swale filling lies on clays of bed h, which are penetrated by rootlets. Gravel of bed a overlies the organic horizon, with sand of bed d above it.

Plate 3
Section WRBCb at West Runton, 408 m west of Woman Hithe, 1964 (Fig. 11). Clay penetrated by rootlets, related to the Cromerian land surface, overlies bedded sand of pre-Cromerian age. The 'rootlet-bed' is overlain by laminated tidal clay of bed b, of Cromerian IIIa age.

Plate 4
Section WRAW at West Runton, 44 m west of Woman Hithe, 1964 (Fig. 7). The clay in front of the spade is the tidal clay of bed d i, Pastonian II in age. Behind the spade are cross-bedded sands of bed f, showing their characteristic westward apparent dip. Above this are beach gravels and shelly sands of bed i (Beestonian). The beach gravels of bed a of the post-West Runton Freshwater Bed succession overlie them, and contain *Mya* in position of life. Between this and the till (bed e) are tidal sediments of bed b, sandier at the base, the more massive silt and clay of bed c, and the sand of bed d.

Plate 5
West Runton, 320 m east of Woman Hithe, 1974. Laminated tidal sediments of bed b, Cromerian III in age, overlie the black detritus mud of the West Runton Freshwater Bed (bed f).

Plate 6
West Runton, looking west, 225 to 245 m west of Woman Hithe, 1972 (Fig. 10). A sloping dark surface of Cromerian organic swale filling thickens westward and is overlapped by laminated tidal sediments of bed b, Cromerian III in age. Massive silt and clay of bed c overlies the tidal sediments.

Plate 7
West Runton, about 550 m east of Woman Hithe, 1972. Ice-wedge cast in upper part of the bedded sand of bed d in the post-West Runton Freshwater Bed succession, underlying Anglian till.

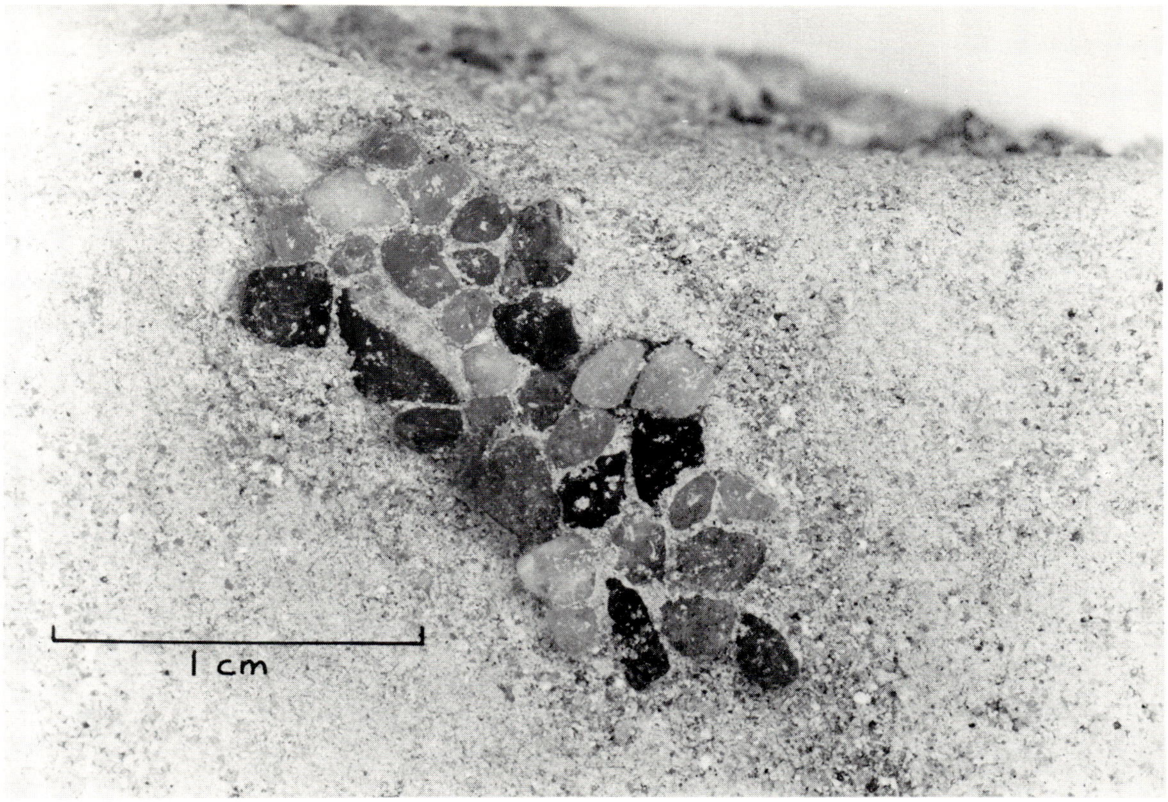

Plate 8
Caddis-fly larval case from laminated silty clay of bed h at section WRAR, 344 m east of Woman Hithe, 1964 (Fig. 6). Beestonian.

Plate 9
West Runton, view down from cliff top, at 35 m west of Woman Hithe, 1965, showing exhumed ice-wedge polygon system of Beestonian age penetrating tidal sediments of Pastonian II age. At the bottom right of the photo, the two sides of the polygon met to form the cast seen in the section WRBD, 32 m west of Woman Hithe (Fig. 7).

Plate 10
West Runton, foreshore, about 1 km west of Woman Hithe, showing stone bed (bed b) on Chalk and containing grey silt of Pre-Pastonian a age.

Plate 11
Beeston, on foreshore opposite section BTA, 137 m east of datum, 1968 (Fig. 17), showing brecciated Chalk (bed a) below stone bed with marine shells.

Plate 12
Section BR–S at Beeston, 23 m east of datum, 1966 (Fig. 18). Black silty mud of bed i (Pre-Pastonian d to Pastonian I) underlies laminated silty clay of the Pastonian II marine transgression (bed k). This clay is overlain by Pastonian IVa black wood peat of bed l, subsequently broken up, and by sand with silt laminae (bed m) of Pastonian IVc age. The wood peat horizon is *c*. 20 cm thick.

Plate 13
Section BW at Beeston, 307 m east of datum, 1974 (Fig. 16). An ice-wedge cast penetrates silty clay of the Pastonian marine transgression (bed k), the underlying freshwater silty mud (bed i), and sands (bed g) of Pastonian I age. The cast filling is sealed by beach gravels of bed o, Beestonian in age.

Plate 14 (*above*)
Western slope of the western Chalk raft at Sidestrand, 1961 (Fig. 22). Section SSB is at the left margin of the photo, SSH near where the Chalk disappears below the beach and SSD towards the right margin. A stone bed (bed a) and the bedded shelly sands of Pre-Pa a age (bed b, Sidestrand Sands) lie on the Chalk.

Plate 15 (*below*)
Eastern margin of the western Chalk raft at Sidestrand, 1968 (Fig. 22). The raft over-rides Cromerian tidal sediments near its base, at the west end of section SSV (Fig. 23).

Plate 16
Eastern end of section SSV at Sidestrand, 1974. Pastonian II cross-bedded sands (bed g) overlie gravel (bed c), sand (bed b) and the stone bed (bed a). A horizon with drift mud and cakes of detritus mud is at sampling site 2 (Fig. 23).

Plate 17
Section SSV at Sidestrand, 1974 (Fig. 23, 27 m), Tidal sediments of Cromerian III age overlie unconformably Pastonian II sands with a ferrugineous gravel horizon (bed j) at the unconformity.

Plate 18
Section SSV at Sidestrand, 1974 (Fig. 23, 27 m), showing a cross-bedded sand set (pollen sample 3a) and overlying drift mud (pollen sample 3b).

Plate 19
Section SSF at Sidestrand, 1962 (Fig. 22). The section is at the eastern end of a Chalk raft, with the sequence sampled horizontally in the vertical strata. The stone bed (bed a) and sand (bed b) lie on the Chalk.

Plate 20
Section SST at Sidestrand, 1966 (Fig. 22). Cromerian I mud (bed i) lies in a channel cut in tidal sediments of Pastonian age (bed f). The arrow marks an unconformity, with a gravel horizon, between Pastonian II and Pastonian IVa sediments. Sediments (beds k, l) of Cromerian III and IV age seal the channel.

Plate 21
Excavated section PG in front of section PB, 1695 m south-east of datum 1 at Paston–Mundesley, 1966 (Fig. 30). At the base of the hole, Pastonian laminated marine clays and bedded sands of bed e occur, with the ferrugineous gravels and clay conglomerate of bed f above. In the cliff section behind, Cromerian IVa laminated tidal silty clays (bed i (lc)) are seen.

Plate 22
Section MW at Paston–Mundesley, 93 m south-east of datum, 1963 (Fig. 32). An ice-wedge cast penetrates sand and gravel of bed f (Pastonian II) and is overlain by beach gravel (bed i (b)) of Cromerian III age.

Plate 23
Section MU at Paston–Mundesley, 74 m south-east of datum 1, 1963 (Fig. 32). The Cromerian II mud (bed h) has been cut away at the right-hand side, with subsequent deposition of marine gravels. Estuarine clay of Cromerian IIIa age predates the erosion and lies on the mud at X.

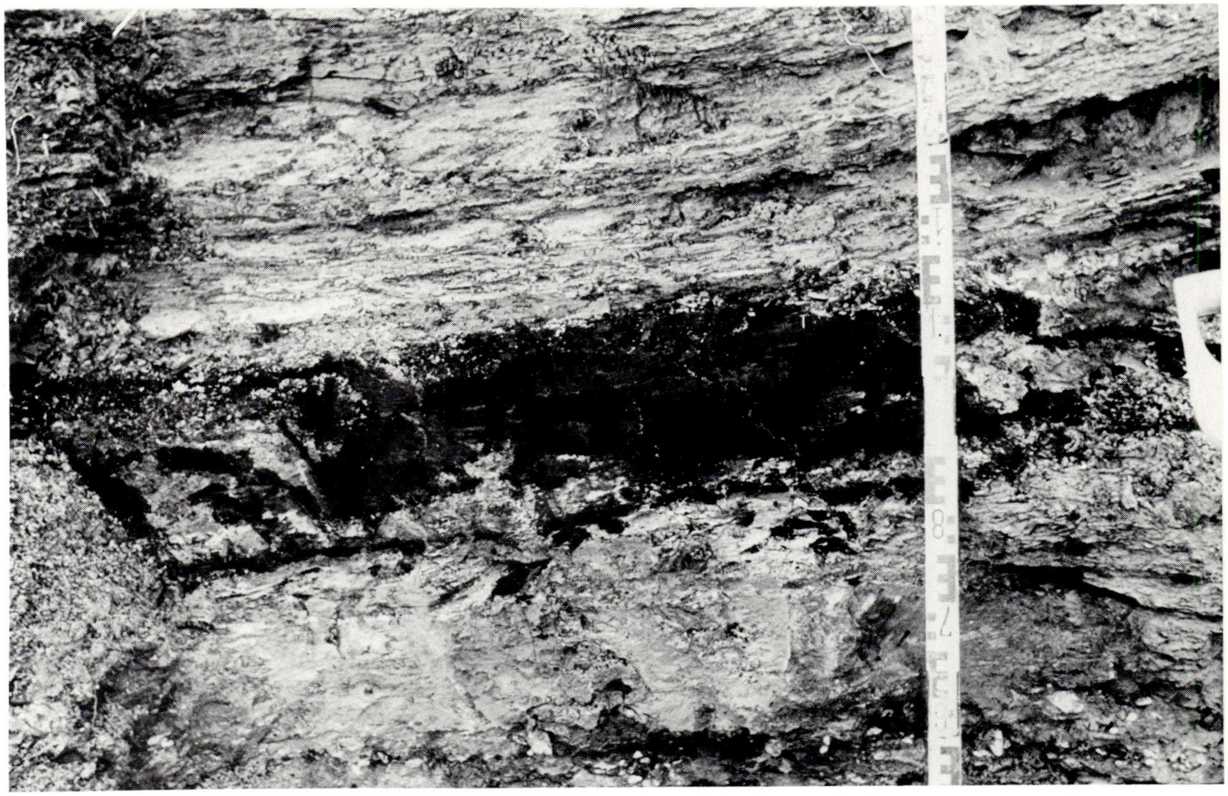

Plate 24
Section MAM at Paston–Mundesley, 49 m north-west of datum 2, 1964 (Fig. 33). Cromerian IVa detritus mud of bed i (o) overlies Cromerian III laminated tidal clays of bed i (lc) and is overlain by a similar Cromerian IVa tidal sediment (bed i (lc)).

Plate 25
Section MAU at Paston–Mundesley, 128 m north-west of datum 1, 1976 (Fig. 34). Laminated tidal clays of bed i (lc), Cromerian IIIb in age, are overlain unconformably by freshwater shelly clay-mud of bed k, Cromerian IVc in age.

Plate 26
Section MAA at Paston–Mundesley, 165 m south-east of datum 1, 1964 (Fig. 32). The base of the section, behind R. W. Hey and Sally, shows Cromerian beach gravels (bed i (b)). An ice-wedge cast penetrates their surface on the right of the section. Early Anglian cross-bedded sands (bed j i) overlie this and contain at the base a large *Taxus* stump. The sands are sealed by laminated clay of bed m.

Plate 27
Section MAG at Paston–Mundesley, 464 m south-east of datum 1, 1964 (Fig. 32). Cross-bedded sands of bed j, of Cromerian IVa age, contain seams of clay and drift mud. They are overlain by laminated clay of bed m, early Anglian in age. An ice-wedge cast penetrates the sands (bed n) above the clay, the clay and the lower sands.

Plate 28
Section MAH at Paston–Mundesley, 460 m south-east of datum 1, 1964 (Fig. 32). Early Anglian silty mud of bed l lies in a channel cut in bedded sands of bed j, Cromerian IVa in age. The channel filling is sealed by the laminated clay of bed m.

Plate 29
Section ME at Paston–Mundesley, 45 m south-east of datum 1, 1961 (Fig. 33). A channel is cut into laminated clay of bed m (early Anglian) and is filled with silt and clay with drift-mud seams, containing a good flora. This is the Arctic Freshwater Bed of Reid. At the base are bedded sands of bed j.

Plate 30
Section PC at Paston–Mundesley, 1675 m south-east of datum 1, 1964 (Fig. 30). The basal part of the section, behind D. G. Wilson, shows Cromerian laminated tidal clays of bed i (lc). Above are sands of bed n, with a channel cut into them and filled with laminated silty clay of bed o. The channel filling, early Anglian in age, is sealed by Anglian till.

Plate 31
Section OD at Ostend, 1961 (Fig. 39). Clay with race (bed b) occupies the basal part of the section grading up into clay with rootlets and branches (bed c). Till seals the 'rootlet bed'.

Plate 32
Foreshore near OD at Ostend, 1961. Rootlets (Cromerian) protrude from the clay of bed c.

Plate 33
The site of the borehole HC at Happisburgh, 1966 (Fig. 39), at the datum point.

Plate 34
The foreshore at low tide at HE, Happisburgh, 1966 (Fig. 39). Logs of wood lie in detritus mud (bed j (o)) overlying tidal clays (bed j). This is the 'forest-bed' at Happisburgh recorded by Gunn and by Reid.

Plate 35
Foreshore at Corton Cliffs south of the bluff opposite the radio station, 1971 (Fig. 43). The Rootlet Bed (bed e) with protruding rootlets is seen in the foregound, and is covered by black mud (bed f) of Cromerian II and IIIa age (photo J. Craik).

Plate 36
View from cliff top at Corton, south of bluff opposite radio station, 1971 (Fig. 43). The Rootlet Bed (bed e) is seen next to the recent beach sand, with the black Cromerian II and IIIa mud (bed f) overlying it nearer the cliff. The bluff shows North Sea Drift overlain by Corton Sands. A crack associated with an ice-wedge polygon system of early Anglian age (*see* Gardner & West, 1975) is marked by an arrow (photo J. Craik).